Lecture Notes in Logistics

For further volumes:
http://www.springer.com/series/11220

Klaus Erlach

Value Stream Design

The Way Towards a Lean Factory

Springer

Author
Dr. Klaus Erlach
Fraunhofer Institute for Manufacturing Engineering
and Automation IPA
70569 Stuttgart
Germany
E-mail: Klaus.Erlach@gmx.de
www.valuestreamdesign.eu

ISSN 2194-8917 e-ISSN 2194-8925
ISBN 978-3-642-12568-3 e-ISBN 978-3-642-12569-0
DOI 10.1007/978-3-642-12569-0
Springer Heidelberg New York Dordrecht London

Library of Congress Control Number: 2012945295

Printed on acid-free paper

Springer is part of Springer Science+Business Media (www.springer.com)

Where is the knowledge we have lost in information?

T. S. ELIOT *The Rock*

Preface to English Edition

'When the course of economy takes an unexpected turn, when instead of the continuous growth which we have become accustomed to, we find ourselves threatened by economic crises associated by us with past ages of laissez-faire capitalism, we naturally blame anything but economic order. Have we not all striven according to our best lights, and have not many of our noblest price winners worked to make this a prosperous world? Have not all our deregulations and beliefs been directed towards greater liberalism, fairness, and privatization? If the outcome is so different from our aims, if, instead of liberalism and privatization, nationalization and indebtedness stare us in the face, is it not clear that speculative forces must have foiled our intentions, that we are the victims of some avaricious power which must be defeated before we can resume the road to prospering things? However much we may differ when we name the culprit, whether it is the greedy investment banker or the stubborn attitude of an ideologized labor union, the incompetence of our politicians, or a welfare state not yet sufficiently downsized, although we have been reforming it for half a century – we all are, or at least were until recently, certain of one thing: that the economic policies which during the last decades have become a reality to most countries of goodwill and have determined the major changes in our business life cannot have been wrong. We are ready to accept almost any explanation of the present economic crisis except one: that the present economic situation may be the result of fundamentally misleading economics and that the releasing of some of our most profitable businesses has apparently caused transformations utterly different from those which we intended.'

After all, Adam Smith's *invisible hand* has been leading our market economy towards the best of all possible economic worlds since 1776, and ever since orthodox economic theory has known full well that man – *homo oeconomicus* – by pursuing his own profit furthers general wealth, too. As a persistent egoist – to quote Goethe's Mephistopheles – man is 'part of the power that would always wish evil, and always works the good'. With the recent financial crisis, however, a new and very surprising insight has hit the liberal and idyllic well-functioning market: Apparently, there are two separate markets – the so-called 'real economy' and the financial world. This of course insinuates that the financial world has nothing to do with real goods – and the underlying perceived truth tells us that is has nothing to do with reality (any more) either. However, as in accordance with the Efficient Market Hypothesis the stock exchange as an ideal financial market will sooner or later correctly, i.e. sensibly, evaluate the intrinsic value of companies, the old Hegel dictum will finally come true: Should the theory of efficient markets not fit the same, more's the pity for reality …

Meanwhile, the USA have taken the physical separation of the two East and West Coast markets one step further by strict functional task sharing: Silicon Valley, representing the so-called real economy, creates values which Wall Street, symbol of the financial world, then speculates away, at least those (still) owned by investing 'muppets' and not directly paid out in the form of bonuses. While German 'stupid money' was looking for bad investment world-wide, the German industry implemented some reforms and improvements, which were hardly noticed at the beginning, but after the crisis are now being followed by a specific 'German Next Economic Miracle'. In hard times, added value once again becomes a matter of importance, without which there could be no value absorption in the financial markets in the first place. This somewhat explains the rediscovery of manufacturing industries in countries which formerly thought of themselves as developing from industrial to service societies. All in all, these are excellent frame conditions for a book dealing with production optimization.

Globally active industrial enterprises need to adjust their production strategies to many different countries and cultures; accordingly, a specialized book dealing with production optimization in a generally accepted manner cannot remain limited to its own local language territory for long.

The design guidelines introduced herein expressly lay claim to universal validity, though factories of course differ in accordance with their respective locational factors. Standardization of factories does not mean the simple copying and transferring of standard elements but a selective differentiation of uniformly defined production principles in accordance with the respective prevailing local circumstances. Encouraged by enquiries from industrial customers world-wide as well as the Springer publishing company, I finally ventured on the project of translating the book into English. With respect to future methodological development, I very much hope to find out more about the actual scope of the expected universal applicability through feedback from readers from academical and industrial backgrounds.

The result is an almost unaltered translation of the second edition from 2010, solely supplemented by a few phrases concerning the hierarchy of goals in lean production systems and a few corrections of earlier minor oversights. To prevent publishing this book in what Prince Charles calls our common world language of bad English, I recruited some professional help and I would like to take the opportunity to sincerely thank the translator Sabine Saaro for her thorough work. With the exception of some intended deviations, I kept the translation of the technical terms defined in connection with figures and equations close to those used in the respective English specialist literature and I very much hope to have succeeded in preserving the character of my book. I wish my readers much pleasure and many helpful insights when working with this book, and above all, I wish you all much success with the application of the value stream method.

Stuttgart, June 2012 Klaus Erlach

Preface to Second German Edition

'Someone must have provoked a deep crisis for the Producing Economy of G., it knew it had done nothing wrong but, one day, its financing ran dry. Every day for years, orders had been placed by customers – purchasers from all over the world – but today they didn't come. That had never happened before. The producers in G. waited a little while, looked with incredulity at the banking houses who conducted their speculations and who were cashing in on government support with a shamelessness quite usual for them, and finally, both alienated and still without orders, called for help. There was immediately some financial aid and the politicians, who hadn't meddled in business for a long time, began with the nationalisations.'

A story about our past two years of economy could well begin like this. Yet, in spite of the apparent stagnation underlying any crisis, even a profound economic crisis like the present one seems to provide opportunities for radical technical innovation – such as the suddenly highly intensified debate on electromobility, or initiatives for large-scale alternative energy production projects. Should this turn into a new social trend, should for instance our old beloved concept of cars becoming ever more powerful be abandoned all of a sudden or may be dated power plant technology finally be discontinued, then the fundamental innovation of the new – though not yet positively identified – economic cycle as per Kondratieff could well consist of new types of energy technology in these times of climate change. This will present new challenges for factory planning and production optimization. But of course, the application of the highly estimated value stream method in *factory planning*, for *factory optimization* and in *factory operations* will continue to be highly rewarding, not only for these new product lines.

The success of a book is summed up by its readers. Not only sales figures, but also numerous comments from managers, corporate consultants, professors and students have proved the first edition of this book to have been considered a real asset by many. This positive feedback as well as detailed involvement with the numerous possible applications of the value stream method made the revision of the book a rewarding and appealing task in spite of the required long hours invested during holidays and on weekends.

The second edition at hand includes some new *explanatory supplements* as well as a certain *contentual expansion*. With respect to topics embraced by other production management methods as well (such as REFA time and motion studies, overall equipment effectiveness OEE), similarities and differences were determined. As an excellent labour-saving device, the book now has an appendix including indices of formulae, symbols and technical terms. Apart from some eliminations of minor errors, the book contains only one significant change: In the operator balance chart, the depiction of quality defects was shifted from the customer takt time to the cycle time, because this had meanwhile proved clearly more practicable and methodically sound (cf. Sect. 2.3.1). As a result of the entailing intensive involvement with time portions and their respective possible ways of representation in the operator balance chart, the industrial sample projects were amended accordingly.

With a view to additional content, several equations were supplemented, making the index of formulae a mathematically simple layout framework for value stream design. In the section on continuous flow production, the methodically significant aspect of technological process integration was illustrated. Besides, the highly consequential application of the flow principle in fixed station assembly – with time and space swapping places so to speak – was also newly included. In addition, the second edition was the perfect opportunity to incorporate some comments on appropriate production automation levels which had been disregarded in the first edition.

The book includes an additional industrial example, mainly to also illustrate the technology-oriented aspects of the value stream method. The newly introduced procedure of campaign formation clearly enables better observance of restrictions with a view to order sequencing in complex productions and is thus governed by the seventh design guideline (cf. Sect. 3.4.2). Finally, previously not considered planning and control tasks in factory operations are described in a 'value stream management' section of their own (cf. Sect. 3.6), thus expanding the applicability of the value stream method from mere factory planning to factory operations.

The two latter amendments – campaign formation and value stream management – are due to Michael Lickefett's highly productive cooperation over the past few years. In joint industrial projects, the idea of creating *lean order management* – initially formulated in the first edition in connection with value stream-oriented factory planning – was adapted and realized as a synthesis of the principles of 'lean production' and traditional order management. Suitable industrial projects and well-attended seminars instrumentally enabled and aided the further methodical development. I would therefore particularly like to thank all those companies who with their orders directly or indirectly enabled the fundamental work required for the expansion of the book.

However, the conceptual synthesis of lean order management has since largely come into inappropriate use and lost not only much of its appeal but also much of its edge as regards contents. It was therefore necessary to come up with a new description for the successful adaption and suitable adjustment of renowned PPC solution approaches to the specific conceptual and methodological frame of value stream design, i.e. *value stream management.* Accordingly, the traditional picture of a value stream manager was equipped with some new and extended aspects resulting from additional tasks partially equivalent to those formerly associated with production planners and methods engineers (cf. Sect. 3.6.1).

After the first edition was published, the new *VDI-Guideline 5200 on factory planning* (guideline published by the German Association of Engineers, *Verband deutscher Ingenieure*, VDI) was completed in cooperation with the author, longstanding spokesman of the VDI expert committee on factory planning. Concisely formulated, it sets new standards and provides a clear overview of the general planning process. This procedural factory planning approach is briefly dealt with in the chapter on value stream-oriented factory planning (cf. Sect. 4.1).

In July 2009, a workbook on the newly developed method of *energy value stream design* (cf. Erlach 2012) was published on the occasion of the fiftieth anniversary of the Fraunhofer IPA Institute for Manufacturing Engineering and Automation, Stuttgart. First application experiences have proved the inclusion of the energy aspect in the value stream analysis and the analogue application of the eight value stream design guidelines for improved energy efficiency to be unexpectedly fruitful. Once again, the value stream perspective has changed the outlook on factories. As the focus is on the product rather than the respective resources, the energy requirement appears in an entirely different context – a sure indication of the value stream method's potential for further development even beyond its current already excellent level.

I would sincerely like to thank all those who contributed to the creation of this book. In particular, I thank Michael Lickefett, who was not only an inspiring interlocutor but always encouraged me to continue with the further development of the value stream method and its possible applications in spite of all obstacles. I am also very grateful to Susanne Ramsthaler for her diligent proofreading, whose attentive eyes eliminated many mistakes and whose insistent questions frequently coerced me into quite cumbersome clarity both with a view to wording and structure.

Stuttgart, Easter 2010 Klaus Erlach

Preface to First German Edition

A spectre is haunting Germany – the spectre of Globalization. Numerous emerging markets seem to be conspiring to wipe out the social market economy coined by Germany together with its dumping wages, slack environmental laws and investment friendly tax system. Accordingly, the German employment trend paints a rather gruesome picture – with many jobs in the manufacturing industry being threatened by impending relocation to low cost countries. Is this picture correct? Is there anything we can do about it?

Current globalization discussions limit the outlook of global communication, acting and travel to the purely economic aspect of the commodity and money markets. *Globalization* itself is presented as an inherent necessity, highly approved of by some as a 'free market' of opportunities, but equally enthusiastically condemned by others as a ruthless profit maximization strategy. Both sides seem to agree, though, that our entire ecology-, tax- and welfare-related consciousness is exclusively governed by our economic existence and its inherent practical constraints. Affirmation or disaffirmation aside – production relocation seems to be the last resort for manufacturing companies to secure their competitiveness and thus their own survival. But how can we add further options to their possible course of action?

Productions of largely standardized, work-intensive, easily transportable products are most likely to be affected by possible relocation. Many companies decide to relocate solely based on possible wage cost savings assessed on the grounds of their current wage hour requirements and the respective differences in pay rates, thus substituting the question of 'if' by the question of 'where to' – all in all a very tempting approach. However, there will always be a company somewhere producing the same product at an even lower price. By trend, this applies more to high-wage countries, though there will always be the possibility of finding an even cheaper location wage-wise – and relocate there, at least until the wages rise at the new location and another, better production is discovered. As a result, *factory nomadism* never comes to an end.

Location decisions are of immense strategic significance: On the one hand, a later correction of the inherent commitment and orientation of the relocating company requires immense effort; on the other hand, cost minimization is not the only target value of a production. Not only the price, but also customer-oriented flexibility, consistent quality as well as sufficiently short and reliable delivery times are equally relevant market success factors. Similar to the question of society's correct response to the effects of globalization, location decisions should never be taken with a view to wage costs alone, but should always keep an eye on the other market objectives as well.

Relocation decisions are often taken with no regard to possible optimization potential at an existing German location – in particular, when factory structures grown over time are compared with newly constructed 'greenfield' projects abroad. A comparison at such unequal preconditions results in an inappropriately poorer evaluation of the production to be relocated, though. What is more, ignoring the potential for optimization leads to a disadvantageous transfer of the respective inefficient production and business processes from the existing to the new location. Together with entailing start-up costs and other relocation risks, this may even worsen the company's circumstances rather than improve the same. Accordingly, the *location optimization* potential should always be investigated before the risk of relocation with all its pertaining cost and time-related perils is taken.

To this end, the book at hand introduces a highly effective and extremely helpful method: value stream design. Originally developed by Toyota for application in the car industry, this method has meanwhile widened out to other industries with great success. The Fraunhofer IPA Institute for Manufacturing Engineering and Automation, Stuttgart, succeeded in integrating the *value stream method* in factory planning projects, thus developing the original approach with its segment-related workshop character into a highly powerful planning tool. As a result, today's planners find it hard to imagine how only a few years ago industrial projects were realized at all without the aid of the value stream method.

The potential of locations in Germany has meanwhile been recognized by all those who in earlier years bought companies with previously neglected production optimization but who were able to sell the same at a handsome profit following reorganization. The time has clearly come to provide managements with a suitable method to review their factory goals and optimize their production processes. On a factory level at least, this approach could take an adequate stand against all this globalization talk.

For the author, a book project is always a pleasure as well as a burden. Special thanks is therefore due to all those whose support and advice helped ease the burden and improve the result of my efforts. I would particularly like to thank Hans-Hermann Wiendahl, Michael Lickefett und Alexander Stamm for their professional suggestions. Susanne Ramsthaler and Karin Erlach's diligent proofreading gave the text its clear structure. I also very much thank Minou Friele, Siegfried Reusch und Karin Mutter. Not least, I thank all those who supplied the foundations for my work. Mike Rother, who introduced me to the value stream method, the Fraunhofer Institute who provided me with such an inspiring working environment and all the manufacturing companies who allowed me to advise them on factory-organizational matters.

Stuttgart, February 2007 Klaus Erlach

Contents

1 Production Optimization 1
1.1 The Organisation of Production 3
1.1.1 The Factory 4
1.1.2 The Value Stream 6
1.2 The Four Goals of Production 10
1.2.1 The Dilemma of Production Process Planning 13
1.2.2 The Logical Square of Goals 19

2 Value Stream Analysis 25
2.1 Product Families 31
2.1.1 Product Family Matrix 32
2.1.2 Production Procedure and Family Likeness 34
2.2 Customer Demand 37
2.2.1 Customer Takt Time 38
2.2.2 Fluctuations in Customer Demand 41
2.3 Production 44
2.3.1 Production Processes 46
2.3.2 Material Flow 66
2.3.3 Order Processing 72
2.4 Potential for Improvement 83
2.4.1 Production Lead Time 85
2.4.2 Line Balancing 90

3 Value Stream Design 97
3.1 Production Structuring 105
3.1.1 Business Types 108
3.1.2 Allocation of Resources 114
3.2 Production Process Design 117
3.2.1 Capacitive Dimensioning of Resources 118
3.2.2 Process Integration and Continuous Flow Production 121
3.3 Production Control 139
3.3.1 Direct Coupling of Production Processes 144
3.3.2 Consumption-Oriented Kanban-Regulation 156
3.3.3 Production Control at the Pacemaker Process 183
3.4 Production Planning 190
3.4.1 Smoothing the Production Volume 191
3.4.2 Levelling the Production Mix 196

3.4.3 Capacitive and Restrictive Bottleneck Control 206
3.4.4 Characteristics of Lean Planning und Controlling 211
3.5 Conception and Implementation 213
3.6 Value Stream Management 218
3.6.1 Value Stream Manager 219
3.6.2 Value Stream Management Tasks 220

4 Towards a Lean Factory 231
4.1 Value Stream-Oriented Factory Planning 233
4.2 Production System 247

5 Case Studies 251
5.1 Comfort Ltd. – Transparency by Continuous Flow 252
5.2 Spring LLC – Lot Sizing in Variant Production 263
5.3 Glass PLC – Short Delivery Times in Single Piece Production 276
5.4 The Seco ULC – Flow Orientation by Technology 287

6 References 305

7 Appendix 307
7.1 Index of Formulae 308
7.1.1 Calculation Formulae for Value Stream Analysis 308
7.1.2 Specification Formulae for Value Stream Design 312
7.2 Symbols 317

Subject Index 319

Chapter 1
Production Optimization

The design of optimal production procedures is a factory planner's most important and central duty. It is accompanied by various obstacles such as factory-specific restrictions and conflicting partial production goals which need to be taken into account. Besides, as soon as an ideal status seems to have been finally achieved, the goal posts are moved again: Customers' wishes, production technologies, legal requirements and/or other significant influencing factors are subject to constant change. Accordingly, the never-ending task of production optimization and its requirements on factory planning are ever present in all factory operations.

Method

Since there's nothing as practical as a good theory, the targeted increase of a production's effectiveness must be based on a sound and sophisticated method. In order to really achieve sustainable improvement, all previous habitual production procedures must be systematically challenged and developed in a suitably target-oriented manner. The book at hand aims to conclusively illustrate the respective outstanding effectiveness of the *value stream method*. Not only does it enable the observance of the entire value added chain by way of excellent visualization, it also supports optimal production design through targeted application of design guidelines.

In this book, the main features of the renowned value stream method (Rother 1999) will not only be presented accurately and systematically but also significantly expanded. The traditional manner of representation and symbolism will be supplemented, in particular to enable a more detailed representation of the respective information processing procedures. The familiar guidelines will be expanded by solution principles for production organization and further developed and elucidated with a view to their coherent correlations. Subsequently, the resulting *eight consecutive design guidelines* for value stream design will be introduced. In addition, the traditional formation of product families will be expanded with the aid of production-relevant criteria and business types to enable product family-oriented segmentation. This comprehensive structuring approach forms the basis for *value stream-oriented factory planning*, combining the project-specific approach of traditional factory planning with the more consistency-related value stream approach.

The book at hand includes experiences from numerous *industrial projects*; individual cases were suitably adapted to serve as realistic examples. Initially, the

K. Erlach: Value Stream Design: The Way Towards a Lean Factory, LNL, pp. 1–24.
springerlink.com

value stream method is going to be introduced using the example of an easily comprehensible serial production, moving on to rather more complex case examples of small batch productions of high-variance as well as customer-specific productions to illustrate the successful expansion of the value stream method's application range to include piece productions with complex multi-step production structures and high product variance.

As the value stream method was originally developed by Toyota for their own production systems, its *application range* initially seemed to be limited to the car industry; however, numerous different examples of value stream optimization and factory planning have since proved this wrong. Thanks to a few methodical amendments and variations concerning the application of the design guidelines, its implementation in other industries was enabled and positive experiences have since been gained in machine construction as well as in the electrical and electronics industries, in medical engineering and optics, sanitary and line technology as well as the customer goods industry. The value stream method has thus proved highly suitable for the analysis and new design of all types of piece productions. Besides, experience has shown the method to be perfectly suited for multinational large scale manufacturers and their suppliers as well as small and medium-sized businesses.

This book is addressed to corporate management as well as to present and future production managers. The value stream method provides the management with a clear picture of the production and thus enables a precise assessment of the inherent production requirements. Simple standardized symbolism allows the illustration and comprehensible corporation-wide communication of different types of improvement potential and their respective solution approaches. The detailed description of the value stream method and the numerous practical examples illustrate its many benefits and wide range of possible applications. The approach is directly applicable and allows the targeted conception and implementation of a improved productive efficiency.

A book on methods will never be able to make up for practical project experience – no matter how many examples are given. However, it facilitates the targeted contextual collection of the respective experiences. This is greatly supported by the objective and neutral observations of an external advisor who is not yet used to habitual restrictions and is thus able to disregard the same in his search for unrestrained, innovative solution approaches. In order keep the initial dynamics of beginning improvements moving and prevent the same petering out halfway, it is recommended to make one person responsible for the intended implementation to avoid prevarication on the grounds of day-to-day business. External monitoring may keep up constant communication and thus also ensure the conceptual focus is kept.

Contents

Though the book at hand concentrates on details of production, production-related cost effectiveness is not the central aspect of our investigations. Instead, the focus lies on the purpose of all production: The actual product to be produced – though this is not entirely correct either, as the purpose of a product is not really fulfilled until it has been sold. Accordingly, the basic idea of assuming the *customers' point of view* in the evaluation is of central importance for the method introduced herein (Chap. 2).

Viewed from the point of corporate controlling, things usually look different, though: Development, production and distribution of products are merely means to achieve financial profit. Accordingly, costs – inclusively those of production – become significant factors for the evaluation of a production. However, without a successful business idea, without marketable products, hardly any profit will be made. Cost assessment alone will therefore not be much help when it comes to appropriately improving the design for the technical activities connected with a production.

Accordingly, there is no point in planning how to increase the efficiency of processes devised to aid the purpose of production until the usefulness of the respective act of producing has been ascertained. *Production optimization* – if understood correctly – observes both aspects and thus begins with a critical analysis of all customer-specific and product-related production processes. This is comprehensively reflected in the basic concept of a production design primarily striving to prevent all waste (Chap. 3). For low-waste, i.e. *lean production*, it is not enough to keep production costs low, the right products must be manufactured in accordance with the customers' wishes – the only way to ensure sustainable profitability. However, the term 'lean production' should not be misunderstood – as often happens – to mean 'slim': The goal is to keep the muscles but get rid of unnecessary fat.

The purpose behind the act of producing is to satisfy the customers' wishes. Customers' wishes are satisfied by products featuring certain characteristics and functionalities. The basic aspects are determined during the product development stage and in the case of customer-specific products are later varied during order processing. However, the customers' specific desires comprise more than a basic adaptation of product characteristics. Customers want products of a certain quality, available in certain quantities and within certain periods of time – and at certain prices, too. The production must therefore be aligned with different goals, the interdependencies of which are going to be preparatively explained in the following (Sect. 1.2). This detailed reflection on the *goals of production* is a vital prerequisite to conclusively achieve the required degree of usefulness in a production.

A purposeful production needs to be organized accordingly. A lean production procedure is realized in the form of a customer-oriented, highly efficient value stream. Seen from this value stream perspective, the entire production is focused on the creation of value. The structure of the production is defined by its spatial and social layout within the factory, where each value stream has its expedient location. The *organization of production* as a value stream is going to be prefatorily outlined in the following (Sect. 1.1).

1.1 The Organisation of Production

The organization of industrial production happens in factories. Not only does the factory building provide the physical environment, the factory itself could also be considered a sociotechnical production system (Sect. 1.1.1). In order to best holistically regard the production procedure with all its correlations within the factory, the value stream perspective is assumed, showing the production as a value-creating flow (Sect. 1.1.2).

1.1.1 The Factory

As a rule, a society's wealth is usually based on its self-acquired material foundations. Today, we quite aptly refer to industrial societies, in spite of information and/or service societies, which are hardly ever fit to exist on their own. Etymologically, 'industry' means 'diligence' and 'activity'. The central location of this activity is the *factory*, in which necessary as well as not so necessary things are produced. This is not done to secure sheer survival, but also to provide 'objectively superfluous' goods, which simply make life more diversified and interesting. The term 'factory' generally refers to a commercial business and/or production facility for the mechanical and work-sharing production of goods – in contrast to a traditional, manually operated manufacture. The *VDI-Guideline 5200* on factory planning (guideline published 2010 by the German Association of Engineers, *Verband deutscher Ingenieure*, VDI) defines a factory 'as a location where value is created by way of work-sharing production of industrial goods using factors of production'. In practice, factories result from historically grown changes and adaptations applied in factory operations as well as planned interventions in factory planning.

A detailed depiction of a factory may be developed in five dimensions: the location, the structure of the entire factory site, the factory building incl. factory structure and layout, the production logistics as well as the work organization, all of which are going to be discussed in further detail in the following.

Location. Initially, a factory is defined by its *location*: On a local level, the choice of a location mainly depends on criteria such as price, size and topography, as well as infrastructure, i.e. water and energy supply, roads, railways and waterways etc. On a regional level, characteristics like human resources, labour market, wage levels, taxes and subsidies as well as climate and cultural environment are taken into consideration. All these location factors greatly influence the factory operations. Conversely, both a factory and its production processes must be suited for their respective location and inherent characteristics. Thus, a factory is entrenched in its environment.

Factory site. The *factory site* is situated at said location. The first boundaries are fixed in connection with the initial structural planning. The existing terrain is shaped, levelled and the parcel of land fenced in. Construction plots are defined for future production, storage, and/or administrative buildings or those for social purposes (such as canteens, company kindergardens). These buildings are logistically connected by paths and streets; space permitting, parking spaces and outdoor facilities may also be planned. The resulting ideal structure of the premises illustrates the connection between the buildings within the factory compound as well as external infrastructural connections. Basically, this is a small town of its own with highly differentiated functions, though. Accordingly, the structural planning of factory sites is actually part of an urban planner's field of work.

Factory. At least one of the *buildings* on the site will be allocated to the factory intended for production in the true sense of the word. Each factory is composed of three basic components, i.e. the outer building shell, the structure of the internal areas and the spatial arrangement of the resources within the factory layout. The factory building comprises three elements: the building shell including supporting structures, HVACR incl. media supply as well as all permanent support elements, such as walls and apertures (windows, doors). Once the factory has been built, it will still be empty from the production's point of view. Within the factory building, different production areas will be created by way of segmentation in order to create a transparent and efficient *factory structure*. In a manner of speaking, once the rooms of a house have been defined, the matching furniture may be planned. Technologies are chosen, appropriate resources such as machines and plants as well as conveyor and storage techniques are devised, developed, constructed, built, planned with a view to layout and finally installed. Among others, the decisive layout criteria concern the material flows and their correlations, defining the transport and storage of material between resources. The final result is the *factory layout*, which describes the space requirements and the physical arrangement of the production facilities in the narrower sense.

Logistics. Now that the construction of the factory has essentially been completed, the real work – the action – begins: Machines and material must be set in motion. This requires information on what needs to be done where and when. The information flow organizes order processing-related tasks such as the recording, compiling, processing, storing and distributing of data and instructions required for production planning and control. The factory is controlled and managed by way of specially designed *production logistics* with the aid of production plans – converting the act of producing into systematic technical action. This systematic action is guided by three logistical goals, namely a high degree of machine utilization, low inventories and high delivery reliability. As these goals will hardly ever be able to be met to identical degrees, the production planner is faced with the so-called dilemma of production scheduling (Sect. 1.2.1). As a result, poorly planned technical actions will frequently be encountered factory operations, though these are usually only ever revealed retrospectively.

Work organization. The fifth factory dimension is still missing – employees with specific skills and qualifications whose production-related obligations and rights are determined by the *work organization*. This allocation of tasks and persons organizes the distribution of work. The social factory dimension also includes working time models, shift rosters as well as pay scales. In addition, the employees are organized in a structural organization which in turn determines the management structure.

All five dimensions described above – location, site structure, factory building incl. factory and layout, production logistics and work organization – are designed and prepared for the actual factory operations during the factory planning stage.

The actual factory, i.e. the fully equipped building, could be regarded as the turnkey product of a factory outfitter, turned over to the user ready for use. Equally important though is the entrenchment of the factory in its geographical and social environment as well as the installation of the target-oriented production logistics to create the factory's efficient and customer-related working order. However, the work organization defining the social rules pertaining to the factory can only be roughly planned – it cannot be sold like a product, it needs to be 'lived' in the factory. The latter is a vital prerequisite for the achievement of the production goals and the success of the developed production concepts. Accordingly, a factory is a highly complex *sociotechncial system* (Ropohl 1982). This does not make a factory a complicated and large technical device, but a technically shaped technomorphically designed and lived-in living space for living people.

A factory, seen as a space for living, could be referred to as a *technotope* (Erlach 2000) – in line with the term 'biotope'. The human existence takes place – and particularly so in a factory – in a technical environment. Human beings do not simply surround themselves with technology – they *are* technology. This anthropological constant seems to be highly influential with a view to our lifestyle. A factory is a place for living rather than a mere building or a simple production facility. This also explains why the construction of factories with an attractive working environment is a vital luxury, it's the only possible way to allow and enable employees to fully dedicate themselves to their best possible performance. It also explains why the design of successful factories often matches that of their products.

1.1.2 The Value Stream

The question of what a factory looks like is generally answered by way of a photograph of the building, or factory site, respectively, as well as the traditional overall result of the factory planning, i.e. the factory layout. However, this only reflects some aspects of a factory, as explained above. More often than not denominations, sizes and spatial arrangement of resources are stated, which are of little informative value with respect to the production procedure and tell us nothing at all about the quality of the factory. The presentation can be greatly improved by charting the material flows. This illustrates the sequences in which certain products pass resources, allowing an evaluation of how well the sequences match the layouts of resources. In cases of complex production procedures, though, the depiction of overlapping material flows quickly becomes confusing and an important question arises: Though we can see that something is moving with the material flow we have no way of knowing why, i.e. which information flow triggers which production processes.

Value Stream Method

This is exactly where the value stream method and its advantages come in. It not only visualizes production processes and material flows, but also information flows – all within one and the same illustration. Additional flow charts of business processes are merely required for extra detail. The process chain diagram found in order processing is firmly rooted in value stream mapping; it visualizes the interfaces between production processes and business processes. Resources with similar functions are merged into production processes in the value stream map. This abstracted value stream depiction increases the transparency of the production processes as opposed to the numerous single machines individually shown in the factory layout. The spatial layout of the resources is not shown, though. Accordingly, a factory will always initially be devised, planned and depicted as a value stream and subsequently as a layout (Sect. 4.1).

With its clear *visualization* of the production processes, the value stream method provides an excellent internal communication platform for mutual interchange concerning the current state as well as the desired future state of a factory. The comprehensive depiction of the production processes allows a transparent factory analysis for the targeted identification of weaknesses as well as the determination of potential improvement (Chap. 2). Based on the aforesaid, the possible effects of changed production procedures may be conceptionally examined with the aid of design guidelines indicating the path towards an improved production and a value stream-optimized factory (Chap. 3).

The term 'value stream method' includes two descriptive elements – 'value' and 'stream'. 'Value' refers to the *value creation* inherent in the production of goods. This describes the general intention of production: namely the conversion of a source material into a product considered to be of higher value. The term 'stream' refers to the fact that an essential characteristic of said production lies in the spatial movement and qualitative alteration of the parts and products in the *flow of production*. Due to machine utilization and work sharing specialization, though, the various process steps cannot all be conducted at the same place – nor at the same time.

Value

During the course of the production, the process of value creation furnishes the source material with additional value, thus the finished products are appropriately referred to as 'goods'. The *value* of such goods results from their evaluation by certain criteria. This leads to the fundamental problem of value arising from product characteristics as well as from conformance with certain criteria. From the viewpoint of the producing company, an evaluation of the goods in line with the respective production effort seems expedient – i.e. production costs are used as a basis. This results in the type of calculation typically found in a supplier market: The price is calculated as the total production costs plus a profit margin determined by the producer (Fig. 1.1, case 1). The customers are allocated the products in the sequence of order receipt or other criteria. As the production costs invariably rise over time due to wage increases, replacement investments and increased purchase prices, frequent increases in sales prices are also inevitable.

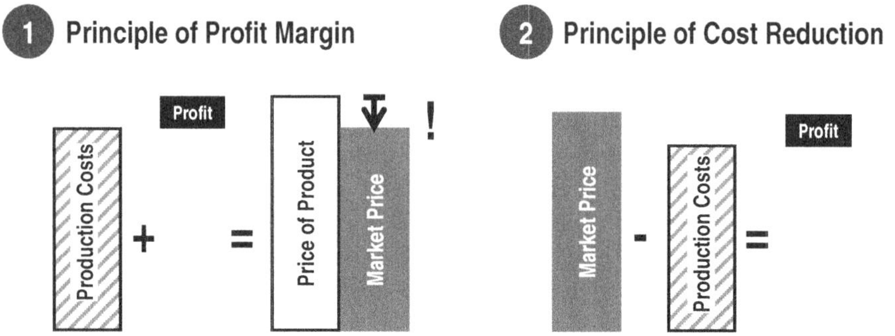

Fig. 1.1 The principle of cost reduction

While a supplier market offers hardly any incentive for *cost reduction,* competitive environments may well result in product prices higher than market prices. In global competition with free market access, the corporate success depends much more on the achievable market prices, where the profit is then conversely determined by subtracting the production costs from the market price (Fig. 1.1, case 2). Accordingly, the profit may also be increased by way of consistent cost reduction. As far as the production is concerned, however, this requires constant improvement in order to ensure a consistent decrease of production costs and thus cushion inevitably rising factor costs through lower production effort. From the production's point of view, this is the only viable leverage to sustainably ensure the competitiveness of a location.

Depending on the various possible location-specific combinations of applicable factor costs, though, sustainable global competitiveness may be a hopeless venture for some products. Therefore the corporate point of view should bear in mind that the price also reflects a product's *value in use* – based on the subjective criteria of the customers' expectations with a view to its actual usefulness. Said usefulness results from the product's ability to satisfy the respective customer's wishes. Accordingly, a lamp may be better or less suited for reading, regardless of its price, whereas representative features may by more dependent on the material quality and may therefore directly influence the price. Regardless of the actual production effort, the benefits of a product may well be much higher – or in the worst case also lower – than the production costs. A production location's success thus largely depends on its concentration on the respective customer benefits generated by the products. It is on the one hand influenced by concrete product characteristics, which shall be disregarded herein, though; on the other hand it depends on the customer-specific performance features of a production. The latter in particular are also reflected by delivery reliability and delivery quality and largely depend on the customer-oriented design of the entire production procedure.

Stream/Flow

The metaphor of a flow has widely established itself for the description of a factory's production. Whenever something changes and/or moves in the production, such as material flowing from station to station, the respective production is regarded to be positive. In a negative production, the production process is said to have come to a halt, with material piling up and backing up. The *production flow* is considered a decisive characteristic of a positive production process, a possible key figure for its evaluation is the degree of flow.

However, change and movement within a factory are no guarantee for value-adding properties of the respective activities, i.e. any real creation of value. This aspect is emphasized by the term 'value stream'. The *value stream* includes all tasks and activities required for the conversion of raw material into a finished product.

- Firstly, these tasks include all directly producing tasks, i.e. activities which change the properties of the respective materials. They are broken down into six main groups of manufacturing procedures as per German industrial standard DIN 8580: primary shaping, shaping, separating, joining, coating and the alteration of substance properties.
- Secondly, they also comprise all logistical tasks concerning handling, transport, storage, supply and commissioning.
- Thirdly, they include tasks generally referred to as indirect tasks, such as preliminary planning and control as well as maintenance, servicing and repair of machines, plants and workplaces.

There are also numerous ancillary tasks which may be required due to specific circumstances, local boundary conditions and technological requirements.

Assuming the *value stream perspective* means taking a comprehensive look at all these tasks as a whole – with a focus on the creation of value. Not all of the tasks mentioned above are considered *value adding* ones, though – the decisive criterion is whether or not the respective tasks issue the material and later product with any value-enhancing characteristics from a customer's point of view – which is not always easily determined with certainty. In addition, there are numerous non-value adding tasks which seem vital for the production procedures due to different boundary conditions. Thus, one of the goals of production optimization is the reduction or elimination of non-value adding tasks.

Improvement of Production

Accordingly, the effectual logic of a lean factory lies in consistent production improvement (Fig. 1.2).The presentation of this correlation is derived from the so-called Deming Cycle known from quality management systems – the PDCA (Plan-Do-Check-Act) Cycle. Continuous improvement is the force that pulls the rotating wheel of the value stream upwards on the steep slope of production quality with the aid of an appropriate work organization. Waste of any kind is the obstacle which must be overcome through increased effort in the realization of improvement measures. Once implemented, improvements are secured through determination of binding standards, because positive production procedures are characterized by repetitive execution of procedures in an identical manner. The illustration clearly highlights the vital significance of the respective *standardization*

– without which the gravity of bad habits would soon drag the level of production back down the hill again. Besides, certain disturbances and a great part of the waste occurring in production can only be made visible against the background of such standards. All design guidelines for a value stream optimized factory therefore aim at standardization or even presuppose the same. Whenever new production technologies are introduced – usually in connection with new or significantly changed products – the process of continuous improvement begins from scratch. Those promoting the changes will have learned from previous experience, though, and may thus be better equipped for a faster qualitative ascent this time.

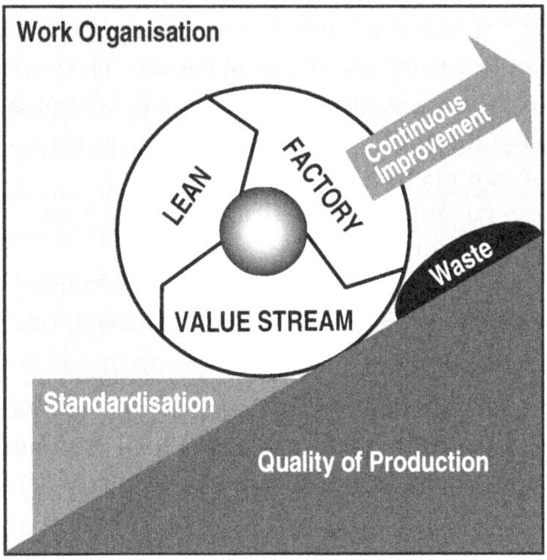

Fig. 1.2 The effectual logic of continuous production improvement

However, before a new production can be designed, or an existing production be improved, respectively, the respective factory goals and their priorities must be clarified and weighted accordingly. Specific individual goals may be hierarchically arranged in a specific system of goals. On the highest level of abstraction, there are four independent goal dimensions which are not mutually reducible, though. These goal dimensions as well as their correlations and interdependencies are going to be discussed in greater detail in the following.

1.2 The Four Goals of Production

The first question that arises in the optimization of a production asks which goals the intended improvements should measure up to. The achievement of the respective goals then indicates the quality and efficiency of a production, enabling the respective factors indicating the success of such factory products in the market to be defined accordingly. Said success factors indicate which customers' desires need to be met by which performance criteria in order for a factory to successfully establish itself in the market.

The primary objectives of production are known as the 'holy trinity' of cost, quality, and time – the sequence varying according to the 'felt' importance. Most important is to point out the right direction of goal achievement – low production costs, high quality of products, as well as short lead times in production and order processing. Recently, though rarely, the product variety on offer is added to these production goals. Gone are the times of mass production, characterised by Henry Ford's statement, "You can have any colour you want, as long as it is black." Products produced in series and not positioned within the lowest price segment can only be distinguished from competition by the 'long tail' of innumerable possible variants.

Goal Dimensions

Accordingly, a production's efficiency is always basically determined by four independent goal dimensions (Fig. 1.3). On a general level, the four dimensions of variability, quality, speed and economy span the entire width of a production's applicable *system of goals*. Each goal dimension comprises several typical partial goals, which in their reciprocal correlations define a production's system of goals and accordingly also the *factory goals*. These highly interdependent factory goals of the four dimensions may sometimes conflict with each other, though (Sect. 1.2.1). They are not interchangeable, i.e. goals allocated to a specific goal dimension cannot be converted into or reduced to any of the goals pertaining to other goal dimensions. Quality targets for instance will not automatically be met along with necessary changeover reductions; nor may they be regarded merely with a view to profitability when designing a factory, though they may be allocated different levels of relevance through appropriate weighting.

Fig. 1.3 The four goal dimensions of production

The four goal dimensions shall be illustrated in further detail by way of examples of their respective partial goals in the following:

1. The *variability* of production indicates how wide the attainable production range is. This dimension determines how many variants will be produced for what product – and whether there are customised products. The flexibility of a production system indicates to what extent short-term variations of market demand can be met. Mutability describes the ability

of production to respond to requirements changing in the short to medium term by adapting structures.
2. On the one hand, the *quality* of production indicates the rate of yield of production processes. On the other, it describes how well the tolerance levels are complied with, as well as how reliable any of the production processes works. This is closely connected to the question whether the exact number of pieces can be produced to obtain the necessary quantities. Also ergonomics and occupational safety of the production processes are a quality measure within production. In addition, schedule reliability – both internal and external – is a quality feature of production.
3. The *speed* of production indicates how time-consuming the value-adding steps are, as well as the auxiliary operations such as changeover activities. The length of production lead time also falls under this goal dimension. The availability of machines and other facilities expresses the qualitative reliability of production processes in terms of time. Frequency and duration of breakdown greatly impact the efficiency of production.
4. The *economy* of production indicates productivity in relation to all production factors. Here, all factor costs arise that are influenced by the requirements of variability, quality, and speed. Directly relevant are personnel productivity, capacity utilisation, and material utilisation; i.e. sustainability in a broader sense.

The above given sequence of goal dimensions does not reflect their significance. The importance of any goal and goal dimension depends on the specific conditions of production. The sequence, however, indicates the logical succession for defining the required goals in each goal dimension. Therefore, one begins with the definition of the product range to be produced to achieve market success. Then, the desired product quality and accompanying production quality have to be determined. If, in consequence, the technologies for the production processes are established, operation times and process duration can be calculated. Finally, economy can be checked after designing and selecting the facilities to achieve the other above-stated goals. At last, this economy view has repercussions on the other three goal dimensions. Often, some of the aspired features of production design are too expensive or inefficient. As a rule, the repercussions lead to mutual restrictions in goal accomplishment. For cost reasons as well as for other restrictions, variability, quality or speed of production will be limited to certain extent.

Production optimization aims to consistently increase the *efficiency* of the production in spite of the four conflicting goal dimensions, bearing in mind, though, that the relative weighting of the various goals has changed over time – mainly as a result of technological developments and increased social wealth. With increasing wealth and technological experience, the original desire to simply own a certain product gives way to certain expectations with a view to desired product qualities, such as functionality and quality. Such rising expectations as to product efficiency are often accompanied by ready acceptance of (slightly) higher prices, but sometimes lead to immense pressure with a view to pricing and lower expectations. This experience is based on the fact that increasing social wealth is only possible with consistently decreasing prices or increasing efficiency at only moderately increasing prices. Accordingly, cost pressure in production is the inevitable downside of social wealth, enabled through technological progress, which allows decidedly higher profitability in the production and at the same time offers various ways to meet more of the performance-related goals with only minimal loss of profitability.

Success Factors

These rather broad boundaries allow companies the specific positioning of their products in the market. The concrete goals within the four goal dimensions depend on the specific *success factors* of the company in the market. And related to these success factors, a product can be more or less costly. The respective price level can be justified by material quality, workmanship, durability and reliability, design and image of brand, availability, product innovation, consequential charges, environmental friendliness and other more (Fig. 1.4). The analogous structure of the goal dimensions of production and the success factors of products shows that both should be adjusted to each other. The objectives at a specific production site must comply with the respective specifications of success factors. The strengths of each production site should meet the specific demands of the products to be produced there.

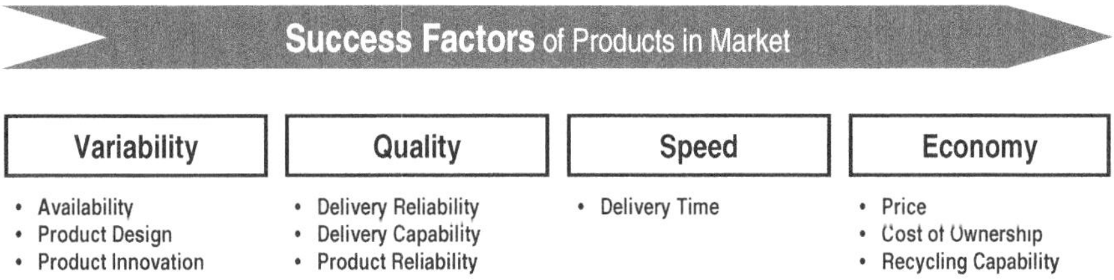

Fig. 1.4 The four success factors of products in market

> **Goal dimensions**
> Production optimization aims at the constant improvement of production *efficiency* with respect to four conflicting goal dimensions. These four dimensions – variability, quality, speed and cost-effectiveness – are reflected in the market as the success factors of the respective products. The essential strategic parameters of a production are comprehensively defined with the aid of the market goals of availability, delivery capability, delivery reliability and delivery time.

1.2.1 The Dilemma of Production Process Planning

The challenge for production management is to find the best way to achieve the different and conflicting goals. Typical measures to improve the achievement of one goal lead to a deterioration of goal achievement in the other goal dimensions.

These goal contradictions appear in scheduling and material planning with regard to three target criteria for both cases. They are known in literature as the 'dilemmas' of material planning and scheduling (Gutenberg) or as the *goal conflicts* of production logistics (Wiendahl 1995) and are going to be slightly modified and illustrated in the form of a triangle diagram below. Based on this preliminary work and supplemented by the fourth goal dimension of 'variability', a general diagram of goal dimensions is going to be developed (Sect. 1.2.2).

Dilemma of Production Scheduling

The *dilemma of production scheduling* usually occurs during production scheduling, or production-related time scheduling, respectively (Fig. 1.5), which aim to complete production orders and, above all, customer orders on schedule. Delivery reliability is a measure of logistical process reliability in factory, which has to be maximised (goal dimension 'quality'). If customer demand is varying, this goal can be achieved quite simply with more or less unlimited production capacities. Due to restricted investment resources, capacities are always limited. That's why scheduling also aims to ensure high capacity utilisation. It is a precondition to achieving low respectively marketable production costs (goal dimension 'economy'). For that it is necessary to level production quantity.

The realisation of a levelled production volume leads inevitably to variations in the throughput time of both customer orders and production orders, if customer demand is varying. This jeopardizes the goal of delivery reliability, because the compliance of schedules requires foreseeable throughput times. In general, short throughput times are easier to predict than longer ones. This is firstly because there is less time for schedule-related postponements. So to speak, production orders are already executed before any schedule deviations can be planned. Therefore, and secondly, production systems with short throughput times can be more easily controlled, because laborious changes of schedules and priorities can be avoided. Thirdly, higher percentage variations of throughput time are of slighter importance. A fifty percent delay for a two day delivery time would be one day, which can be compensated by express delivery, whereas the same delay for a one month delivery time would be painful two weeks (goal dimension ‚speed'). Consequently it is the main objective of production scheduling to achieve high schedule reliability, high reliability of throughput time as well as low production costs. Accordingly, the maximisation of capacity utilisation contradicts the maximisation of schedule reliability as well as the minimisation of throughput time, whereas the latter both are to some extent compatible as also shown in the illustration (Fig. 1.5).

Limited production capacity is additionally restricted by the technical *availability* of the resources. Machine utilisation, throughput time and, above all, schedule reliability decline in an unpredictable way. Further restrictions are caused especially by setup requirements. On the one hand, setups diminish the available capacity and thereby increase resource costs, resulting in higher capacity requirements. On the other hand, setups necessitate lot-sizing, which usually does not correspond with customer demand. *Lot sizes* and their sequence, the *setup sequence*, are both parameters to optimise production scheduling with respect to conflicting goals (Fig. 1.5). The setups generate a goal conflict between the necessity of producing a variant desired by the customer and the necessity to create appropriate lot sizes and maintain the setup sequence in order to have enough capacity available. Thus, production scheduling is internally influenced by three determining factors – availability, lot sizes, and setup sequence.

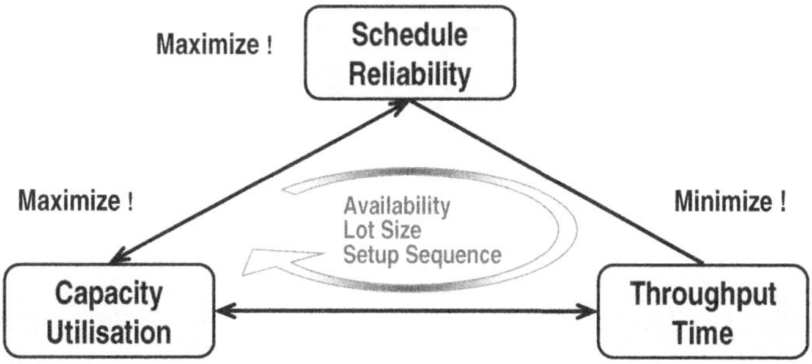

Fig. 1.5 Dilemma of production scheduling with conflict of goals and determining factors

Dilemma of Material Planning

The *dilemma of material planning* appears both in planning the material of semi-finished parts and final products as well as in the materials management for raw material procurement (Fig. 1.6). The objective of material planning is to reach high delivery capability, which means to be able to meet the customer's requested delivery date. Delivery capability is a measure of logistical process capability in factory, which has to be maximised (goal dimension 'quality'). This could be easily reached if all products would be held available as finished goods in the maximum quantity of customer orders. According to the multiplicity of final product variety, the capital tie-up costs and the need for storage space would be exorbitant. To avoid this, it is necessary to plan raw material and semi-finished parts. The required minimisation of inventory presumes a shortened replenishment lead time for suppliers as well as internal prefabrication. It is an important additional condition that delivery time must be short enough to meet market demand (goal dimension, speed').

Destocking as a result of shortened replenishment lead time requires more *frequent orders* in smaller order quantities with more frequent transports. More frequent orders, lead to increasing material costs. This is a result of higher administration expenses, smaller order quantities in smaller packaging, as well as more frequent deliveries. The task of procurement is to achieve low costs of material planning and procurement while order transactions become more frequent with smaller quantities. In the same way, placing more frequent production orders increases the internal costs of prefabrication. These costs have to be minimised by reducing controlling expenses, transporting distances, and changeover times (goal dimension ,economy'). Accordingly, scheduling aims to secure a factory's process

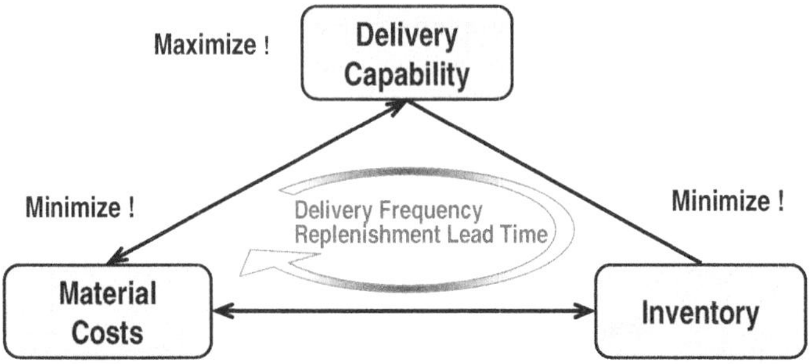

Fig. 1.6 Dilemma of material planning with conflict of goals and determining factors

capability with a view to material availability and/or delivery capability, it keeps procurement costs low and minimizes raw material and semi-finished goods inventories. Accordingly, the maximisation of delivery capability contradicts the minimisation of inventory as well as the minimisation of material costs (Fig. 1.6).

Material planning and management can be optimised by adjusting *delivery frequency* and *replenishment lead time* (Fig. 1.6). Increasing delivery frequency allows for placing smaller order quantities with suppliers and prefabrication, this being prerequisite for reducing raw material and semi-finished parts. The maximum level of delivery frequency is confined by order costs. Cutting down on costs incurred by an operation makes it possible to ease the goal conflict and improves the production process as a whole. Lowering replenishment time shortens the planning horizon so that procurement can respond quickly to varying customer demand. However, reducing replenishment time by increasing inventories at the supplier's to enable him to supply from stock at short notice is not sufficient, because sooner or later this will lead to higher purchase prices. Nor is it helpful, with lengthy replenishment times, to simply increase delivery frequency, as inventories would be increased by wrong forecasts, bearing in mind that there are usually multiple deliveries on the road.

Goal Dilemma of Production

The above investigation of the scheduling and material planning dilemma proves that the generally known goal conflicts in production are confined to the three major goals of quality, speed and economy. In sum, this can be visualised by a triangle scheme indicating the dilemma of the three traditional production goals (Fig. 1.7). In the following, the goal dimensions will be further analysed taking the example of a number of selected sub-goals of key significance, depicted below with their respective direction.

Quality. Here, a distinction has to be made between *technological* quality goals and *logistic* quality goals. Technological quality becomes noticeable in the output, i.e. the *product quality*, as well as in the processes, i.e. in *production quality*.

Fig. 1.7 Overview displaying the dilemma of the three goals of production

The difference refers mainly to when and where an error is detected. If this happens in the factory, it results in scrap or rework. Quality defects are handled directly on the shop floor. If a defect is detected after being delivered, then this impacts the customer's evaluation of product quality. This affects one of the success factors.

The logistic quality performance is mainly measured in terms of schedule adherence. Basically, four different dates can be distinguished: The *customer requested due date* stated in the order, the *delivery date* agreed in the order confirmation, the internal *completion date* planned on the shop floor, and finally the actual *shipping date*. Ideally, all of these four dates refer to the same point in time. If the dates deviate from each other, the deviations from two dates will result in separate quality performance measures. In industrial practice it makes sense to measure the following three performance indicators:

- *Delivery capability* indicating to what degree the customer requested due date corresponds with the delivery date confirmed for the desired delivery quantity.
- *Delivery quality* indicating the share of customer orders delivered in the right quantity without product quality defects (product quality) at the confirmed delivery date (delivery reliability).
- *Delivery reliability* indicating the share of parts and products manufactured in the right quantity and on schedule by the planned completion date.

Delivery capability is related to the logistic *process capability*, whereas delivery quality is related to the logistic *process reliability* (Wiendahl 2002). Both are market goals pursued by companies. Internal delivery reliability, however, is usually concealed from the customer, being only relevant to the scheduling of internal operations. In general, there is a difference between market and company goals. If possible, all quality performance measures relating to schedule adherence and absence of defects should be maximised.

Speed. For time-based production indicators, a similar distinction can be made between market and company goals. The former are indicated or measured as delivery times, the latter as lead times. With regard to market, the aim is to come up with competitive *delivery times*. Internally, this is accomplished through rapid order fulfilment, extending from order entry or order capturing to shipment.

This is to be distinguished from internal *production lead time*. It correlates with inventories, which accordingly, despite being cost factors, function more or less as time-based production measures. Delivery time and production lead time are to some extent independent of each other. While delivery time in comparison to lead time exceeds the latter because of the additional steps of administrative order handling, it is shorter than the latter due to the missing part of customer-independent prefabrication.

Another time factor is the *replenishment lead time*. It may refer to the length of internal customer-independent prefabrication as well as to the external procurement of raw material from suppliers. The latter directly affects the delivery time of customer-specific parts. In case of a sudden and strong increase in demand, even order-independent material may run out, with negative effects on delivery time. The internal replenishment time does not only depend on the length of the prefabrication time, i.e. the corresponding lead time, but also on the time required for scheduling and dispatching the relevant production orders.

The success factor of shorter delivery times can be supported by reducing the time-based measures mentioned above, especially by *speeding up* production, procurement and order fulfilment processes. If the time-based measures are minimised so that they fall below the value needed to achieve the requested delivery time, even the quality goal of delivery capability can be fostered despite limited capacity. This correlation points to the fact that the goals are interdependent and mutually affect each other, as described below.

Economy. The key market success factor – provided you could agree on the services to be ordered – is *price*. To make profits with competitive prices in global competition, it is vital to minimise all *costs*. Economy calculation, however, is a rather complicated affair and therefore will not be dealt with here. Only the three main factors of personnel, operating resources and material are briefly highlighted. It should be noted that not the absolute cost rates but the corresponding relative performance measures are key.

- Essential for an assessment of the economic efficiency of personnel are not the labour costs per man-hour but *personnel productivity*, e.g. measured as unit labour costs. It should be noted that investments also affect personnel productivity. Thus, fixed costs overlap with variable costs, though they cannot be influenced in the same way.
- Machine-hour rates are part of fixed costs but virtually depend on *utilisation*, which in turn depends on demand and the selected lot sizes.
- *Material costs* do not only depend on external factors, such as purchase price and transport costs, but also on internal procedures including ordering costs, which are related to order frequency. Production technologies also have an effect on the material cost per unit via the degree of material utilisation, e.g. in case of scrap.

Performance figures

Each of the three goal dimensions of quality, speed and economy can be distinguished in terms of *market goals* and *factory goals*. Market goals include delivery quality and delivery capability, delivery time and delivery price. Factory goals basically cover delivery reliability, lead time, replenishment time, productivity, utilisation, and material costs. These are some of the common key figures used to measure production performance, which can be extended according to the respective company circumstances.

A *performance measurement system* for a production system must set out from the factory goals with parameters being adjusted to market goals, and needs to be structured, of course, in line with the goal dimensions.

It has not been dealt with the mutual dependencies of the three classic goal dimensions. Moreover, the fourth goal dimension of variability needs to be added to the triangle scheme. This should be achieved in the following by constructing the logical square of goals.

1.2.2 The Logical Square of Goals

The fourth goal, which is about an increased variability of production systems, is rarely discussed in the technical literature on production organisation or explicitly mentioned in passing only. Thus another dimension is added to the goal dilemma of production described in greater detail above. In setting out the factory goals it is therefore necessary to consider the specific goal conflicts of the four independent and not interchangeable dimensions, and to take appropriate decisions. The following section will not only deal with the fourth goal dimension, but also with the course the individual conflict lines take between the goal dimensions in the expanded goal dilemma of production.

Variability

The market goal here is *deliverability*. This parameter indicates whether a product is deliverable at all, regardless of its momentary availability. While delivery capability informs whether an item is available on a specified delivery date or not, deliverability indicates whether, in principle, a customer need can be met and supplied, i.e. what the product range on offer looks like.

Accordingly, one factory goal must be the ability to cover the variety of products on offer. *Flexibility* and *mutability* ensure that the entire product range in its variety and extent of customisation can be produced. In comparison with the other goals, variability is of increasing importance to the business success. In many markets, companies can only maintain their competitiveness by attaching particular weight to this goal. Top quality is often considered a matter of course and only the ability to make tailor-made products that are made quickly available to the customer allows a company to distinguish itself from the product standard available all over the world with, for the most part, low profit margins.

Goal Conflicts

As already indicated above, the four goal dimension conflict with each other at varying degrees. The goal conflicts are more or less severe. Some goals can be more easily achieved than others; some goals are compatible to a certain extent; the attainment of others is completely incompatible. In sum, the possible relationships between the four goal dimensions can be distinguished into four types:

1. The *contradictory antagonism of goals* describes the strongest type of conflict. Here, an improved level of goal achievement for the one goal deteriorates that of the other goal.
2. With the *contrary antagonism of goals,* the goal achievement of the two opposing goals is irreconcilable. This somewhat alleviated conflict means that the attainment of the two goals cannot be improved at the same time; anyhow, the fulfilment of the one goal can be improved without negatively affecting the fulfilment of the other goal.
3. Some goals are basically easier to accomplish than others thanks to their lower implementational requirements. This *subordination of goals* is to be understood in a logical sense, not in terms of practical relevance.
4. *Compatibility of goals* exists if the levels of goal achievement for two goals can not be deteriorated simultaneously, if not explicitly aimed at. In this case, the two goals can be better accomplished independently.

Logical Square of Goals

Now the four goal dimensions can be arranged into a square. The four sides and the two diagonals add up to a total of six oppositions for the four goal dimensions. Each of these oppositions can be related to one of the goal conflicts described above. The relationships between the four goal dimensions and the relevant goal conflicts can be depicted by the *logical square of goals* (Fig. 1.8). As the arrangement of the four goal dimensions into a square depends on the respective type of mutual goal conflict, it is not arbitrary. This schematic depiction illustrates the different logical features of the goal conflicts in production. The general representation of the relationships makes it easier to find out what type of antagonism lies behind currently realised goal conflicts and to achieve at least partial compatibility. All six goal conflicts will be described in greater detail in the following section.

1. Quality & Economy: contradictory antagonism of goals
This conflict line indicates that the improvement of quality will lead to higher costs. Thus, enhanced product quality requires more precise and thus more expensive resources, more highly skilled and therefore better paid employees, and maybe additional quality assurance processes that are not for free. If one refrains from these additional expenses, direct scrap costs will increase for the supposedly higher quality requirements, as now fewer parts will be within the tolerance range. The frequently pursued goal of increasing economy with larger lot sizes to achieve better utilisation, however, will reduce logistic quality. Since larger lots lead to longer and more variable waiting times in front of the resources, the times spent

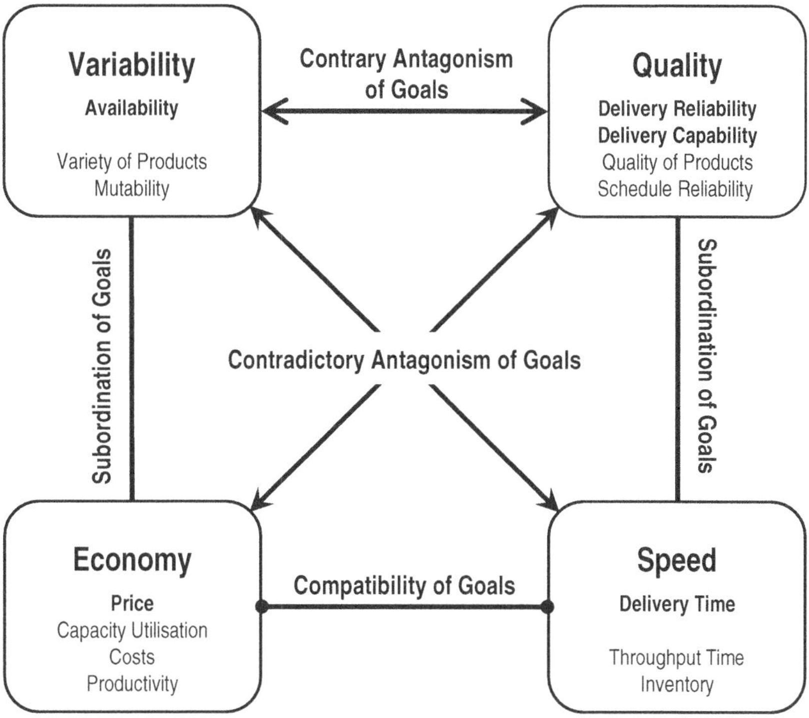

Fig. 1.8 The logical square of production goals with its six conflict lines

waiting for urgently needed parts will rise, thus reducing delivery reliability. The goal conflict can be resolved to some extent by trying to achieve a quality level in line with the market price.

2. Variability & Speed: contradictory antagonism of goals
This conflict line indicates that an increase in variability will lead to longer delivery times or higher stock levels. A simple example of this effect would be the Kanban supermarket warehouse in which all variants must be stored. Stock levels will rise with the number of variants and so will production lead time (Sect. 3.3.2). The shortest possible delivery time can be achieved by holding products on stock, since stock withdrawal is usually faster than manufacturing. Make-to-stock, however, is only possible for a limited product range; for make-to-order products this is utterly impossible. In general the following is true: The higher the number of variants and the more customised a product, the earlier in the manufacturing process chain must customer-oriented production start and the longer the customer must wait. The goal conflict can be resolved to some extent by moving the point of variant creation to the latest possible point in the manufacturing process. Similarly, if flexible machines are used for processing, then the whole process often slows down. The more tools are required and the more different work fixtures are used, the more difficult it will be to increase processing speed.

3. Variability & Quality: contrary antagonism of goals
This conflict line indicates that, on the one hand, an increase in variability makes it more difficult to meet quality goals and, on the other hand, an increase in quality requirements lead to restricted variety and flexibility. Thus, an increase in product variety due to customer-specific design adaptations, for example, brings about a completely new type of risk associated with unplanned delays. Accordingly, delivery reliability may decrease, but need not. Becoming better in this field than in series production is a rather implausible goal. By definition, quality problems due to slow or incorrect design adaptations cannot occur with catalogue products. In general, an increase in variety places higher demands on production systems, leading to poor parts quality, poor schedule adherence or poor delivery capability. The goal conflict can be resolved to some extent by slowly increasing requirements on the one side to enable the other side to improve alongside, i.e. not to fall behind or, if so, only marginally.

4. Variability & Economy: Subordination of goals
This conflict line indicates that most often it is easier to improve productivity and utilisation than to expand the product range. In general, it is easier to cut the manufacturing cost of a standard product than to make an existing production system more flexible in order to be able to manufacture a diversified product range. Increasing the adaptability of an existing production system is an even greater challenge, since it requires construction as well as changing the design of manufacturing resources. Conversely, a more flexible machine can also be better utilised than an inflexible machine. Accordingly, both goal dimensions can be fulfilled to greater extent, though being located on a completely different level of production design. At the same time, they will conflict with each other, as too much flexibility will, as a rule, make it inefficient.

5. Quality & Speed: Subordination of goals
This conflict line indicates that, as a rule, it is more difficult to raise quality than speed. Thus, by holding goods in stock it is easy to minimise the delivery time, the real challenge lying in the ability to deliver punctually in any case, typically having anything in stock except for the desired variant. Likewise, it is also easier to manufacture quickly than in good product quality; or to reduce production lead time by decreasing stocks than being capable to deliver at all times. If one develops a good strategy to manage the manufacturing process, it is also possible to improve delivery reliability by cutting down on lead time. Similarly, the quality of some production processes will rise, if execution is accelerated.

6. Economy & Speed: Compatibility of goals
This conflict line indicates that, as a rule, economy and speed can both be improved at the same time. For example, the reduction of setup time along with slightly reduced lot sizes, resulting in decreasing inventories would at the same time reduce lead time, increase machine utilisation and cut down on setup costs. Decreasing lead time inevitably saves associated inventory costs. To some extent both goal dimensions do positively correlate.

Logical Square of Goals

The logical square of production goals presented here shows how the four goal dimensions of production are arranged in six different types of *goal conflict* relationships. Improving the individual goals does not necessarily mean that another goal is affected to the same degree; some goals can actually be improved simultaneously.

The objective of production optimization is to counterbalance operation of production and the product range at a specific production site with the four goal dimensions in order to achieve the best level of goal achievement.

A closer look at the traditional methods of lean production and their objectives reveals their underlying goal systems. The respective goal hierarchies are going to be described in the following, broken down into logistical and production-related goals (Fig. 1.9).

The goal dimension of quality is easily the utmost priority of lean production. Quite rightly, rejections are considered the most severe technical kind of waste, as with one single process step they wipe out all previously generated value-adding activities as well as the utilized raw material. In cases of defects only detected in later process steps, subsequently created value is also lost. In addition, the production is disrupted as rejections are unavailable for customer orders at short notice. *Zero defect production* is thus of superordinate significance for a reliable low-waste production procedure. From a logistical point of view, another qualitative goal, namely delivery reliability is attached the highest possible priority. The consistent assumption of the customer's point of view as well as the entailing strict alignment of the production with the customer demand is one of the basic concepts of lean production (cf. Sect. 2.2). All other specifications are developed on this basis.

The speed-related goal of *short production lead times* typically associated with lean production comes second. Accordingly, inventories are considered the most severe logistical kind of waste, provided customer deadlines are met. From a production-related aspect, this aims less at an acceleration of machining processes in order to shorten operation times than more at a maximization of machine running times and entailing improved uptimes. The method of total productive maintenance (TPM) is of central importance in the context of lean production to reduce breakdown-related time loss and loss of speed. The thus desired increase in machine efficiency illustrates the importance of time management.

The goal dimension of economic efficiency ranks third. This may seem surprising, as even in a lean production, costs are no insignificant factor, and should definitely be kept at a minimum. Initially, though, this is indirectly achieved through prevention of non-monetary waste such as rejections, machine breakdowns and inventories, not with the aid of directly reduced cost rates. The third goal dimension aims to keep resource-related investment (and thus machine-related hourly rates) low in accordance with the principle of *low cost automation*. From a logistical viewpoint, this allows a lower weighting of the utilization-related targets in production planning, i.e. the forgoing of high machine utilization for the benefit of lower inventories and conformance with delivery dates. With respect to the workers, lean production similarly does not aim at lower wages, but rather wants to achieve high worker productivity on the basis of high employee qualifications and staff motivation. Contrary to resources, maximum utilization is desired here – not logistically, though, through personnel planning, but in a production process-specific manner by way of temporal alignment of all work contents, ensuring *optimal employee productivity* for each and every workplace. At the same time this enables a production-wide application of the operator balance chart known from assembly processes.

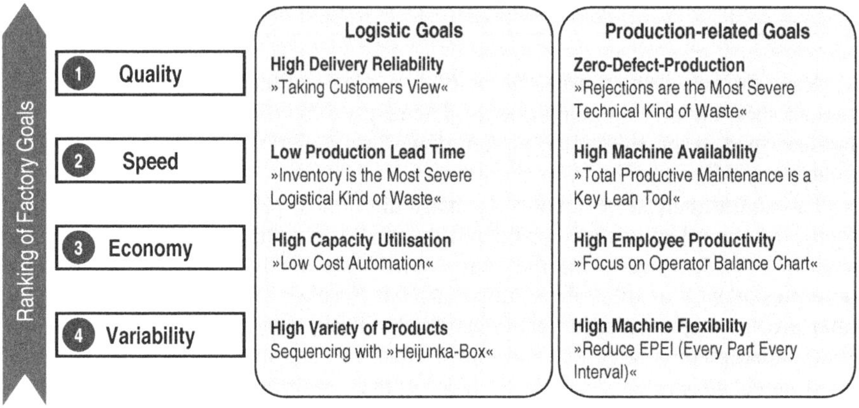

Fig. 1.9 The ranking of factory goals in a conventional lean production system

From a logistical point of view, the goal dimension of variability is implemented with the aid of the simple heijunka principle which enables easy sequencing of production orders of different product variants. However, this is only of limited benefit in productions of high variance – a weakness which leads to its rather *low prioritization* in lean production. As a result, this method for the levelling of the production mix has been significantly advanced in the book at hand (cf. Sect. 3.4.2). With respect to production-related goals, high machine flexibility is of vital importance for the achievement of a high product variance, as an indicator of which the so-called EPEI value (cf. Sect. 2.3.1) lends itself, though its low degree of popularity also somewhat reflects the low priority of this aspect.

The book at hand aims to illustrate how to design an optimal production. With the aid of value stream design, design guidelines will be suggested which shall enable the design of a production in such a way as to consistently operate the pertaining factory at a specific goal-related operating point. The traditional solution approaches of lean production were expanded in the design guidelines introduced herein – in particular with a view to the previously neglected goal dimension of variability (Chap. 3). First of all, though, the production needs to be mapped in its current state in order to clearly and appropriately visualize the starting point for an intended redesign and/or new design of the production, for the purpose of which the value stream analysis introduced in the following is the perfect tool (Chap. 2).

Chapter 2
Value Stream Analysis

The most popular field in German engineering is the in-depth optimization of individual production processes with a view to quality, reliability and output – vital factors of sustainable corporate success, as this is where specific technological know-how is generated and developed. However, being among the leaders – or indeed in the vanguard – of technological core competencies is not always enough to defy competition; the joint correlation of the different optimized processes is an equally important success factor. Accordingly, companies need not only technical skills but also workflow management competencies. Improvements achieved in individual production processes may easily go up in smoke unless planned and implemented with reference to the entire production process. Only when all production procedures are finely coordinated and appropriately linked logistically can a company be truly successful. The *value stream perspective* therefore concentrates on this superordinate aspect of correlation between the various individual production processes, thus moving the focus of attention towards certain crucial success factors in production, which otherwise might be overlooked.

Since a purely technological viewpoint only allows for highly selective inspection of a production, i.e. of individual production procedures, the examination of a production on the whole requires a more production procedure-related approach. The analytical findings of the former will achieve merely isolated improvements; only a holistic approach will enable an improvement of the entire production procedure. The latter, though, is the decisive factor for success, as isolated top performances will hardly improve a mediocre overall performance. On the contrary, the perspective has to be inverted. Based on the overall procedure of production the *requirements* for individual processes must be developed. This is very convincingly enabled by the value stream method.

Target-oriented current state mapping is an indispensable prerequisite for production improvement. Many analysis methods aim to collect data *as exhaustively as possible*. Unfortunately, however, these approaches are generally rather laborious with results of a seemingly higher precision than their actual informative value really lives up to. Besides, rather than giving a well-arranged overview of the overall production, they merely produce individual results condensed into graphs.

Another method for the correct current state mapping of a production is to depict all production procedures *as comprehensively as possible*. This, however, requires a systematic simplification of the recording process, summarizing differentiated data analyses in estimated empirical and/or mean values, thus considerably decreasing the relevant analytical effort and at the same time improving both

K. Erlach: Value Stream Design: The Way Towards a Lean Factory, LNL, pp. 25–96.
springerlink.com

clarity and informative value of the respective results. In addition, analytical findings are available much faster thanks to the reduced effort – it will take only days to complete analyses that would normally have taken weeks. With all these qualities, the *value stream analysis* is a truly outstanding analytical tool, the application of which is described in detail in this chapter.

The value stream analysis is *the* perfect method for clear and comprehensive current state mapping. This is mainly enabled by taking into consideration production processes as well as material and information flows, but also by visualization with the aid of simple symbols. The book at hand expands the renowned method of graphic symbols (Rother 1999), in particular with respect to the information flow, and supplements the same with easy methods of data analysis. This considerably increases the application range as well as the informative value without any significant increase in effort or difficulty.

Value Stream Depiction

In order to be able to map the various productive activities of a factory as comprehensively as possible, a suitable *modelling* is necessary. A suitable model is a model that depicts an item in a simplified and functional way. This is exactly what the value stream method accomplishes in a highly effective manner. The modelling of a factory's value stream is based on *six basic elements*, which may each be described by specific parameters and further differentiated by type (Fig. 2.1):

1. 'Production Process' stands for directly productive activities within the factory as well as external processing activities,
2. 'Business Process' describes order processing tasks incl. production planning and production control,
3. 'Material Flow' is the movement of materials between production processes incl. materials on hand,
4. ‚Information Flow' stands for the transmission of data and documents between individual business processes and towards production processes incl. data frequencies,
5. The 'Customer' reflects the customer demand to be met by the production, thus modelling the system load.
6. The 'Supplier' represents the production's supply with raw materials and parts.

The value stream flows from the suppliers through the factory to the customers, i.e. from the left to the right in the illustration below (cf. Fig. 2.1). Accordingly, downstream production processes are closer to the customer than upstream ones. The order processing-related business processes, the actual material flow in the factory and the entire information flow to all production processes together form the complete production logistics of a factory. The *logistic linkage* between two production processes reflects the material flow between both processes and their respective control logic. On a superordinate level we can therefore differentiate between production and logistics.

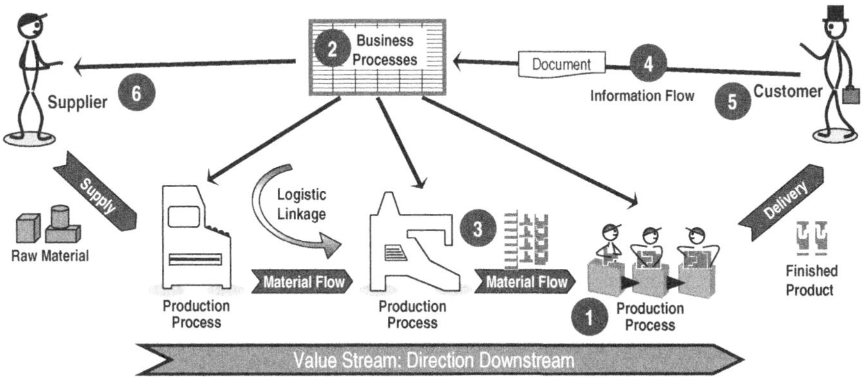

Fig. 2.1 Value stream in a factory

Based on said modellation, the value stream method enables a concise *visualisation* of the value stream. Each of the six model elements mentioned above is represented by a simple *symbols*. The value stream perspective concentrates on logistic linkages which are depicted by different symbols depending on their particular characteristics – thus making the typical specificities of a production procedure immediately identifiable. Significant properties can be made visible by adding further self-explanatory icons. The *value stream depiction* of an entire production thus consists of a drawing made up of symbols.

Depicting a production with the aid of simple symbols has two significant advantages: On the one hand, it allows for a very condensed depiction of the complete analytical findings; on the other hand the method of *value stream mapping* has several merits in its own right, which are described in detail in the following:

1. The simplicity of the symbols allows the creation of a value stream depiction on-site in the form of a pencilled sketch, so that the analysis actually maps the situation *prevailing* on the shop floor – the de facto *current state* – a significant difference to a data analysis conducted from outside the factory.
2. The visualisation accomplishes a well-arranged depiction of the entire production. Is it compact – and as a rule enables the depiction of all production processes from the supplier to the customer on a single DIN A3 (297mm x 420mm, or 11.69in x 16.54in, respectively) or similar sized sheet, providing much greater user *transparency* with a view to the conditions surrounding the production. At the same time, possibly unnecessarily high *levels of complexity* of individual production procedures are identified.
3. Depending on the type of modelling, the attention is drawn to different – usually averaged – significant *parameters*, such as operation times, setup times, and inventories. The structural approach of value stream mapping prevents losing focus in view of overly complex and detailed masses of data.
4. A numeric analytical result is generated by comparing the total operation time with the respective production lead time. The difference between the two points out great *potential for improvement* with respect to the overall procedure.

5. The intuitively understandable symbolism facilitates in-house communication with respect to current production procedures and their weaknesses as well as concerning approaches to the future layout of the production. It can also serve as a basis for *communication* with external partners as to a production's efficiency and implementational leeway the respective requirements thereon.

The Customer's Point of View

The basic concept underlying any value stream analysis is the consideration of the *customer's point of view, b*ecause it is the customer's point of view that determines the requirements placed on the overall production as well as on each individual production process. Starting from the shipping end, the customer's point of view should be carried upstream along the material flow up to the production processes. A value stream analysis is thus not conducted in the traditional direction, i.e. starting from the delivery, or receiving dock, respectively. Proceeding downstream intuitively seems more logical, probably because it reflects the principle of cause and effect. Yet, proceeding in the other direction, starting with the customer, the actual final cause for all production, has numerous clear advantages:

1. Any production only exists for the purpose of satisfying the customer's desires. Adopting the customer's point of view in an analysis and approaching the material flow from the back end allows us to better include the customers' requirements in our production.
2. Each inspected production process is itself a customer of an upstream delivery process. The analysis thus carries the customers' expectations through the entire production – every single step of the way. While the reverse method of approach gives rise to the question of how the customer process is going to deal with the delivered material, this approach asks what the delivery process needs to be like in order to ideally serve the customer process.
3. With multi-part or assembled products, the value stream branches out as we go upstream. The value stream analysis of any chosen manufactured product has its clear starting point at the shipping end, but at the receiving dock it is rather more difficult to decide which component or material to start the analysis with. At upstream forks, priorities can deliberately be set, but all loose ends will finally finish up at the receiving dock.
4. In a value stream analysis, the technical *'How?'* question, i.e. 'How is a certain part produced?' leads through the production like Araidne's thread. This makes things easier to understand for external visitors, as the individual results of the production processes make ‚How'-questions easier, since the 'Why' of the different technical steps is already known at this stage.

Approach

The *execution* of a value stream analysis for a manufacturing company consists of four steps (cf. Fig. 2.2):

1. **Deduction of product families.** A value stream analysis begins with the methodological classification of the entire range of products into product families. Based on differences with respect to the production process, product families are grouped together to be mapped in the same value stream. Companies using this method for the first time must then choose a suitable product family (Sect. 2.1).
2. **Customer demand analysis.** Next, the customer demand is modelled, ideally based on the previous business year's sales quantity. With this, all preparatory activities are complete (Sect. 2.2).
3. **Value stream mapping.** The actual value stream mapping is done walking round the factory. The value stream is depicted in its current state, mapping all significant parameters of the production process and logistics (Sect. 2.3).
4. **Potential for improvement.** The relation of operation times and production lead times and the coherence between individual production processes enable an evaluation of the quality of the respective value stream thus mapped , pinpointing potential for its improvement (Sect. 2.4).

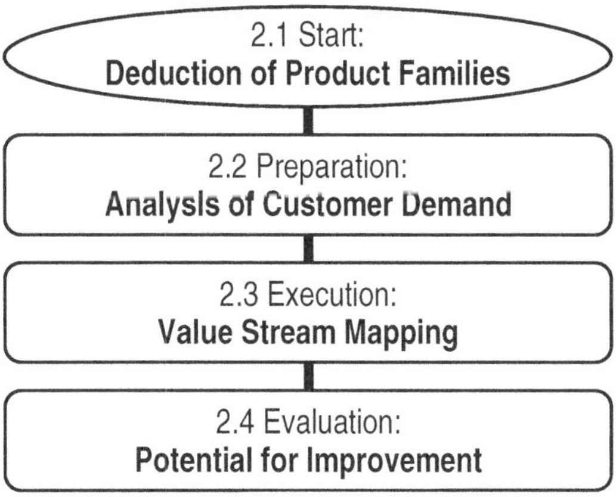

Fig. 2.2. Value stream analysis procedure

Case study – Liquipur plc. Many aspects of this method are better clarified by way of a practical example. The methodological explanations in this book shall therefore be accompanied by the value stream analysis of a relatively simple case study, based on a fictitious company called *Liquipur*, a name borrowed from Latin. At their Stuttgart location, the stock company develops and produces filter systems for both stationary and mobile applications, while a staff of approx. 1,800 generate a turnover of almost 300 million Euro.

Value stream analysis for the mapping of a production

Goal is the efficient mapping and clear depiction of the currently existing situation in a factory in order to gain a better understanding of the current production processes and to point out potential for improvement. An important factor in this context is the consistent assumption of the customer's point of view.

Approach The analysis is conducted via snap-reading method to represent typical factory conditions. In two consecutive runs, starting from the customer end, first the production flow and then the order flow are mapped by way of interviews, measuring and counting. The value stream with the most important production parameters is sketched on-site by hand. This method requires very little time effort.

The *result* is a transparent and well-arranged depiction of a complete value stream incl. production processes and material and information flows on a single sheet. This changes our perspective from the observation of individual processes and resources to the logistic linkage of the different production processes.

Advantages:

- The depiction of an entire production – almost like a bird's view – shows the overall context of all processes and prevents a reduction on individual isolated problem areas.
- The transparent depiction of technical processes and material and information flows with all significant parameters enables a comprehensive understanding of the production without fading out important aspects. No other method shows the production processes and their underlying information flows together.
- The fairly low level of detail shows the essential without any risk of getting lost in the data tangle of traditional analyses.
- Use of clearly defined symbols improves in-house communication concerning suggestions for improvement and objectives through visualisation of the effects of present conditions and intended improvements.
- The value stream depiction shows how well the different processes match each other and the customer demand, both with a view to layout and the factory as a whole, rather than reflecting seemingly positive appearances based on efficiency evaluations of isolated individual parameters without relation to the overall context.

2.1 Product Families

Before we begin our actual value stream analysis, we need to determine which product it is going to refer to. As a rule, a value stream applies to a combination of all production processes for one particular product. A different product may well follow a different path through production or may place different requirements on the same production processes and will thus create a value stream of its own. If we tried to map complex productions of products with different properties in one single value stream, very little would be discernible. The result would show a web of overlapping connections between production processes of different relevance, close to a factory material flow but without a factory layout's floor space information.

In the case of mass producers with only very few different products, each product may be viewed individually in a value stream analysis. In the case of assembly-to-order, however, mapping individual products would not be the best possible solution, similar products may be produced in similar production processes from mostly similar raw materials and the production procedures of such similar products are better joined in one value stream. Thus, any value stream analysis begins with the breakdown of the entire *product range* produced in a factory according to production-relevant criteria. This results in the classification of *product families*– not to be confused with the product groups defined in sales, though. Each of these product families is then mapped in a different value stream depiction.

Thus the first step of a value stream analysis is the *deduction of product families*. In a manner of speaking, a product family is a segment isolated from the rest of the factory to be observed on its own. A skilled analyser knows to place the dividing lines in such a way as to reduce the complexity of the interrelated, overlapping production processes in a factory without separating processes, parts and/or products that belong together. This is usually recommended to be done by a simple method called *product family matrix* (Sect. 2.1.1). This matrix regards each individual product, it is basically a mass data analysis and therefore only suitable for mass productions. The book at hand shall introduce a different method which requires very little effort and is much better suited for value stream analyses, because it does not start with individual product details (Sect. 2.1.2). Rather, it explicitly determines properties of *family likeness* based on general schemes of production procedure, based on which individual products may then be allocated to their respective product families.

Selection of product families

To conduct a value stream analysis for the first time, a suitable product family must be chosen. For this pilot project, we should not pick a very difficult one, but it shouldn't be too unimportant either. Ideally we recommend a product family with high lot sizes but few variants of not too small shares of sales. This offers the added advantage of a relatively easy initial application of the method paired with a substantial leverage effect of the achieved improvements.

2.1.1 Product Family Matrix

A seemingly simple and safe method to select product families is based on the so-called *product family matrix*. This matrix shows all products together with their applicable process steps. From there, all products which are subject to the same or similar process steps are grouped into families. In detail, the procedure is as follows:

1. All products, identified by their individual item numbers, are entered in the matrix one below the other at the beginning of the lines.
2. All production steps required for the entire range of products are entered next to each other at the top of the different columns.
3. All product steps required for the production of the respective items are marked by crosses in the relevant matrix cells.
4. All products with the same markings, i.e. which require the same production steps, belong to the same product family.
5. The principle of *similarity* allows for products with similar process sequences to be grouped into the same product family.

However, the definition of similarity remains rather vague– the decision as to which products are similar enough to be grouped together into one family must remain at the user's discretion and the determination of which process steps should be grouped together to form a production process is equally discretionary. If the level of detail is chosen in the same manner as for products, then all machines must be observed individually, only almost identical machines may be grouped together. For the determination of product families, however, the requirements on the respective machines are decisive. For instance, all machining centres with similar working areas will be grouped together, because they are best equipped to handle geometrically similar parts. Production steps on the other hand, need to be defined in greater detail than general production process descriptions such as 'milling', as otherwise the different requirements on the machines would be lost in grouping. Consequently, a *product family* in the sense of the matrix at hand includes all products which are produced in similar production steps *and* on mostly identical machines.

Item number	Drilling	Milling (vertical)	Milling (horizontal)	Milling (small working area)	Washing	Hardening	Pre-assembling 1	Pre-assembling 2	Pre-assembling 3	Assembling 1	Assembling 2	Assembling 3	
612680-20	x		x		x		x			x			PF 1
612842-20	x		x		x		x			x			
612682-20	x		x		x		x			x			
612682-23	x		x		x		x			x			
612861-10			x		x		x			x			
612684-10			x		x		x			x			
612685-10			x		x		x			x			
612839-20	x		x		x		x			x			
612234-10			x		x		x			x			
612688-10			x		x		x			x			
615386-10		x			x			x			x		PF 2 bus oil filter
615387-10		x			x			x			x		
615387-23		x			x			x			x		
615722-10		x			x			x			x		
615389-10			x		x	x			x			x	PF 3
612871-20			x		x	x			x			x	
615502-10			x		x	x			x			x	
615608-05			x		x	x						x	
615608-10			x		x	x						x	
615719-10				x	x				x			x	PF 4
612866-20				x	x				x			x	
615736-05				x	x				x			x	
...													

PF = Product family

Fig. 2.3 Excerpt from the *Liquipur* product family matrix

Case study *Liquipur* currently uses well over a thousand existing item numbers. For the sake of clarity, an excerpt with only 22 items is shown here (Fig. 2.3). In all, twelve different production steps are listed. The machining centres and milling machines are broken down into three groups, namely 'milling horizontal', 'milling vertical' and 'milling with small working area'. Pre-assembly and assembly each include three different work spaces with different appliances and different members of staff. The numbering of the assembly areas only serves for better differentiation; it does not reflect any sequential arrangement. The items depicted are grouped together into four product families. In-house personnel will know that product family no. 2 includes the four different types of oil filters for busses produced by *Liquipur*. Product family no. 1 consists of items with and without drilling, here we find similar process sequences. What is not apparent, though, is the fact that in product family no. 3 hardening is followed by another milling process using the same machine as the one used before hardening.

In the case of complex product ranges with many item numbers, though, above described method is rather laborious – and with several thousand items it is easy to lose track very early on. After all, similarities with respect to process sequences are not always clear from the beginning and what is more, the actual sequential arrangement of the different process steps can only be depicted in a very limited manner. Not all combinations in a table will be equally suited for a production-oriented sub-categorization of the entire product range. We are thus faced with the major disadvantage of not being able to explicitly define set similarity criteria in a product family matrix. In a way, said criteria are hidden behind the definitions of the various process steps and the possibility of differently marked cells within one product family. Imprecise criteria, however, may easily lead to mistakes and in order to avoid such mistakes, another method for the deduction of product families is introduced in the following:

2.1.2 Production Procedure and Family Likeness

Above described matrix creates groups *bottom-up*, i.e. from individual products and machines up to product families. The following family likeness method derives product families in the reverse order, i.e. *top down*, starting the first step with the functional modelling of production sequences and production procedure schemes. These are then specified and broken down with the aid of product- and part-related characteristics which serve to substantiate the requirements placed on the machines; the level of detail applied depends on the respective scope of products.

Production Procedure

As a first step, all production process sequences within the entire product range are each modelled as individual *production procedure schemes*. At this point we are still looking at the production process level, leaving aside resource-related aspects, as these specifications are not considered until we launch the second step. The individual elements of the production procedure schemes are thus kept more general than the production steps of the matrix. All existing production procedure schemes of the factory are determined. This first step aims at outlining the main differences between the various production procedures. In cases where only one production procedure of particular significance is depicted initially, items pertaining to other procedures may be left out at this point.

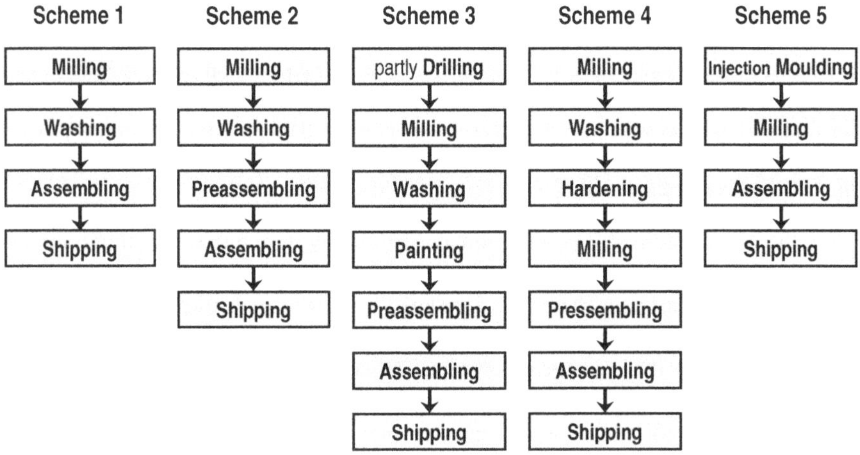

Fig. 2.4 *Liquipur* scheme of production procedures

Case study At *Liquipur*, five production procedure schemes of very different complexity can be identified (Fig. 2.4). The first – very simple – procedure concerns oil filters for passenger cars, the second one includes oil filters with casing for busses and trucks. The third procedure for instance applies to hydraulic filters with the partial drilling in the first production process being of little significance; therefore the two procedures are not differentiated. The fourth procedure includes two milling sequences. The fifth procedure finally pertains to filter systems with plastic casing.

Family Likeness

The production procedure schemes generally determine which technologies are required. However, the relevant technological requirements of a production procedure may differ considerably depending on the product to be manufactured. The *criteria* for the subdivision of the various production procedures may be chosen from below select list of relevant properties (Fig. 2.5).

The subdivision of production procedure schemes into product families is carried out with the aid of said production-relevant product criteria, or the requirements determined by the raw materials and/or parts used. Accordingly, we need to differentiate between aluminium, steel and carbon fibre reinforced compounds when turning, even in otherwise identical production procedures. The specific requirements, in particular concerning assembly, caused by the different levels of manageability of the various parts can justify the classification into different product families; further typical categorization criteria result from the geometry, complexity and/or functionality of the various products, where product families are then grouped together according to measurements as a size requirement for milling machines, volumes of injection moulding parts as shot size-related criteria for injection moulding machines, the number of mounted parts as a decisive factor with respect to logistic-oriented assembly, and so forth.

Main feature	Criteria
Raw material of parts	Process requirements (lubricant, tools, contamination, ...)
	Durability
Geometry of products	Weight
	Volume or greatest dimension
	Degree of customization (standardized, parameterized, free form)
Complexity of products	Number of mounted parts
	Sum of work content
	Change frequency of parts
Functionality of products	Degree of customization (standard, options, design modification)
	Number of possible variants
	Testing requirements
Handling	Managebility of products and parts
	Protection requirements of some parts (surface, cleanliness, ...)
	Value of parts (inventory management, anti-theft protection)

Fig. 2.5 Typical product and part criteria for the description of family likeness

Very often, production procedure schemes will vary very little within a given product range. Electronic manufacturing, for instance, includes repeated sequences of component placement, soldered or bonded, either on automatic placement machines or manual workstations. In this case, a differentiation into product families by production processes and/or machines is very difficult; the matrix method cannot really find a hold here. However, product families may very well be grouped together by product characteristics such as type or number of the required subassemblies, circuit board size, or possibly the overall size of the assembled end device.

Case study In the subdivision of the *Liquipur* production procedure no. 2, the criteria of raw material and product complexity play a significant role (Fig. 2.6). The first differentiation concerns the material of the processed parts: In order to keep lubricants and shavings separate and also for reasons of tool economy, cast steel parts should be processed on other machines than the usual aluminium die-casting parts. The latter even include parts which due to their geometry and the existing machines need to be mounted twice. Once again, product family no. 2 is the product family for oil filters for busses. However, the deduction of identical product families in different approaches is not the rule.

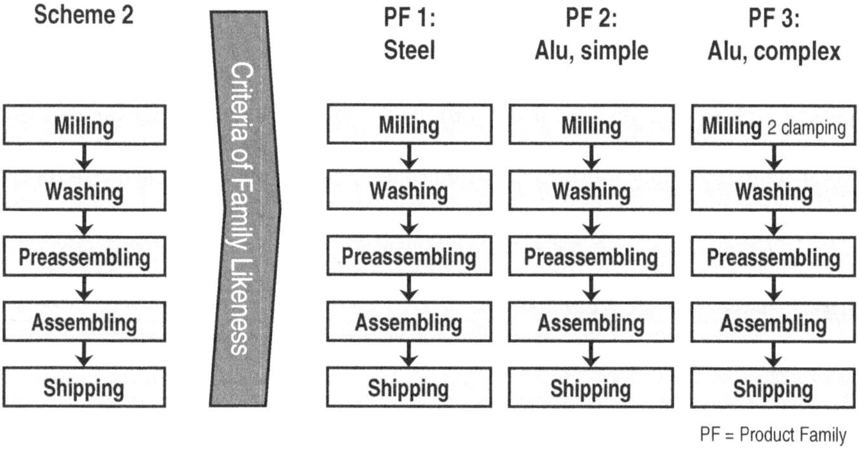

Fig. 2.6 Breakdown of a *Liquipur* scheme of production procedures into three product families

The *family likeness* within a product family is determined by production relevant criteria of products and parts. These classification criteria should be defined absolutely unambiguously, so that new products may easily be allocated an existing family or logically constitute a new product family.

Within any product family, we find so-called *variants*, which in the context of a value stream analysis are largely treated identically. In value stream mapping we concentrate on one particular item, which then represents the product family on the whole. Production processes and their pertaining times and other parameters are described on the basis of said *representative*. Each representative should feature the typical properties of the product family in question as well as a reasonable lot size. Should the variants differ considerably in certain aspects, however, it is also possible to quote the relevant scope of variant-specific parameters. Depending on the quality of the respective data, the calculation of a mean value from the previous year's product may generate a *virtual* representative of the product family.

Significant aspects for the deduction of product families

Goal is the transparent classification of a product range by product requirements.

Approach The deduction of a product family is conducted top down in two steps: Firstly, the relevant production procedure schemes are determined in accordance with their required production processes and sequences. Secondly, these production processes are broken down by specific properties of the processed raw materials, parts and products to identify uniform requirements on the machines. Please also note the following:

- The definition of any product family must be recorded in writing.
- Within one product family, all variants must be treated as one product.
- Within a product family, one product serves as a *representative* and as such supplies the specific data concerning times, measurements and other product-related parameters. The respective findings are then transferred to the other products in that family.

The *result* produces product families with homogenous technological requirements. Thus, a production is equipped with serial product characteristics which it does not normally seem to comprise.

Once a product family has been defined as to its constituting characteristics, we can focus our attention on the customer demand.

2.2 Customer Demand

Value stream design aims at customer-oriented production. The customer's perspective is therefore assumed even at the very start of the analysis and the *customer* is also the first element to be observed once the product family has been defined. As a consumer he is allocated the symbol of a house (Fig. 2.7). Underneath the customer symbol a data box shows all the relevant information needed to determine the stresses and strains placed on the production, calculated in customer takt times (Sect. 2.2.1). In isolated cases it may be a good idea to supplement the thus calculated mean stresses by an analysis of the fluctuation margin in customer demand (Sect. 2.2.2).

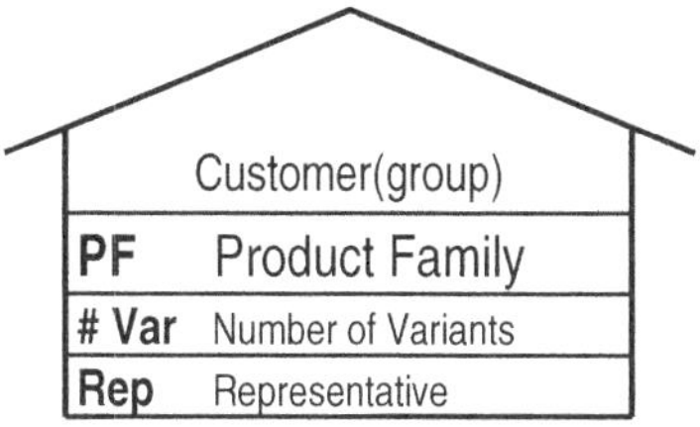

Fig. 2.7 Customer symbol

First of all, the customer(s) is/are entered in the *customer symbol* (Fig. 2.7). The customer in turn may be another factory where the products under observation are processed further. This is usually the case with the supplying industry, whereas the consumer goods industry generally directly supplies retailers or crafts enterprises. In the most straightforward case, i.e. mass production, there will only be one major customer whose name and delivery address are recorded. With growing variety or customer specificity of the products, however, the number of customers will also increase, so that the number of customers or the name of the respective customer group will be noted rather than the name.

In addition, the description of the respective product family (PF) is entered in the customer symbol. The number of pertaining variants (# Var) indicates a possible heterogeneity of the product family. The informative value of the value stream analysis depends very much on well chosen simplifications. Therefore not every single item number is observed, but the diverse variants of a product family are represented by a concrete product typical for the respective family. Unless stated otherwise, all later findings established in the value stream mapping then refer to that particular representative (Rep).

Different customers have different *delivery requirements* with a view to mode of dispatch, mode of transport, packaging, package sizes, delivery times or qualities (tolerated delay, permitted underdelivery). Much – such as the mode of transport and/or packaging – depends on the geographic location, i.e. wooden crates would be used for oversees shipping or a recycling system for containers could be introduced for nearby customers. Modes of transport like 'ex works', 'free domicile' as well as delivery sizes determine whether the transport is carried out by self-collectors, internal or external carriers, parcel services or company trucks with set routes. Finally, delivery requirements very much depend on how the customer demand arises – for instance whether a production is supplied just in sequence of if the stocks of a sales subsidiary need to be replenished.

All delivery-related logistical customer requirements described above can strongly affect the correct design of the production procedure. Within existing product families there may thus be a need to define additional *customer groups* ordering similar products at different conditions. Each of these customers must then be allocated a customer symbol of his own.

2.2.1 Customer Takt Time

The customer demand of a product family is annotated in a separate data box underneath the customer symbol (Fig. 2.8). As a rule, this reflects the *annual sales volume* of the previous business year, i.e. the annual number of items (Pcs.). Of course, other reference periods may be chosen, but observing an entire year automatically covers seasonal fluctuations, too. With next year's target quantity inserted here, the analytical findings indicate whether the production is well-equipped for future requirements.

In general, daily or monthly production quantities are used to gain information on the efficiency of a production. These figures are easily obtained from annual production numbers by dividing the same by the number of working days per year

Pcs.	Annual Piece Number
FD	Factory Days
WT	Working Time
TT	Customer Takt Time
DT	Delivery Time
DR	Delivery Reliability

Fig. 2.8 Data box for customer symbol

according to the factory calendar (factory days FD) as well as the daily effective number of working hours per production process – in the case of manual processes, rest times must be deducted (working time WT).

This approach, however, has an important disadvantage: It does not immediately supply any information as to the optimal working speed in which the work should be carried out in order to meet the mean customer demand. But the reciprocal value of the production volume per time unit supplies us with just that. It shows the rate of production by stating the time per item produced, based directly on the mean sales rate. This rate of production is referred to as *customer takt time* (TT) and is calculated as follows (Eq. 2.1). It is always stated spanning various customer-groups.

$$TT = \frac{\text{available working time per year}}{\text{customer demand per year}} = \frac{FD \times WT}{Pcs} = \frac{WT}{DD}$$

with: TT takt time [time unit/pcs.]

FD factory days [d/a] (2.1)

WT daily working time [time unit/d]

Pcs annual piece number [pcs./a]

DD daily demand [pcs./d]

Though mathematically very simple, this calculation is hardly ever carried out in practice. Knowing about customer takt times, though, helps balance the rhythm of production with the sales rhythm. Customer takt is the beat an ideal production follows. Once all company processes involved in order processing are tuned to this rhythm, the company fully meets the market requirements and its production is completely customer-oriented. On the one hand, this of course only refers to mean values. i.e. absolutely consistent demand, on the other hand it is hardly feasible on a consistent basis in view of prevailing technical and organizational restrictions. As a guideline for production design and layout as well as order processing, though, customer takt should not be underestimated. The customer model is thus completed by entering the takt time in the data box underneath the customer symbol.

The customer takt time adds a little 'market feeling' to the production by making the customer demand directly visible in relation to the production rhythm. Contrary to this, prognosis-based make-to-stock production with changing time optimized lot sizes hardly offers any conclusions as to customer demand and is generally highly intransparent. As a result of high inventories with different ranges of coverage and planned orders of different volumes, it is impossible to tell from the shop floor whether the production is too fast or too slow in relation to the customer demand. In addition, long pauses between lots are not visible on the shop floor; the respective demand seems to just 'disappear'. Solely an ERP enquiry can tell whether or not a demand is actually met – providing the data have been well maintained.

The customer takt time reflects the quantitative performance requirements on the production. The *time-related performance requirements* placed on the production by the customers are quoted in required delivery time (DT). This could for instance be a guaranteed 24h service or a delivery time as per catalogue with delivery confirmation on a daily or calendar weekly basis. Compliance with the delivery requirements is measured by the key figure of delivery reliability (DR). Depending on the reference parameters chosen (delivery date desired by customer or confirmed date of delivery) and the company-related and/or product family-specific range of tolerance, these figures will of course differ. Thus, the correct delivery date, but also the correct delivery week or an asymmetrical tolerance, such as three days early until one day late, may be considered as on-time delivery. In this context, comparability of actual and target values is of immense importance. All other delivery-related logistical customer requirements (shipping, packaging) are quoted in connection with the external material flow (Sect. 2.3.2).

Case study *Liquipur AG* would initially like to conduct a value stream analysis for their product family no. 2, i.e. oil filters for busses. Said oil filters are produced in four variants for twelve factories of different customers at an annual volume of 192,000. The individual variants' product volumes are 96,000; 58,000; 30,000 and 8,000 per year. The entire production runs three shifts on 240 days per year with an hour's break per shift. The milling machines continue to run during break times. In accordance with Eq. 2.1, this results in two different customer takt times (TT) for manual and automatic machines (for simplicity's sake the customer takt time is generally quoted without the reference parameter 'per piece'):

$$TT_{man} = \frac{FD \times WT}{Pcs} = \frac{240\,\text{d} \times 21\,\text{h}}{192.000} = 94{,}5\,\text{sec}. \quad TT_{aut} = \frac{240\,\text{d} \times 24\,\text{h}}{192.000} = 108\,\text{sec}. \tag{2.2}$$

Here the maximum output of the milling machines may be slightly lower than in cases where workers have to continuously feed and remove parts. Taking into consideration the manual tasks, one oil filter must be completed every one-and-a-half minutes in order to meet the respective demand.

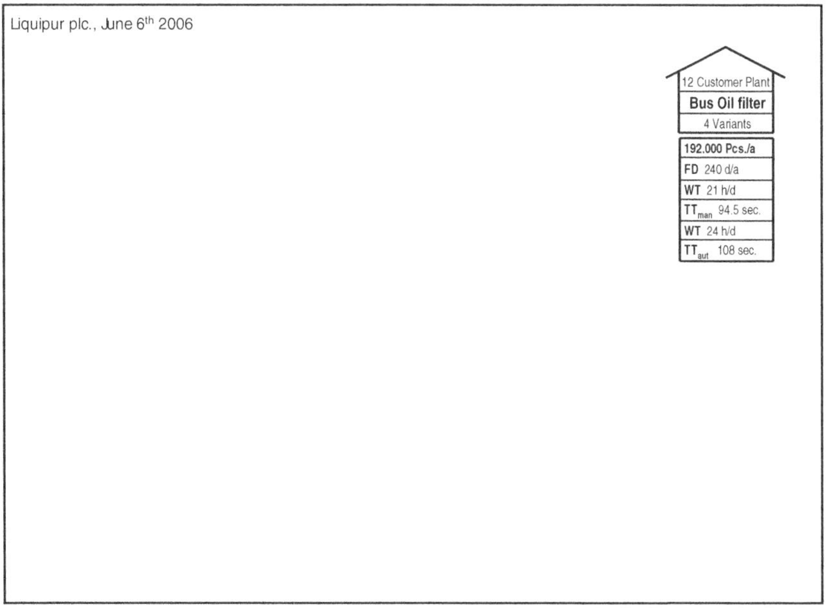

Fig. 2.9 Value stream mapping at *Liquipur* (1): Customer demand

As a rule, value stream mapping starts with entering the customer symbol in the right top corner of a *dated* DIN A3 or similar sheet (Fig. 2.9), thus ending the preparations in the conference room so that the walk around the shop floor to map the production processes may begin (Sect. 2.3). First of all, however, let us have a look at the fluctuations in customer demand.

2.2.2 Fluctuations in Customer Demand

Takt times offer a simple mean value analysis with a view to customer demand, supplying a fast and qualified overview without drowning the viewer in a flood of mass data. The performance takt to be kept by a factory is thus defined. At times, though, the order inflow from customer on the other hand can vary considerably. Based on the customer takt time, these *fluctuations in customer demand* determine the flexibility range a production should be able to cover. The following considerations show how to determine additional flexibility requirements placed on the production in a relatively effortless yet highly reliable manner. Flexibility requirements may be eased by stocks to cushion some of the fluctuating demand.

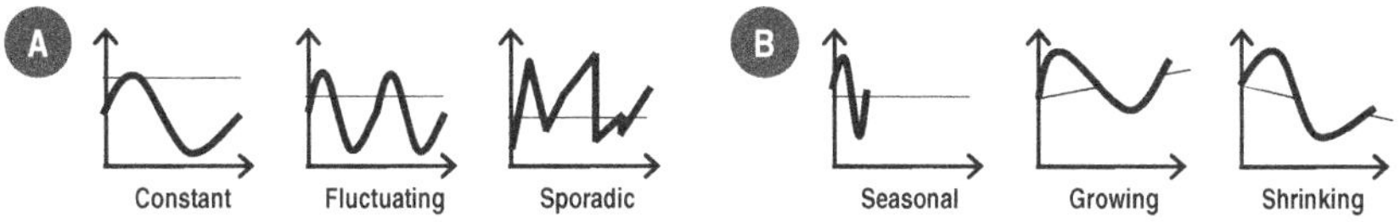

Fig. 2.10 Six types of fluctuation in customer demand

Fluctuations in customer demand are empirical values which depend on various different factors. The sales department for instance may influence customers' order patterns through temporary or volume-related discounts – and thus negatively increase the fluctuation range from the production-related point of view. On the other hand, using the degree of freedom in contractual negotiations sales department may consider the actual capacity situation in factory. Overall result is a *short-time,* more or less fluctuating customer demand somewhere between constant and sporadic conduct (Fig. 2.10, cases A). In addition, this may be overlaid by *medium-term* tendencies, such as regular seasonal fluctuations, which generally place very high demands on production flexibility, or by tendential developments with increasing or decreasing lot sizes according to related market tendencies (Fig. 2.10, cases B).

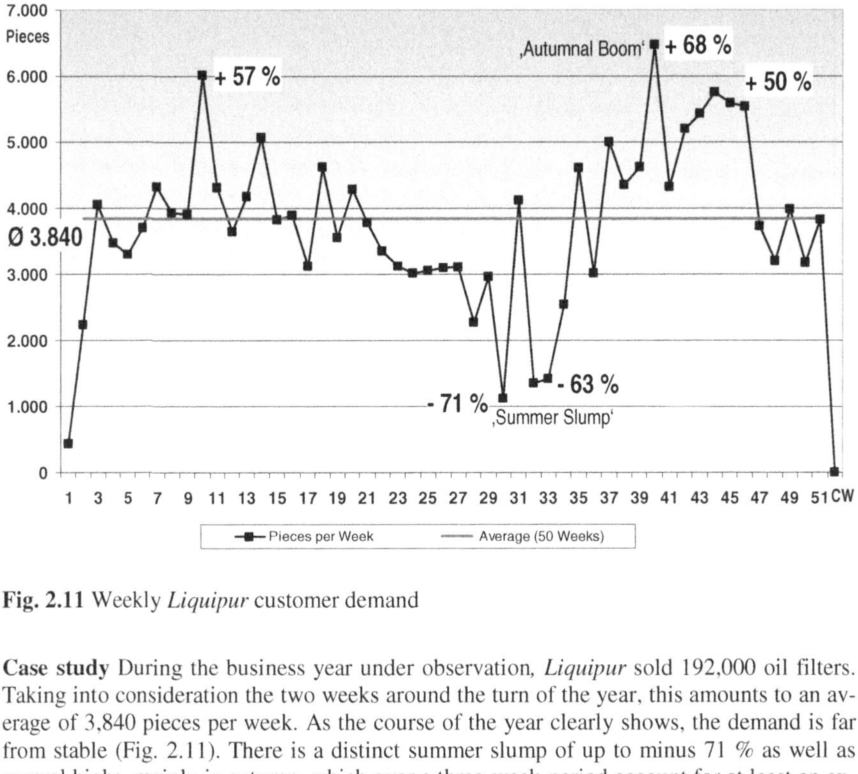

Fig. 2.11 Weekly *Liquipur* customer demand

Case study During the business year under observation, *Liquipur* sold 192,000 oil filters. Taking into consideration the two weeks around the turn of the year, this amounts to an average of 3,840 pieces per week. As the course of the year clearly shows, the demand is far from stable (Fig. 2.11). There is a distinct summer slump of up to minus 71 % as well as several highs, mainly in autumn, which over a three-week-period account for at least an extra 50% and for a good extra 40 % over another three weeks. The fluctuation margin to be covered by production here is thus between +68 % (6,468 pieces) and –71 % (1,116 pieces) on a weekly basis. Trying to meet these requirements solely through capacity flexibility would constitute immense implementational strain and excessive surplus capacity.

To supplement the above, let us have a look at the cumulative *theoretical inventory log*, which is based on the fictitious assumption of weekly production exactly equalling the mean customer demand and all customer demand being met and delivered exactly on time while the production volumes remain completely unchanged. As a result, each week's volume shows a positive or negative balance, which in turn is set off against the following week's result, depicting the fluctuations in customer demand over the course of the year. This inventory, calculated for stable mean production, may well – such is the freedom of mathematics – turn out negative results, indicating that the production would be lagging behind.

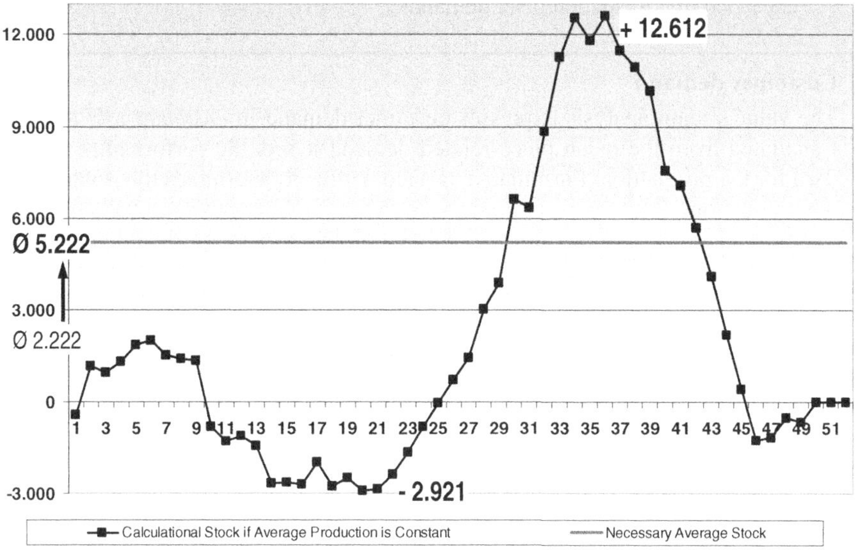

Fig. 2.12 Theoretical *Liquipur* inventory log

Case study At *Liquipur,* the excess demand of the first half year leads to a shortfall quantity of max. 2,921 pieces (Fig. 2.12). During the summer slump, a peak inventory is created to cover autumn's peak demand. The mean calculative inventory is 2,222 pieces. If we now – as in real life – consider negative inventories as not permissible, the opening stock for instance needs to be at approx. 3,000 pieces; the required mean inventory to fulfil all delivery obligations without capacity flexibility requires 5,222 pieces, equalling a mean range of coverage of 6.5 working days. Accordingly, the maximum inventory amounts to approx. 15,600 pieces; this corresponds to a maximum range of coverage of roughly one month. However, this does not take into account any possible product variety-related inventory increases.

In all, we can differentiate three approaches – and combinations thereof – to flexibly cover fluctuations in customer demand:

1. Firstly, an increase in demand may be made up for through inventories. In a manner of speaking, the production is separated from the actual customer demand and works on the assumption of a smoothed out demand. This is the typical ideal solution for productions with few variants, such as *Liquipur*.

2. Secondly, we may alter the capacities available, for instance through overtime. Workers from other, less busy areas may be transferred, or temporary staff may be employed short-term.
3. Thirdly, delivery dates may be deferred, delivery times may be linked to the respective demand. With customer-specific productions, this approach is sometimes inevitable, as stocks are not an option. Sometimes, dates must be changed if the planned capacity or inventory-related flexibility margin is exceeded due to sudden peak demands.

Customer demand

The value stream analysis expresses customer demand in *customer takt times*. Calculated from the mean piece-related demand it sets the performance standard for the production. Formulated as a clock rate, it determines the rhythm of the production.

Fluctuations in customer demand call for stocks or capacity-flexible production. The range of fluctuation with respect to mean customer takt times is concluded from a data analysis or determined as a target value, the production-related *flexibility range* is adjusted accordingly. Higher fluctuations in customer demand lead to delays in the fulfilment of the customer demand.

2.3 Production

Once the product family to be observed has been determined (Sect. 2.1) and the customer demand has been duly analysed (Sect. 2.2), the actual value stream mapping of the production may commence. The current-state mapping is always conducted *on-site* in the factory, i.e. on the shop floor at the machines and in the foremen's and the production planning offices. The required tools are DIN A3 or similar paper, pencil, eraser and stop watch – and a pocket calculator afterwards. Value stream mapping is conducted in two individual runs:

1. During the first run, we map the production processes and their connecting material flow. We begin at the interface between production and customer, usually at the shipping end, so that the value stream may be traced from the customer's point of view all the way to its source at the receiving dock.
2. In the second run, we record order processing-related business processes and the pertaining information flow. We begin at the interface between order processing and the customer, usually at the point of order acceptance, from where all work places generating and handling production documents are accessible step by step.

Once all employees concerned and the works council (if applicable) have been informed, you work your way through production and order processing step by step, asking the following *four key questions*: In order to remain as close to reality as possible and to reflect the actual circumstances prevailing at the various places of work as precisely as possible, it is vital to speak directly with the foremen, masters and employees involved in the production processes and working in the production planning offices.

1. What are your tasks and duties?
This question aims at process descriptions in production and order processing. With production processes it is advisable to always map entire work cycles and to time them yourself. Standard times on file do not always coincide with the actual conditions on-site. They may be too short, having been taken under the pressure of daily business, or they may be too long with additional planned times and/or partial changeover times included; the actual production procedure may meanwhile have changed and been improved. The times determined by our stop watch are random samples to give an overview of the way the production processes work; they cannot substitute well-established time recordings for the purpose of wage determination. A digital camera serves well to record other impressions of the production conditions, such as tidiness and cleanliness, the visualisation of allocated areas, processes and key figures, or possibly the transparency of the shop floor. As far as the business processes are concerned, the time aspect is not so important at this point, rather the logic sequence is observed.

2. How do you know what to do and when to do it?
This question aims at the control information available in production, to be found for instance in order lists, production orders, work process sheets, stock withdrawal documents, commission lists, or maybe online in the form of planning and control software or spreadsheet programmes. Copies of the various production documents used and attached to the value stream depiction can be very helpful. Often we find informal flows of information, which partially serve to smooth out deficiencies in the defined procedure. These are also to be included in the value stream analysis. With business processes, the trigger mechanisms for individual steps must be enquired.

3. How large is your inventory, or work basket, respectively?
This question determines inventories in the material flow by counting the number of pieces stored or buffered between two production processes. Please note that there may be different storage locations for identical stocking levels. Physically walking to all these storage locations will give a better overview than merely checking inventory management data. Besides, discrepancies between actual and electronically recorded inventory are quite frequent. Accordingly, the number of ongoing operations per business process needs to be determined for the information flow.

4. Where do you get your required material and/or work orders from?
For this question an in-house guide will take the value stream mapper (who may not be familiar with the factory) to the next work area, upstream production process or downstream business process. Please note that this question may have to be asked rather insistently, as some side arms of the flow in question may be known to few persons only and will only surface upon explicit enquiry.

Value stream mapping is like taking a *snapshot*, i.e. the questions aim at depicting the concrete production and order processing conditions at a certain point in time

– which could turn out poorly chosen if the situation proved to be an exceptional one. Experience shows, though, that this is rarely the case, even if people like to speak of exceptional situations, maintaining "it's never normally like this' etc. A repetition at another time generally confirms the findings, the so-called exceptions then usually occur somewhere else. The answers to our four key questions supply the information needed for a snapshot mapping of the value stream.

Our DIN A3 or similar sheet is usually big enough to record the information thus gathered in a structured and well-arranged manner with the aid of the symbols described in the following. During the course of the mapping process, repeated corrections of omitted special cases or misinterpreted details will become necessary. Certain things will need to be analysed in greater detail than initially expected. This is why the corrigible sketch developed on-site in pencil has well proved its worth – a good eraser will sometimes prove your most important tool in value stream mapping. The central result of a value stream analysis therefore is a *value stream drawing* with symbols. For the sake of documentation, each value stream analysis should be dated. During the course of a restructuring project, this piece of paper with the value stream depiction will remain your constant companion, as an aid to memory as well as to record additional details which only turn out to be of importance during the course of your project.

The symbols required for a value stream depiction and their pertaining parameters are successively explained in the following sections by way of the six basic elements of value stream depiction, of which the 'customer' has already been explained (Sect. 2.2):

1. **Production processes.** A production mainly consists of machines, equipment and work stations which convert the raw materials to products. In the value stream analysis, these machines are functionally modelled as production processes (Sect. 2.3.1).
2. **Material flow & suppliers.** In order to process any material at all, this must be transported to and from the various work stations. The entire transport and movement-related linkage of all machines is functionally depicted as a linkage of the production processes. The external material flow from the supplier needs to be recorded here as well (Sect. 2.3.2).
3. **Information flow & business processes.** To ensure that the material moves in a controlled manner rather than randomly, information is needed as to which material needs to be taken where and how then to process it. Contrary to the classic approach of product visualisation – layout and material flow depiction – here we also extensively map information flows and their pertaining business processes for order entry, production planning/control and procurement, here summarized as order processing (Sect. 2.3.3).

2.3.1 Production Processes

In the value stream analysis, productions are depicted as successions of production processes. The symbol used to depict a *production process* is a simple rectangle, at the top of which the process denomination is entered (Fig. 2.13, case 1). For the sake of better differentiation, though, we should not simply just it 'process'. As a rough capacity indication, the two most important basic parameters of the respective production process are also recorded. Experience has shown that it is a good idea to also mark these two parameters with little icons:

1. The number of operators allocated to this process per shift. These are stylised as a head with arms.
2. The number of alternative machines available for this process, here referred to as resources. Resources are symbolized by a small vertical rectangle.

Strictly speaking, a production process is a *technical process*. Materials and parts are subjected to various types of processing and thus transformed in a value adding manner. In particular, this includes packaging, assembling, coating, thermal treatment and manufacturing. Ancillary activities like sorting, commissioning, marking and/or inspecting do not directly serve the fabrication of the product as desired by the customer. However, these *logistic processes* are also depicted as production processes. They include shipping, quality inspections, functional inspections, commissioning and the receipt of goods. Activities which merely result in a change of location with respect to materials and parts, such as transfer to and removal from stock as well as movement and transport are generally allocated to the material flow and are not modelled as independent process boxes.

An *external production process* constitutes an exception, though (Fig. 2.13, case 2). In this case, the value stream leaves the observed facility for a particular process step. The respective processing is then carried out by a supplier who as a contract manufacturer offers various technologies not available at the factory in question. Acting like an extended workbench, the supplier complements existing technologies. An external process consists of both technical and logistical parts; basically, it's an entire value stream of highly compact depiction of its own, which is treated like a 'black box'. Its key parameters are the name of the supplier and the entire production lead time, measured from readiness for dispatch until completion of receipt of the goods after return delivery. In our symbolic depiction, the process box is supplemented by the stylized shed roof of a factory building. Instead of the in-house production processes' capacity information, the respective lead times are recorded. External production processes resemble material flow elements.

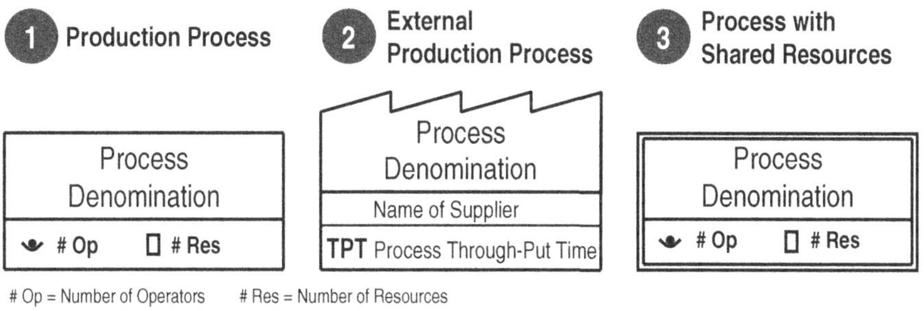

Fig. 2.13 Symbolic depiction of a production process

Certain production processes are not exclusively available to one value stream (Fig. 2.13, case 3). They include resources which also handle other products pertaining to product families not immediately observed at this point in time. This is immensely important in the capacity-related evaluation as well as in the later future-state concept. Therefore process boxes of production processes containing *resources jointly used* by several value streams are marked by double line borders.

When modelling a production process, we are faced with the question of which activities to allocate to a certain process step and which activities to depict in the previous and/or the following process steps. The desired *level of detail* in value mapping is the decisive factor when it comes to demarcating individual production processes. Production processes consist of a succession of activities. The smallest parts of such activities are single movements – such as the taking hold of a screw – which may each be timed individually. Several of these partial activities together form a complete activity unit – for instance 'positioning and screwing a lid onto its casing'. An activity unit corresponds to the finest possible level of detail that can be depicted in value stream mapping. Mostly, however, it makes more sense to map production processes in less detail.

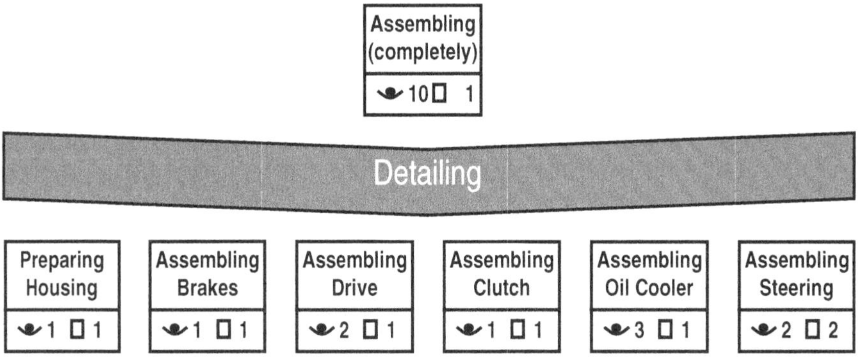

Fig. 2.14 Scaleable level of detail in production processes

Accordingly, an entire final assembly process – confer 'Assembling (completely)' above – could be summarized in one process box (Fig. 2.14). All ten operators working at various successive assembling stations are allocated to this final assembly process. The above example of a gearbox assembly includes six stations, the last of which has been doubled due to excessive workload. Each of the six production processes is shown with their respective work contents and pertaining process characteristics.

Parameters and Figures to Characterize Production Processes

In a value stream analysis, production processes are described by four different time types: operation time, processing time, changeover time and cycle time. The first three are parameters to be measured – or retrieved if measuring would be too time consuming due to the nature of the process in question. Cycle time is the process performance figure derived from the former.

The *operation time* (OT) indicates how long a particular part stays in the production process (Fig. 2.16, case 1a). This includes both the operator's manual work and the machine operating time. In the simple and straightforward case of one person operating one machine, the operation time equals that particular person's basic working time (Fig. 2.15). This basic 'Human Being' time is composed of the actual operation time of the worker plus waiting times resulting from scheduled breaks occurring while the operated machine runs automatically. In a milling process for instance, the basic time results from the time needed to mount the respective part, start the machine via program fetch, mill and retrieve the part and finally debur it. CNC milling machines the operators will incur waiting times – except in cases of multi-machine handling. Said waiting times could of course well be used for deburring parallel to the primary operating time. At conventional milling machines, however, waiting times are virtually non-existent.

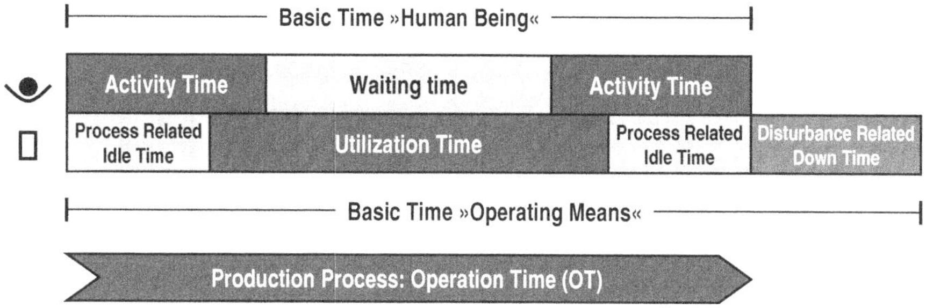

Fig. 2.15 Operation time in relation to different time types (REFA 1992)

The utilization time of a machine, which is divided into main and ancillary utilization time, includes changing of tools, clamping, processing and retrieving. If the machine has to wait for the operator, we speak of process-related idle times. A disturbance-related idle time, however, is determined via reliability data (cf. Fig. 2.19), therefore the machine's utilization time and basic time only play an indirect role in a value stream analysis.

The employee's working time consists of principal and ancillary activities for the indirect or direct performance of work tasks. The actual processing time spent on the transformation of the product into exactly what the customer wants is the time spent on principal activities and is referred to as *value-added time* (VAT) (Fig. 2.16, case 1b). Process-related ancillary activities and waiting times constitute more or less superfluous *ancillary times* (AT). The allocation of individual activities within the production process to said time types depends on detailed time recording with waste analysis, which may be conducted in connection with a value stream analysis for the purpose of process optimization.

The *processing time* (PT) indicates how long individual parts remain in the respective production process. If one part is handled per process, then the operation time equals the processing time – in which case the latter is not even stated. It becomes more interesting, though, as soon as several parts or materials for several products are processed together in one process. We differentiate two types of processes: In the continuous process, parts are fed and retrieved individually (Fig. 2.16, case 2a), so that – other than in sequential lot processing – several parts are in process at any given time. In batch processing, however, parts are grouped together in batches to be processes simultaneously (Fig. 2.16, case 2b). A continuous furnace for heat treatment would be an example for the first case, the washing or hardening of parts in baskets would be an example for the latter.

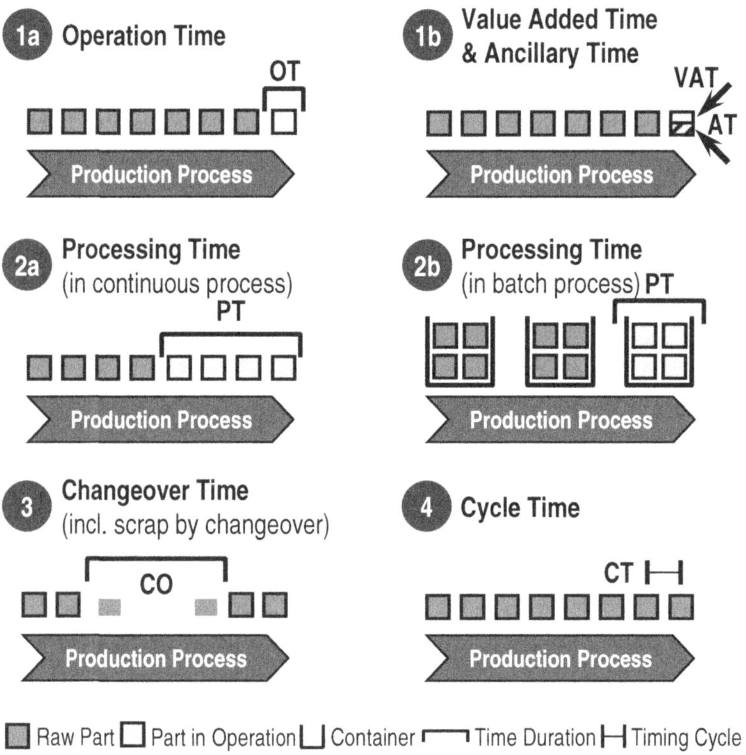

Fig. 2.16 Schematic diagrams of the different time parameters

Changeover time (CO) is the time during which a machine is unavailable for processing due to a change of fixtures, tools or materials (paint, coil, granulate) for a new variant (Fig. 2.16, case 3). i.e. the time is recorded from the last good piece of the previous variant to the first good piece of the following variant.

The *cycle time* (CT) is an important specific figure for the value stream method. It indicates how much time is needed to complete a part or product in a production process (Fig. 2.16, case 4). If there is only one machine, then the cycle time is identical with the operation time. In the case of more than one alternative machine involved in a production process or in cases of continuous or batch processes, the cycle time needs to be calculated (cf. Eq. 2.3). In clocked facilities, such as assembly or placement lines for printed boards, not only operation times but also *machine takt times* can be measured at the individual stations. This value is then incorporated in the cycle time which will only coincide with the measured machine cycle if there is only one single piece of equipment. In addition, it is well worth measuring the individual stations in order to determine the slowest one. The longest operation times of the various pieces of equipment should be in tune with the cycle time.

In a data box similar to the customer symbol, each production process is now provided with its fundamental *parameters* and their respective *key figures* for process description. The general idea behind the determination of parameters is to record the actual conditions prevailing in the factory, which is also why we should never transfer standard times from work sheets to value stream drawings. The value stream mapper personally measures the operation times with the aid of a stop watch. This is not done to aim at maximum precision or even to lay the foundation for a new pay system. Instead, this is done to better evaluate the efficiency of the various production processes, by absolute numbers well as in relation to each other and with a view to customer takt times. This works a lot better with time measurements carried out personally, and contrary to abstract times transferred from elsewhere, here you know exactly which concrete activities are included in the respective production processes. Detailed, methodically verified time recordings are not required until the individual production processes are designed in connection with the future-state concept (Sect. 3.2.2).

The production process data box described in the following sections is considerably more extensive than normally to be found in relevant literature. This is mainly due to the fact that for the first time we also observe the production of variants in standardized data fields. Besides, practice has proved the relation between operation and cycle times of a process to be quite complex at times. Additional production process parameters are therefore recorded for the sake of better data comprehensibility and readability.

All in all, we can divide the *data box* into three sections, each of which summarizes the respective value stream-specific information in a key figure as follows:

1. Information for the determination of cycle times as the key figure for *process capacity*.
2. Information for the determination of the EPEI value as the key figure for the changeover time-related and availability-specific *variant flexibility* of a production process.
3. Key figures for the evaluation of the *quality* of a production process and information for the determination of the process-specific customer takt time.

These three data blocks are going to be explained in more detail in the following. For better in-house communication and also as an aid to memory it is advisable to enter not only the verbal process description but also the name of the respective piece of equipment in the data box (Fig. 2.17, first line).

Cycle Time

The cycle time contains information on the available capacity of a production process. The first figure to be entered in our data box is the basic value of a production process per se, i.e. the operation time (OT), which reflects the *time effort* of a production (Fig. 2.17). When measuring time on-site with a stop watch, we do not differentiate value-added and non-value-added time fractions; we leave that up to a data analysis. Ancillary times not directly related to the processing of an individual piece but conducted jointly for several products are allocated to the changeover time (cf. Fig. 2.19).

For several parts simultaneously being involved in a production process, the operation times are best recorded individually as processing time (PT). As a rule, processes working with *batches* or in *continuous flow* are those which in a broader sense alter the material properties. In production engineering, these are processes which serve to change the material properties, such as coating or cleaning by heat treatment, varnishing or sandblasting. In process engineering, the corresponding processes would be mixing, filtering, distilling and/or alteration of the material in question in reactors. All these processes generally include considerably longer residence times than manufacturing processes like primary shaping, forming, separating and assembling. Accordingly, the processing times of production processes in process engineering are only conditionally comparable to operation times of manufacturing processes and should therefore be recorded separately. Thus, value stream analyses may also be conducted in the process industry. However, in addition to the processing time, the number of parts included in a batch must always be recorded stating the process quantity (PQ). In hardening, for instance, this may be the number of parts per basket, in the case of a continuous furnace the number of parts simultaneously treated in the furnace.

	Name of Equipment
OT	Operation Time
PT	Processing Time
PQ	Process Quantity
# P	Number of Parts per Product
CT	Cycle Time

Fig. 2.17 Data box, section 1, with information for the determination of cycle times

Some appliances (for instance on a milling machine) or tools (e.g. in injection moulding) allow *simultaneous processing* of several parts. This gets even more complicated when there are various alternatives for one part (like double or quad cavities in injection moulding). In this case, it is generally best to record the process volume with respect to the operation time, as this does not constitute a batch production with its pertaining specific characteristics. This type of multi-access machining usually requires rather a lot of space and is thus highly dependent on the respective variant.

Some production processes generate several *identical parts* for each end product (# P), the growing number of which will increase the time required for the respective overall operation. Sometimes various different parts may be fabricated in one production process: In this case we simply add all operation times and record the respective number of parts in the data box.

Based on said production process data, the *cycle time* is determined – an important interim result for the value stream analysis. In the most straightforward case, the cycle time equals the operation time of the process. If, as shown by the respective entries in the process box, several alternative resources are available, the process turnout will be higher, as several parts can be processes simultaneously. This in turn reduces the cycle time, which is calculated by dividing the operation time by the number of resources (Eq. 2.3, left). If one fitter takes ten minutes to assemble a product, then one product will be completed every ten minutes. Doubling the number of resources halves the cycle time. In our example two fitters would then complete one finished product once every five minutes. The cycle time of a continuous or a batch process is the quotient of processing time and process volume (Eq. 2.3, right). In a continuous process, the output takt time exactly equals the cycle time, here we also need to take into consideration a possibly higher number of identical machines. For two continuous furnaces with 100 parts each and a processing time of 200 minutes, the cycle time accordingly amounts to one minute. All in all, the cycle time is calculated as follows:

$$CT = \frac{OP \times \#P}{\#Res} \quad \text{oder:} \quad CT = \frac{PT \times \#P}{PQ \times \#Res} \tag{2.3}$$

with:
- CT cycle time [time unit]
- OT operation time [time unit]
- PT processing time [time unit]
- PQ process quantity for batch or continuous processing [pcs.]
- # P number of identical parts per finished product [pcs.]
- # Res number of same resources [pcs.]

Cycle time

The *cycle time* reflects the efficiency of the production process in time units in continuous operation without changeover- or disturbance-related interruptions. The cycle time describes the *available capacity* of a production process and shows which customer takt time may possibly be undercut yet in order to meet the respective customer demand.

Changeover Time and Lot Size

Changeovers required in the fabrication of variants or customized products reduce the available capacity of the respective production process accordingly. As the cycle time does not make any allowances for such limitations, we shall now determine the same with the aid of the key figures shown in the second section of the data box.

First of all, the time required for a changeover, the changeover time (CO), is entered in the data box (Fig. 2.19). The changeover time, however, varies with different variants. Optimal changeover sequences can be determined by way of a *changeover matrix* (Fig. 2.18). The highlighted and framed areas in the figure indicate which variants are best handled successively. Please note that our matrix is not symmetrical; i.e. a change from the VZ 10 type to VZ 5 may be more complex than the other way round; the first changeover would take 240 minutes, while the latter would be completed in only 140 minutes. The best option is to enter the mean value resulting from both planning and control logic of the respective current state. Depending on the level of optimization with respect to the various lots' mounting sequences, the resulting changeover times to be entered will have been minimized to a higher or lesser degree.

to (columns) — from (rows) — changeover times in minutes

variants	VG 01	VG 02	VG 03	VG 04	VG 05	VG 06	VG 07	VG 08	VG 09	VG 10	VG 11	VG 12	VZ 01	VZ 02	VZ 03	VZ 04	VZ 05	VZ 06	VZ 07	VZ 08	VZ 09	VZ 10	VZ 11
VG 01		130	130	130	130	130	130	130	130	130	130	130	135	135	135	140	140	145	145	160	160	240	240
VG 02	130		130	130	130	130	130	130	130	130	130	130	135	135	135	140	140	145	145	160	160	240	240
VG 03	130	130		**35**	**35**	130	130	130	130	130	130	130	135	135	135	140	140	145	145	160	160	240	240
VG 04	130	130	**35**		**35**	130	130	130	130	130	130	130	135	135	135	140	140	145	145	160	160	240	240
VG 05	130	130	**35**	**35**		130	130	130	130	130	130	130	135	135	135	140	140	145	145	160	160	240	240
VG 06	130	130	130	130	130		**35**	**35**	130	130	130	130	135	135	135	140	140	145	145	160	160	240	240
VG 07	130	130	130	130	130	**35**		**35**	130	130	130	130	135	135	135	140	140	145	145	160	160	240	240
VG 08	130	130	130	130	130	**35**	**35**		130	130	130	130	135	135	135	140	140	145	145	160	160	240	240
VG 09	130	130	130	130	130	130	130	130		**35**	130	130	135	135	135	140	140	145	145	160	160	240	240
VG 10	130	130	130	130	130	130	130	130	**35**		130	130	135	135	135	140	140	145	145	160	160	240	240
VG 11	130	130	130	130	130	130	130	130	130	130		**35**	135	135	135	140	140	145	145	160	160	240	240
VG 12	130	130	130	130	130	130	130	130	130	130	**35**		135	135	135	140	140	145	145	160	160	240	240
VZ 01	130	130	130	130	130	130	130	130	130	130	130	130		**35**	**35**	140	140	145	145	160	160	240	240
VZ 02	130	130	130	130	130	130	130	130	130	130	130	130	**35**		**35**	140	140	145	145	160	160	240	240
VZ 03	130	130	130	130	130	130	130	130	130	130	130	130	**35**	**35**		140	140	145	145	160	160	240	240
VZ 04	130	130	130	130	130	130	130	130	130	130	130	130	135	135	135		**35**	145	145	160	160	240	240
VZ 05	130	130	130	130	130	130	130	130	130	130	130	130	135	135	135	**35**		145	145	160	160	240	240
VZ 06	130	130	130	130	130	130	130	130	130	130	130	130	135	135	135	140	140		**35**	160	160	240	240
VZ 07	130	130	130	130	130	130	130	130	130	130	130	130	135	135	135	140	140	**35**		160	160	240	240
VZ 08	130	130	130	130	130	130	130	130	130	130	130	130	135	135	135	140	140	145	145		**35**	240	240
VZ 09	130	130	130	130	130	130	130	130	130	130	130	130	135	135	135	140	140	145	145	**35**		240	240
VZ 10	130	130	130	130	130	130	130	130	130	130	130	130	135	135	135	140	140	145	145	160	160		**70**
VZ 11	130	130	130	130	130	130	130	130	130	130	130	130	135	135	135	140	140	145	145	160	160	**70**	

Fig. 2.18 Example changeover matrix

The *lot size* (LS) in the data box further indicates how many non-variable parts are processed in direct succession during the course of the respective production process (Fig. 2.19). In contrast to the process quantity, the parts included in a lot size are not processed simultaneously. As a rule, lot sizes differ with variants, they are determined by the mean customer demand. In the data box, we best enter the smallest, the largest and the mean (or most frequent) lot sizes, they will be useful reference values in case of possible alterations when future-state mapping.

CO	Changeover Time
LS	Lot Size
# Var	Number of Variants a Part
UP	Uptime [%]
EPEI	EPEI-Value

Fig. 2.19 Data box, section 2, with information required to determine the EPEI value

Lot sizes determined by schedulers in industrial practice are often based on empirical values, aimed at a compromise between changeover and storage costs without risking shortfall quantities. As a rule, changeover times of more than five percent are considered excessive. From time to time, the mathematical calculation methods offered by PPS and ERP systems are used for the determination of economic lot sizes to verify said empirical values. These are usually based on a method developed by Ford Whitman Harris back in 1913. This lot size calculation compares changeover costs and storage costs based on the following fundamental idea: If lot sizes are chosen too small, the partial changeover costs are too high, as they will need to be shared by too few parts. If the lots are too large, however, storage costs rise, as it takes too long to use all parts. Storage costs are determined by adding an overall storage cost rate to the manufacturing costs. Together with the changeover costs and the annual piece number of variants, we can use to the following square root to determine the *optimal lot size* in accordance with the 'economic lot size formula':

$$LS_{opt} = \sqrt{\frac{200 \times COC \times \# P_{Var}}{MC \times SCR}} \qquad (2.4)$$

with: LS_{opt} optimal lot size of one variant [pcs.]

MC manufacturing costs [$ / €]

SCR storage cost rate [%]

COC changeover costs [$ / €]

$\#P_{Var}$ annual piece number of variant [pcs.]

The development of changeover and storage costs within the required period underlying the economic lot size formula model thus results in a parabolic development of the total costs, the minimum of which coincides with the optimal lot size (Fig. 2.20). This type of cost optimization, however, is rather dearly bought, as various problems result from the model's assumptions. Some of these may be rectified by more complex mathematical modelling, others, however, are of a more fundamental nature.

- The model only considers the storage costs for the entire lot stored. The carrying costs arising during the manufacturing process of the lot are disregarded completely.
- The model assumes consistent consumption – together with complete lots being passed on, this results in an idealized sawtooth graph (cf. Fig. 2.37). Dynamic cost models, on the other hand, are able to consider fluctuations in outward stock movement.
- Recalculation of the lot size for each individual period results in continuously adjusted lot sizes to match the respective variants. This generally applies to all dynamic models. However, it makes controlling more complex and reduces production procedure transparency, as the production is thus forced to handle ever-changing volumes.

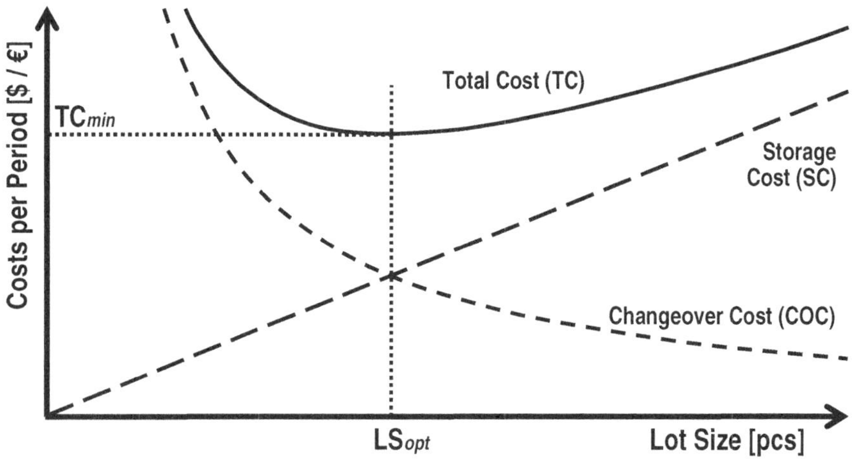

Fig. 2.20 Cost development in the economic lot size formula model

- The model works on the assumption of a single-stage process. In multi-stage productions, however, the optimal lot size must be determined cross-category for several production processes, as due to differences in costs, optimal local values also vary with each production process.
- Storage costs are influenced by capital costs, storage location costs, and stockpiling and stock removal costs. These expenses should not be underestimated – a reasonable empirical value would be approx. 25 % of the production costs. This is more of an estimate than a precise cost unit rate, though, because it suggests that all storage costs develop in line with production costs and may be allocated accordingly. As a result, inventory management costs for instance seem to be lower for low-cost parts than for costly ones.
- Often changeover costs are assessed too high by mistake, as the machine hour rate includes depreciations which would equally accumulate during idle times due to shorter changeover times.

- For exotic variants with small piece numbers, quite plausible minimum numbers are defined, which in turn are higher than the calculated optimal values, though.
- Although the parabolic total cost curve's unmistakeable minimum coincides with the point of optimal lot size, it is usually too flat for any changes in lot size to significantly affect the total costs (Fig. 2.20).

The respective results suggest that this method enables optimal values based on known costs, though weaknesses in production are largely ignored. There is no incentive to try and decrease changeover costs, for instance, mainly because thanks to economic lot size calculation, optimal values seem to have been achieved already. Besides, this approach is based solely on cost unit rates and does not consider the logical efficiency of a production. Accordingly, changed lead times do not affect the evaluation. Flexibility-orientated procedures would easily enable the determination of lot sizes to take place as late and thus as demand-specific as possible, at the same time increasing costs as little as possible, yet require too much effort to be really practicable. Another procedure for the determination of economic lot sizes, solely based on the logistical conditions, shall be introduced in the following in connection with the future state concept (cf. Sect. 3.3.2).

Uptime

So far we have not considered limitations placed on the available capacities by machine downtimes. The remaining production process-related performance ratio entered in the data box is the *technical uptime* (UP, Fig. 2.19). The fact that resources are at times unavailable for work either through scheduled service and maintenance or due to unforeseeable technical malfunctions leading to breakdown or reduced performance is thus accounted for in a summarized figure. Maintenance carried out outside customer takt times as per working time model (Eq. 2.1), i.e. on a Saturday in the case of our fifteen-shift model, has no negative influence on the uptime. The breakdown of technical systems, however, depends on their actual stresses and strains, wear and tear of parts and components, possible fabrication defects of machines, maloperation and/or insufficient or deficient maintenance.

The relative weight of the different influential factors described above is often analysed in greater detail by measuring the efficiency of the entire equipment in order to find out the reasons for machine breakdowns and to be able to develop selective counter measures. Decreases in the performance ratio of equipment are quoted in *Overall Equipment Efficiency* (OEE), a key figure considerably differing from the uptime value determined in a value stream analysis (Fig. 2.21). This OEE figure generally refers to the planned production time, which contrary to the working time has already been reduced by downtimes caused by lack of orders or scheduled service and maintenance.

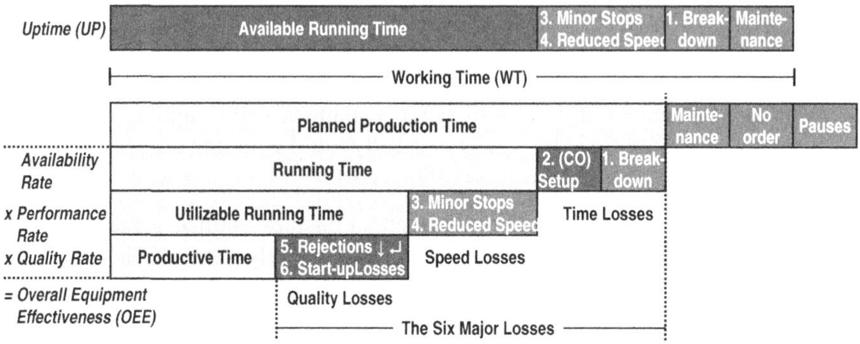

Fig. 2.21 The six major losses of the OEE figure in comparison to uptimes in a value stream analysis

On this basis, the six major losses, broken down into three categories, are applied in the determination of the OEE figure (Fig. 2.21, bottom).

Firstly, unintended *time losses* occur because of equipment breakdown – due to technical malfunctions, lack of material or operating personnel – as well as intended downtimes caused by setups and adjustments. Such changeover procedures are independently mapped in the value stream analysis and recorded as changeover times due to their high logistical significance. The availability rate resulting from the relation of the remaining running time of the respective equipment and the planned production time must be clearly differentiated from the uptime.

Secondly, *speed losses* occur through minor stops or reduced speed of equipment. These minor disturbances are routinely rectified by the operating personnel. Due to their shortness of duration and lack of documentation these are very often hardly noted as performance decreasing factors, though in their entirety they may well reach significant percentages. The running time is thus decreased by the performance rate to result in utilizable running time.

Thirdly, *quality losses* occur through rejections and starting losses, followed by rework and compensatory production. In the value stream analysis this is recorded separately in the third section of the data box.

In an OEE analysis, the multiplication of the utilizable running time and the determined quality rate indicates the actual productive time of the equipment. Assuming absolute top values, which only occur under perfect conditions, the OEE multiplication of an availability rate of 90 %, a performance rate of 95 % and a 99.9 % quality rate only results in a mere 85.4 %. However, if we deduct 5% for changeovers and the quality rate, the resulting availability rate reaches a top value of approx. 90 %.

The availability rate reflects the relation of the utilizable running time of a resource and the working time (Fig. 2.21, top). Scheduled preventive service and maintenance during working times, malfunction-related breakdowns with repairs as well as reduced performance due to minor stops or reduced speed are also factored as performance decreasing criteria.

EPEI Value

With the aid of the ratios shown in the second section of the data box (cf. Fig. 2.19), we can now determine the *variant flexibility* of a production process. This equals the period of time required to complete the entire changeover cycle for all variants and is referred to as *EPEI value*. EPEI is short for 'Every Part – Every Interval'. The number of variants (# Var) entered in the data box also states how many changeover procedures are required to mount one lot of each variant once (Fig. 2.19). The succession of all variants and their pertaining operation times and interim changeover times can be graphically depicted in a pie diagram (Fig. 2.22, right). The availability of several resources reduces the EPEI value – but only if these are used to process different variants simultaneously.

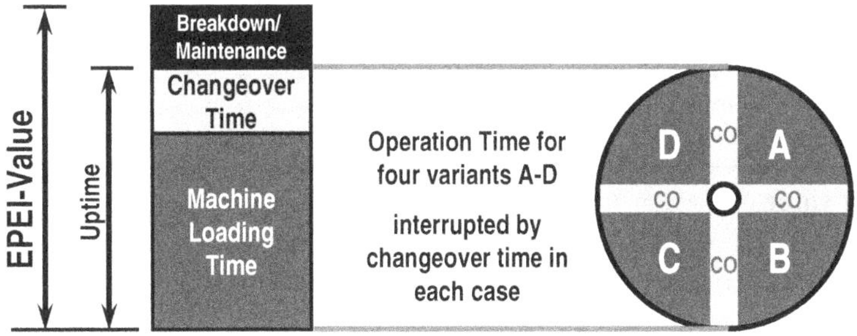

Fig. 2.22 Schematic representation and calculation of the EPEI value

In the simple case of identical lot sizes and changeover times for all variants, the EPEI value is determined in accordance with Eq. 2.5 by adding all operation and changeover times; alternatively, the respective mean values may be used. In cases of variant-specific lot sizes or sequence-related changeover times, the respective individual values are added.

$$EPEI = \frac{\sum OT + \sum CO}{\# Res \times WT} = \frac{\# Var}{\# Res} \times \frac{((LS \times OT) + CO)}{WT} \tag{2.5}$$

with:

EPEI	EPEI Value [d]
# Var	number of variants [pcs.]
# Res	number of same resources [pcs.]
LS	lot size (average) [pcs.]
OT	operation time per piece (average) [time unit]
CO	changeover time [time unit]
WT	daily working time [time unit/d]

The thus achievable ideal EPEI value (Eq. 2.5) then needs to be adjusted by the decrease in performance resulting from resource-specific uptimes (Fig. 2.22, left), and will thus be increased by the respective factor:

$$EPEI_{UP} = \frac{\sum OT + \sum CO}{\# Res \times UP \times WT} = \frac{\# Var}{\# Res \times UP} \times \frac{((LS \times OT) + CO)}{WT} \tag{2.6}$$

with: $EPEI_{UP}$ EPEI Value considering uptime [d]

UP	uptime [%]
# Var	number of variants [pcs.]
# Res	number of same resources [pcs.]
LS	lot size (average) [pcs.]
OT	operation time per piece (average) [time unit]
CO	changeover time [time unit]
WT	daily working time [time unit/d]

> **EPEI value**
> The *EPEI value* of the current state of a production process is determined by adding the operation times of all product variants in the respective lot sizes and the required changeover times as well as planned and unplanned downtimes. This value indicates how much time is needed under the prevailing conditions to produce every single variant once. Other than a mere statement of changeover times and lot sizes, this value gives an indication of the *flexibility level* of a production process.

Quality of the Production Process

Decreases in production performance, however, are not only caused by changeover times and limited availability but also by quality defects. The respective quality data are indicated by the key figures of *first pass yield,* (↑) *rejections* (↓) and *rework rate* (↵) of the respective production processes (Fig. 2.23). An increase in

↑	First Pass Yield [%]
↓	Rejections [%]
↵	Rework [%]
WT_P	Process Working Time a Day
Pcs_P	Process Annual Piece Number
TT_P	Process Takt Time

Fig. 2.23 Data box, section 3, with information to determine process quality and process-specific customer takt times

these percentages also leads to an increase in capacity requirement, as this causes the number of pieces to be manufactured to increase beyond the actual customer demand.

As quality defects count among the worst possible types of waste (Sect. 3), a rejection-related increase in piece numbers should not be expressed as a decrease in customer takt time. Quality problems should not be depicted as a seeming increase in demand but rather as a decrease in process performance. This is calculated by way of a percentaged increase in net cycle time – proportionate to the number of unusable parts in the case of rejections (Eq. 2.7), or proportionate to the number of parts to be reworked in the case of rework required. Should the time needed for the respective rework considerably differ from the regular operation time, it must be weighted accordingly. However, should a separate work station or production area be allocated to rework, said rework will not affect the cycle time of the original production process under observation at all. The rejecting of bad parts unfit for reworking has another negative side effect, though: The resulting higher demand is also *inherited* by all upstream processes. In order to meet the customer demand with a view to the finished product, not only the rejections of the production process under observation need to be fabricated in addition, but also the additional quality-related demand must be produced for all subsequent processes. This increases the cycle time of the respective production process by a quality-related factor as shown in the following equation:

$$CT_Q^m = CT_{net}^m \times QL^m = CT_{net}^m \times \left(\hookleftarrow_m + \sum_{pss=1}^{m} \left(1 - \uparrow_{pss}\right) \right) \tag{2.7}$$

mit: pss index for m-1 subsequent processes after production process m

ZZ_Q time loss by quality defects [time unit]

CT_{net} = CT; cycle time according to eq. 2.3 [time unit]

QL rate of quality losses [%]

$\uparrow$ first past yield [%] *often also as:* $\downarrow$ rejections [%]

$\hookleftarrow$ rework [%]

In the simplest case, rework is carried out immediately during the fabrication of the respective lot or directly afterwards for all parts to be reworked. This raises the lot size by the respective percentaged increase of rework required. Quite possibly, parts will be returned for rework from a subsequent process, though, should the need for rework not be detected until later, thus increasing the lot size of a later order for the respective variant. In both cases, the EPEI value calculated as per Eq. 2.6 needs to be increased by the influence of the rework rate, resulting in the following equation for the determination of the *EPEI value for fixed lot sizes*:

$$EPEI_{fix} = \frac{\sum_{var=1}^{n} LS_{var} \times OT_{var} \times (1 + \hookleftarrow_{var}) + \sum_{var=1}^{n} CO_{var}}{\# Res \times UT \times WT} \qquad (2.8)$$

with: *var* index for *n* variants

$EPEI_{fix}$	EPEI Value for fixed lot sizes [d]
LS	lot size [pcs.]
OT	operation time per piece [time unit]
$\hookleftarrow$	rework [%]
CO	changeover [time unit]
# Res	number of same resources [pcs.]
UP	uptime [%]
WT	daily working time [time unit/d]

The additional rework creates an additional material flow, either flowing back from the subsequent process when the rework requirement is detected, or branching off to a special rework process. In the latter case, however, the EPEI value of the primary process remains unaffected.

Process quality
The key figures for *first pass yield* and *rework rate* measure quality defect-related reductions in production process performance caused by additional operational effort exceeding customer demand, 'passing on' the rejection-related extra demand to all upstream production processes. Quality problems gravely affect the overall *value stream performance*.

Process-Specific Customer Takt Time

Some production processes are not required by all products belonging to the same product family, but are skipped by various variants. Looking at this the other way, we might say that some more complex variants require additional production processes. It is also possible for a value stream to consist of two parallel branches (cf. Sect. 2.3.2) in parts. Different performance requirements placed on the production processes in such cases are also recorded in the data box and lead to the calculation of process-specific customer takt time (TT_P). The respective key figures are entered in the third section of the data box (Fig. 2.23). In the case of *branched value streams* with alternative process sequences for different variants, however, the respective branches are simply allocated percentaged parts of the overall piece numbers (Pcs_p/Pcs), thus increasing the customer takt time by multiplication with the respective percentaged parts as follows:

$$TT_P = TT \times \frac{Pcs}{Pcs_P} \times \frac{WT_P}{WT} = \frac{FD \times WT_P}{Pcs_P} \tag{2.9}$$

with: TT_P process-specific customer takt time [time unit]

WT_P/WT process-specific related to general working time [%]

Pcs_P/Pcs rate of process-specific annual piece number [%]

FD factory days [d/a]

Some production processes, however, conform to a *process-specific working time model*. In these cases, the respective working times (WT_P) differ from those generally integrated in the determination of customer takt times (Eq. 2.1) in a value stream analysis. The thus changed time budget proportionally alters the customer takt time in relation to the two different types of working time (WT_P/WT). This is another factor used in the calculation of process-specific customer takt time as per Eq. 2.9.

Process-specific customer takt time
For individual production processes the *process-specific customer takt time* adjusts the overall customer takt time performance standards formulated for the entire value stream. This is required in the case of process-specific deviations of working time models as well as separating value streams, which both lead to *heterogeneous* production processes with different production speeds and thus pose special challenges for a smooth production flow.

In value stream mapping, all production processes are mapped and their key figures entered in the value stream drawing, starting from the shipping end. The process boxes are drawn in sequence from right to left on the lower half of the drawing sheet, beginning underneath the customer symbol and working upstream (Fig. 2.24). Underneath each process box, the pertaining data box is placed with its respective key figures.

Case study The basic body of a *Liquipur* oil filter consists of a milled aluminium die-casting body housing. In addition, the finished product contains another twenty-four purchased parts, such as gaskets, valves, switches, filter units and plastic housing components. All in all, the product passes through six production processes: The cast piece is received, milled and washed, subsequently assembled in a two-step assembly process and finally shipped as described in the following detailed process description:

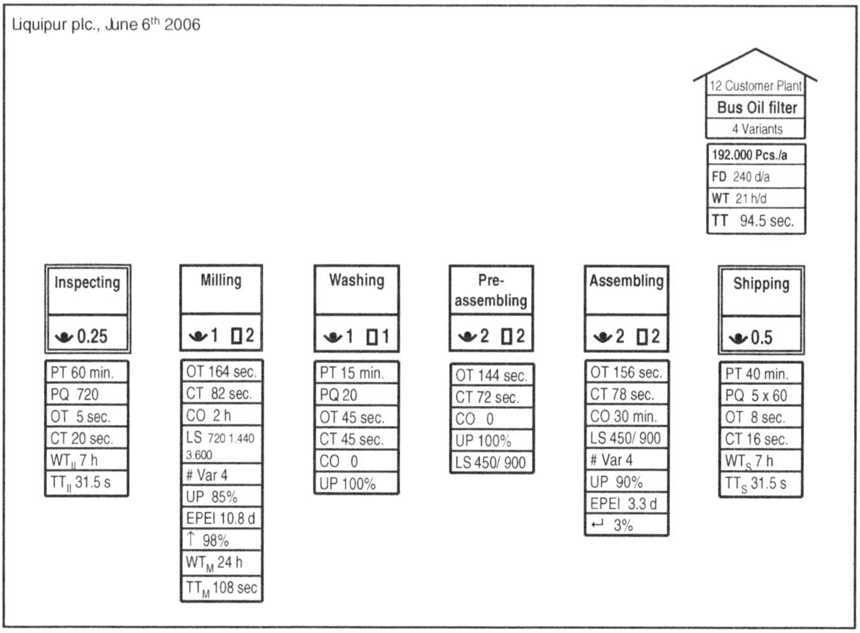

Fig. 2.24 Value stream mapping at *Liquipur* (2): Production processes

1. Shipping The oil filters are shipped in pallets with five layers of 60 pcs. each. On average, the preparing and loading of each pallet as well as the printing and allocating of the respective shipping documents takes 30 minutes. In addition, the receipt and taking into storage of the goods transferred from production approximately take another 10 minutes. This results in an operation time per piece of 40 min. / 300 pcs. = 8 seconds. The process is also used for other products, 1/2 manpower is allocated to the oil filters. This results in a cycle time of 16 seconds. The shipping personnel works in single shifts (7 hours). Based on a customer takt time of 94.5 seconds in a three-shift production (21 hours), Eq. 2.9 thus results in a process-specific customer takt time of 31.5 seconds.

2. Assembly The process finishes with the aligning and mounting of the plastic components of the casing, which takes one worker 156 seconds. In the case of two identical workstations with one worker each, the cycle time is 78 seconds. The adjusting device can be changed for another variant in 30 minutes with both workstations processing the same lot. The lot sizes are 900 pcs. each for the two high-volume variants, 450 pcs. each for the two smaller ones at an availability of only 90 %. As both workstations always process the same variant, the changeover effort for a complete variant cycle increases accordingly. The EPEI value is calculated as follows upon adjustment of Eq. 2.8:

$$EPEI_A = \frac{\sum_{var=1}^{4} LS_{var} \times OT_{var} \times (1 + \lrcorner_{var}) + \sum_{var=1}^{4} \#Res \times CO_{var}}{\#Res \times UP \times WT} \quad (2.10)$$

$$EPEI_A = \frac{[((2 \times 900 + 2 \times 450) \times 156\,sec.) \times (1 + 3\%)] + 4 \times 2 \times 0{,}5\,h}{2 \times 90\% \times 21\,h/d} = \frac{69{,}2\,h}{21\,h/d} \approx 3{,}3\,d$$

with: OT operation time [time unit]

CO changeover time [time unit]

LS lot size [pcs.]

↵ rework [%]

WT daily working time [time unit/d]

Res number of same resources [pcs.]

UP uptime [%]

It takes 3.3 days to mount all variants in above lot sizes once each, always parallel on both workstations, at a rework rate of 3 %.

3. Pre-assembly All small parts are assembled in the cast housing, which takes one worker 144 seconds. As two identical workstations are used, the cycle time is 72 seconds. This is done without appliances; therefore there are no changeovers and/or possible related disturbances. Lot sizes are identical to those in assembly.

4. Washing The milled cast parts in containers of 20 pcs. are automatically washed in continuous processes of 15 minutes each. Division of the processing time by the number of containers results in a cycle time of 45 seconds. In addition, one worker is needed for the handling and moving of the parts from the milling and to the pre-assembly areas; accordingly, an additional operation time of 45 seconds per piece must be factored. The washing machine has no malfunctions, nor does it need any changeovers.

5. Milling The unfinished castings are entirely mechanically processed on two milling machines. One worker sees to the handling and changeover at both machines. The operation time per piece is 164 seconds, the cycle time is therefore 82 seconds. The changeover time per machine is 2 hours; both machines simultaneously process the same lots. The lot sizes are 3,600 pcs. each for the two high-volume variants, 1,440 pcs. for the third variant and 720 pcs. for the exotic variant with extremely low piece numbers. The availability is 85 %. In accordance with Eq. 2.10, the EPEI value is calculated at a little less than 11 days:

$$EPEI_M = \frac{((2 \times 3.600 + 1.440 + 720) \times 164\,sec.) + 4 \times 2 \times 2\,h}{2 \times 85\% \times 24\,h/d} = \frac{260\,h}{24\,h/d} \approx 10{,}8\,d \quad (2.11)$$

The available capacity is higher for the milling process than for the other processes, because the automatic milling machines continue to run during breaks. According to Eq. 2.9 this results in a customer takt time for milling of 108 seconds:

$$TT_F = TT \times \frac{WT_P}{WT} = 94{,}5\,sec. \times \frac{24\,h/d}{21\,h/d} = 108\,sec. \quad (2.12)$$

with: TT_F specific customer takt time of milling [time unit]

WT_P/WT process-specific related to general working time [%]

6. Inspection of goods received Random samples are taken from 3 pallets of 240 pcs. each; incl. handling this equals an effort of 60 minutes for the 3 pallets. The operation time per piece is thus 60 min. / 720 pcs. = 5 seconds. The receiving dock is also available for other cast parts and works in single shifts (7 hours). One worker assigns one quarter of his time to this product family, therefore the cycle time is 20 seconds and – like at the shipping end – the process-specific customer takt time amounts to 31.5 seconds.

2.3.2 Material Flow

Production processes are logistically linked by the *material flow*. The material flow consists of three components: transport, handling and storage. Storage is the temporary placement of materials, parts and/or products in an appropriate storage facility. Transport means the moving of material, parts and/or products to the respective retrieval areas at the subsequent production process or another storage location. Handling describes the manual activities required in stockpiling and removal from stock. Whenever these occur in a significant volume, they need to be considered separately and counted as a logistical process of their own in the value stream analysis (Sect. 2.3.1). Otherwise they may be left out, i.e. the material flow is mapped without belonging operation times. The material flow must therefore depict two aspects: transport function and storage function.

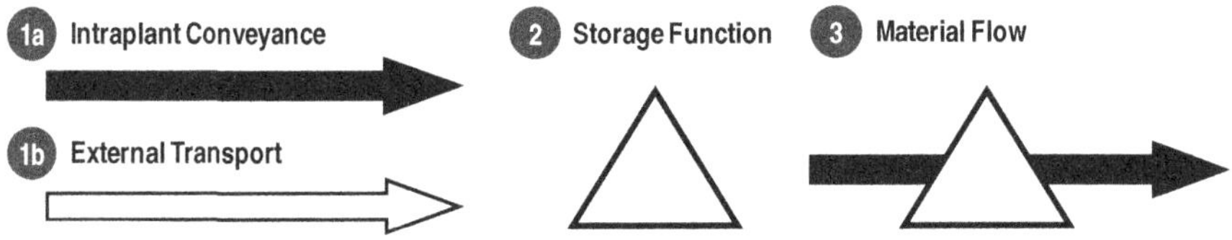

Fig. 2.25 General transport and storage symbols

1. The *transportation* from one production process to the respective following process is indicated by arrows. As a rule, intraplant conveyance – shown as a black arrow – must be differentiated from external transport – symbolized by a broad arrow, white on the inside (Fig. 2.25, cases 1a and 1b). The arrows may supplemented by details on the respective means of transport or loading units.
2. The *storage* between two subsequent production processes can be either short-term buffering or be intended for a longer period of time to build up stocks or interim storage. The storage function is symbolised by an upright equilateral triangle (Fig. 2.25, case 2). The following abbreviations have proved useful for entering in the triangle symbols: FG (finished goods), HRR (high rise rack storage), SFG (semi-finished goods storage), Raw (raw materials storage), WIP (work in process). The storage symbol is supplemented by a data box.
3. The *material flow* is shown by overlapping of the two symbols (Fig. 2.25, case 3). The material flow's specificity is strongly dependent on the pertaining planning and control logic. Information flow and material flow together

constitute the respective logistic linkage. Depending on the type of linkage, storage symbol and transport arrow are depicted differently (Sect. 2.3.3). Material flows may also separate.

If a material flow separates, or branches into different paths, the percentaged piece number is entered next to the *branching symbol* (Fig. 2.26, case 1). When these partial flows join again later, no symbol is required since no loss will have occurred.

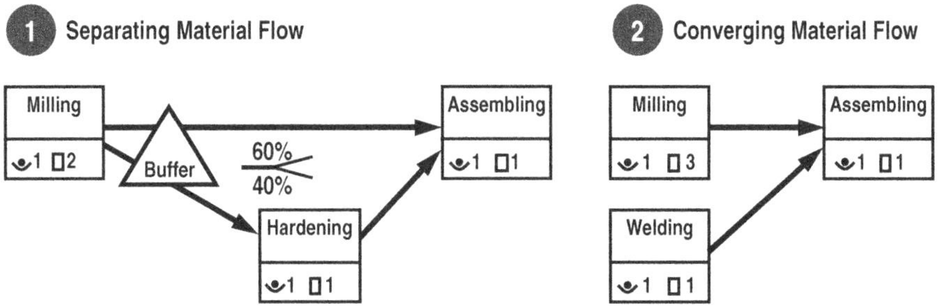

Fig. 2.26 Separating material flow

Those production processes which from a logical point of view run parallel to each other are drawn over one another in a value stream depiction. Assembled products often consist of various complex and valuable parts which are processed in various production processes before they are assembled. In this case, the value stream is made up of several parallel flows which then converge in the last process of the final assembly (Fig. 2.26, case 2). In the case of alternative confluences, for instance for plastic or aluminium casings, the symbol of branching may be mirrored and equipped with the respective partial percentages.

Conveyance and Transportation

The transport symbolized by material flow arrows may be further specified with respect to means of transport. Basically, we differentiate three types: external trucking, internal forklifts and conveyor techniques (Fig. 2.27). Transport is by loading units on load carriers. In the case of defined standards for transport volumes, we may supplement the material flow arrow by a container symbol, an open rectangle, in which we can enter the respective *container quantity*.

1. The usual means of transport outside the factory premises is the *truck*, symbolized by a respective pictogram (Fig. 2.27, case 1). The name of the carrier may be entered in the attached trailer of the symbol; alternatively, self-collector or express service may be entered, depending on the prevailing circumstances. The respective means of transport are mainly characterized by the number of parts or products per load carrier. The numbers of parts transported on pallets, in skeleton containers, cardboard packaging,

wooden crates, returnable containers and/or part-specific special containers is indicated by the container quantity. The *delivery frequency* states how often finished products are transported to the customer or how often raw materials are delivered by the supplier, respectively. Alternatively, the respective weekdays of the shipping route or the delivery frequency, e.g. weekly deliveries may be entered.

2. Internal transport is conducted by different means of transport, typically by forklift (Fig. 2.27, case 2). Alternatively, lift trucks or automated guided vehicles maybe sketched for the depiction of traffic inside a factory hall, trailers for outside traffic. There may possibly be a time table. In cases of a material flow interrupted by a storage function, the realistic transport symbol is usually left out to leave room for the storage symbol (Fig. 2.25, case 3).
3. Production processes can be firmly linked by conveyance technique such as a roller conveyor, conveyor belt or suspension track. This special case can be depicted as an arrow between two lines (Fig. 2.27, case 3). The conveying system usually includes a fixed number of load carriers or accumulating conveyors of a certain length to accommodate a fixed maximum number of load carriers. This maximum buffer quantity (max. BQ) serves as a buffer and is mapped as a stock value between two production processes.

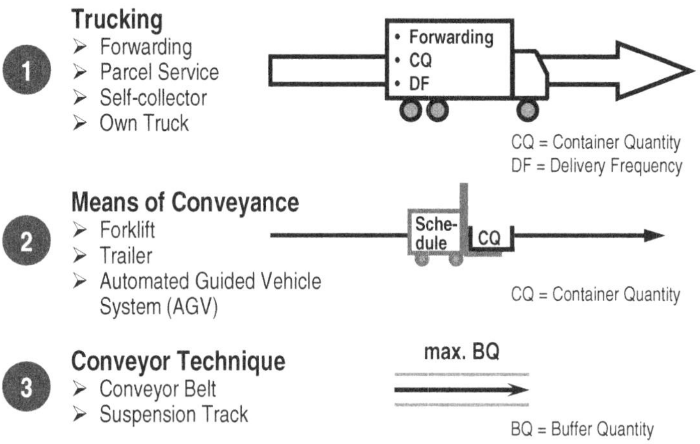

Fig. 2.27 Material flow arrows and means of transport

Inventory – Storage and Stocks

Each storage location needs – similar to a production process – a *data box* (Fig. 2.28). First of all, the denomination of the storage function and its pertaining means of storage are entered, i.e. a specific storage location and the storage technique used, such as automatic high rise rack storage, rack storage, flow rack storage, floor storage, paternoster storage and many more. Underneath the storage location, the alphanumerical EDP identification of the respective store must be recorded as well as the denomination of any in-bought materials, interim products

Denomination Store; Stock Location	
Denomination Material	
# SP	Number of Storage Places
SQ	Stock Quantity
# P	Number of Parts per Product
RC	Range of Coverage

Fig. 2.28 Data box depicting storage functions

and/or parts stored. The respective *storage capacity* is roughly indicated by the number of storage places (# SP).

The most important production procedure-specific characteristic of a store is its *stock*. Stock includes all prefabricated parts required for all variants of the product family under observation. Ideally, we should not enter the value indicated by the EDP system, but the value stream mapper should personally count the stock and – similar to the time recording procedure – capture the prevailing conditions using his very own eyes. On the one hand, this serves to unveil errors, on the other hand, sometimes previously unknown storage places are discovered and stocks become perceptible with all their real-life implications, such as space requirement, arrangement, distribution and accessibility. The counted volumes are quoted as stock quantity (SQ).

The decisive indicator with a view to the value stream analysis here is the period of time covered by the stocks with a view to the mean customer demand. The *range of coverage* (RC) of the respective stocks in days is calculated by simple division of the stock quantity by daily demand (DD). However, the range of coverage decreases due to rejections (↓) resulting from subsequent production processes. Besides, the number of identical parts (# P) per product needs to be considered:

$$RC = \frac{SQ \times \uparrow}{DD \times \# P} \tag{2.13}$$

with: RC range of coverage [d]
SQ stock quantity [pcs.]
DD daily demand [pcs./d]
↑ first pass yield [%]
P number of identical parts per finished product [pcs.]

As a relative value, the range of coverage is a much more significant key figure than the absolute value of stock quantity, in particular in connection with the evaluation of potential for improvement (Sect. 2.4).

Shipping and Supply

The material flow arrow indicating external transport connects the in-house procedures of the observed value stream with external sources and recipients of materials and products. The further away we get from serial mass production, the more diverse is the range of customers, both with a view to the geographical structure of the recipients and the *delivery requirements* resulting from the respective means of transport and the customers' wishes. These logistic requirements influence the entire production procedure. The truck symbol of the external material flow should thus include the following information (Fig. 2.27, 1):

- The respective means of transport determined by the required delivery time (DT, cf. Fig. 2.8) and the distance to the customer's location,
- The required mode of dispatch, for instance self-collector, (externally) contracted carrier, own truck, express delivery,
- The required type of packaging and container quantity (CQ), e.g. wooden crates for overseas delivery, cardboard packaging for the rest of Europe, returnable containers for domestic shipment,
- The usual delivery frequency (DF). In the case of deliveries as per shipping tour, the respective planning tasks are to be included in the information flow (cf. Fig. 2.35).

As a rule, the first production process in a value stream is the receipt of goods. The delivery of the respective goods is depicted by a material flow arrow and truck symbol. Similar to the shipping end, information on container quantity (CQ) and type of packaging as well as type and frequency of the delivery (delivery frequency DF) are to be noted in the truck symbol.

The material flow arrow starts at the supplier symbol – like the subcontractor depicted as a factory with a shed roof (Fig. 2.29). In the *supplier symbol* we enter the name(s) of the respective supplier(s) for a specific material, possibly with the respective location. In addition, we note the denomination of the delivered raw material and/or bought-in parts (RM) and the number of the different types of material (# Typ). As the number of suppliers and supplied materials is sometimes very large, only suppliers of central importance to the value stream in question are recorded. This may refer to primary components or maybe 'A' materials determined via ABC analysis, which due to their high purchase prices must be scheduled very carefully. Besides, critical parts of limited availability due to market shortages and/or quality fluctuations need to be given special attention.

The supplier performance is recorded with the aid of a data box. The availability of purchased parts is mainly determined by their *replenishment lead time* (RLT). This includes the time required for purchase order processing and transport as well as the delivery time guaranteed by the supplier. The volume-related quality of the supplier is determined by the respective error rate (ER) and quantity reliability (QR), the supplier's time-related quality is measured in delivery reliability (DR). As far as quantity reliability is concerned, deviations caused by excess or short deliveries, partial and/or wrong deliveries are weighted differently and recorded accordingly.

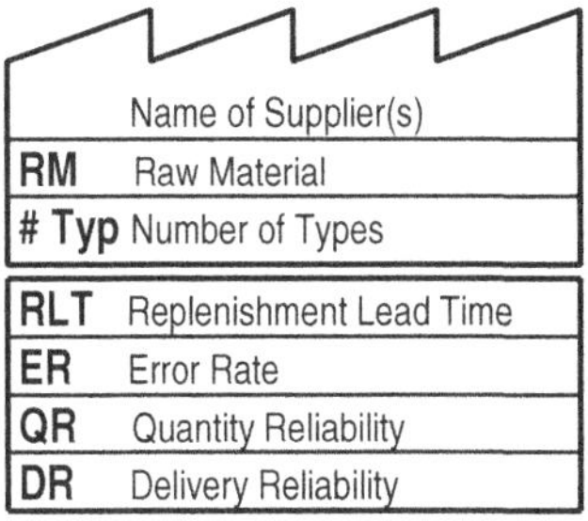

Fig. 2.29 Supplier symbol with data box

In value stream mapping, the material flow between production processes is mapped together with these processes. It is solely for reasons of representation, that all production processes are analysed before the material flows here (Fig. 2.24). In order to depict the material flow, all production processes are connected by storage symbols and dark material flow arrows. Stocks are counted, entered and converted into ranges of coverage. The suppliers are depicted at the top left, on the same vertical level as the customer. Suppliers and customers are linked to the value stream by truck symbols and white material flow arrows (Fig. 2.30). In effect, the external transport now appears in the drawing as a 'vertical' material flow, while the internal material flow runs 'horizontally'.

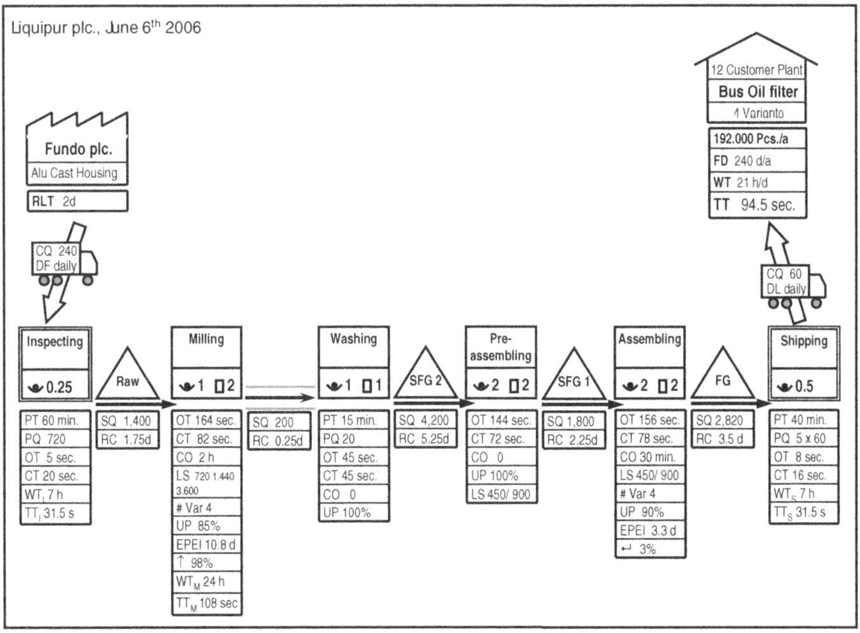

Fig. 2.30 Value stream mapping at *Liquipur* (3): Material flow and suppliers

Case study At *Liquipir*, stocks are counted between all production processes, i.e. at five different locations. For the conversion into range of coverage in days, these figures are divided by the daily demand of 800 oil filters. The mapped data are then entered in the value stream depiction (Fig. 2.30).

1. **Finished goods storage.** There are 47 full pallets with a total of 2,820 oil filters. The range of coverage is 3.5 days.
2. **Unfinished goods 1.** Before and during assembly, two variants with a total of 1,800 pre-assembled oil filters are counted. This equals a range of coverage of 2.25 days.
3. **Unfinished goods 2.** Before and during pre-assembly, all four variants comprise a total of 4,200 processed cast parts. This equals a range of coverage of 5.25 days.
4. **Washing.** The milled cast parts are washed immediately. There is a buffer for 200 parts upstream of the washing machine. This equals a range of coverage of 0.25 days.
5. **Unfinished castings storage.** Thanks to daily deliveries, the raw material inventory was decreased considerably even before commencement of the value stream analysis. On the first day of value stream mapping, there were 1,400 pieces, equalling a range of coverage of 1.75 days.

All external *Liquipur* transport is done by truck. Fundo AG daily delivers aluminium diecast housings in pallets of 240 pieces each. Orders must be placed two days before delivery. Shipping is daily on pallets of 60 oil filters each.

Material flow

Between two production processes we always need to insert a *transport link* to depict the required conveyance of parts, otherwise we would be looking at an integrated production process which could be modelled with just one process box. This rule helps us create coherent value stream depictions. Transport is reflected by two different types of arrows depending on the type of logistic linkage. The branching symbol shows the percentaged distribution for the separating material flows.

In general, intraplant conveyance between two technical processes is interrupted by a *storage process*. Scheduled storage is symbolized by a triangle. Buffers and special storage strategies are symbolized differently (Chap. 3).

The respective stocks are counted during the course of the value stream mapping and converted into *ranges of coverage* in days.

The external material flow links customers and *suppliers* to the observed value stream. *Delivery frequencies* and *replenishment lead times* of raw materials and bought-in parts are central characteristics of external logistics.

2.3.3 Order Processing

To enable the production processes (Sect. 2.3.1) to run smoothly and to enable the material (Sect. 2.3.2) to flow, controlling and/or regulating information is needed. This information is processed and provided by *business processes*.

Well-established value design literature (Rother 1999) uses only one business process, namely production planning and control (PPC), summarizing all information processing tasks in a black box called PPC (Fig. 2.31). This immense simplification, however, has not been able to prove its worth in value stream analyses or value stream mapping, respectively, because order processing is a much more complex process, which at different stages influences the production process with different tasks and activities. A considerably more detailed depiction is therefore presented in the following.

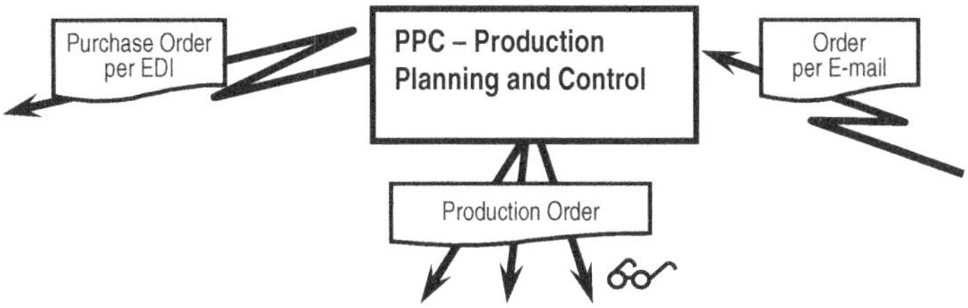

Fig. 2.31 Simplified information flow in traditional value stream analysis

Later efforts in connection with so-called 'lean administration' to depict business processes as value streams of their own – as traditionally done with production processes (Tapping 2003), were not able to overcome this shortfall either, though. The analogue application of the value stream design to administrative processes, e.g. in the service industry, serves to depict and improve the progress of pure service processes. However, tasks arising in connection with *order processing* in the manufacturing industry simply cannot be depicted and/or optimized like this, at least not in relation to the production. Therefore, business processes occurring in the manufacturing industry in connection with order processing are depicted and described here in much greater detail than has so far been done in traditional value stream mapping.

For this purpose, we shall leave the business processes to be observed embedded in their productive value stream; an approach which has already proved its worth in practice and leads to a decidedly improved order processing analysis. We shall not consider its applicability on pure service providers here, though.

We can differentiate three aspects of order processing in a value stream analysis, which shall be regarded in greater detail in the following:

1. Business processes generate, process and store information required for the completion of customers' orders and needed in production planning and control.
2. The information flow transfers data and documents between business processes, to customers and suppliers as well as between production processes.
3. The logistic linkage describes the correlation of information and material flow with respect to the connection of two subsequent production processes.

Business Processes

Business processes are symbolized by a rectangle and identified by their respective business process denominations (Fig. 2.32). The number of employees handling the pertaining business processes is entered behind the symbol – a stylized head with arms. The mapping of business processes is not so much aimed at a detailed representation of the entire work process step by step (order creation, entry and verification of item no., determination of delivery address, checking of availability and recording of delivery date, printing of order confirmation), but rather a clear and well-arranged depiction of the overall procedure. In our order processing example, we could group together several activities in one process box, similar to the depiction of production processes. Each business process should be described by its most important tasks in a *task list*.

Fig. 2.32 Business process symbol

The classification of the respective *business cases* in applicable categories is of central significance for the description of business processes. This allows for orders to be equipped with different order statuses, or priorities, respectively, such as 'urgent', 'regular' or 'top priority', which are also entered in the process box. Another example for an order processing task could be the determination of the order category – such as 'stock replenishment order', 'planned order', 'rework order', 'customer order', 'production order', 'work order' – maybe the determination of the 'mode of dispatch', the 'release status', etc. On the one hand, such differentiation serves to determine which direction the information flow is going to take from the business process in question, on the other hands it serves to map the essential characteristics of the production procedure.

Generally, *electronic data processing systems* are used to support the work process, providing data storing, processing and output functions. EDP systems are symbolized by a cylinder which is also the way they are usually shown in flow diagrams. The denomination of the system in use is entered in the symbol. These could be standard systems such as enterprise resource planning (ERP), manufacturing execution systems (MES) or PPS-systems, or maybe company-specific proprietary developed solutions with similar properties. Often, data bases and/or spreadsheet programs offer special user templates, supporting the work process in a rather rudimentary but possibly quite sufficient way. In addition, EDP aided tools enable product configuration or production process management – such as scrap optimization, control data calculation, quality data recording, key date reporting – or customer loyalty improvement, – for instance through customer relation management-solutions (CRM), programs for dispatch handling with cost estimation and tracking and/or export transactions.

The value stream analysis often shows parallel use of several EDP systems with similar functions in one company, sometimes due to poor functionality of the primary system, deficient training of the respective users or lack of acceptance on the part of the employees. This often causes inconsistencies between multiply stored data.

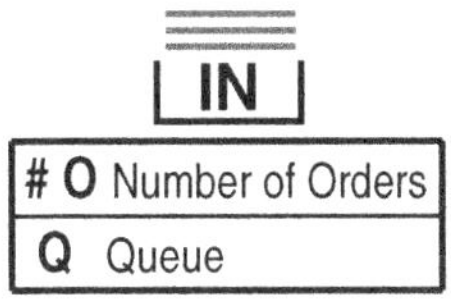

Fig. 2.33 Depiction of order queue

Order processing also accumulates inventories, which in their entirety constitute the opening stock (Fig. 2.33). The first business process – seen from the customer's point of view – is the order entry; but this is actually preceded by the order receipt. Before this first – and all other – business processes, however, varying numbers of orders pile up in the so-called IN box. The symbolic depiction shows a box with a stylized pile of paper. The related data box indicates the number of 'stored', i.e. waiting orders. With the aid of the customer takt time applicable to the shift model of the production in question, we can determine the range of coverage of the respective order backlog. This range of coverage equals the opening stock of the respective value stream, expressed by the length of the *queue* (Q) and is calculated as follows:

$$Q = \#O \times TT$$

with: Q queue [time unit] (2.14)
\# O Number of waiting orders [pcs.]
TT customer takt time [time unit/pcs.]

Information Flow

Just like production processes are linked to each other by the material flow transporting materials and parts, business processes are connected by the information flow which transfers data and documents. In addition, information is passed on from business processes to production processes, controlling not only the production processes, but also the related material flows. The links created by information flows are symbolized by thin arrows (Fig. 2.34a). When observing a production procedure, it is important to note the different transmission formats used for the different types of information, and record the same in symbols of their own. These symbols are drawn over the pertaining *information flow arrows*. The information thus transmitted and the respective formats may be broken up into four categories with a view to processing logic – data, documents, lists and informal coordination procedures, each of which are allocated an *information symbol* of their own.

1. Simple data, or *data sets*, respectively, are exclusively transmitted electronically by way of network cables and/or data storage media such as CD-ROMs or USB sticks. They either serve for data transfer between different data processing systems or as control data for CNC machines, the latter being based on the respective geometrical data of the product to be manufactured. Data sets are symbolized by parallelograms (Fig. 2.34b) – much in the style of flow diagrams – in which denomination and function are recorded. Automatic data transfer with no manual data input in between EDP applications constitutes an *interface*, which is symbolized by a cross in a circle (Fig. 2.34g).

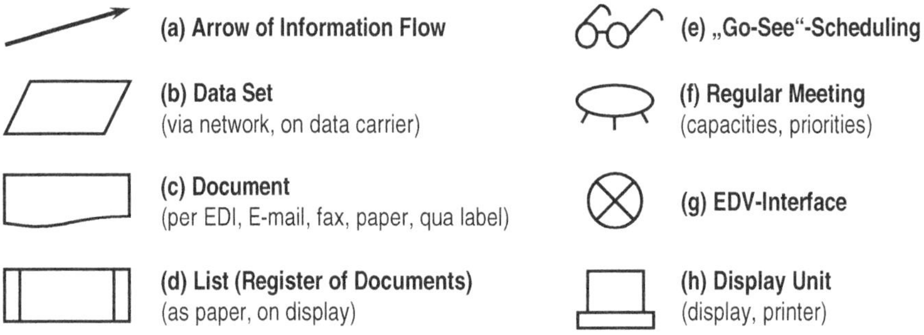

Fig. 2.34 Symbols for the depiction of the information flow

2. By entering order-specific data into forms, we create *documents*. These documents may be transmitted in various ways, via electronic data interface (EDI), email, fax or on conventional paper – either printed or completed by hand. The paper solution very often makes the most sense because it also enables the identification of the pertaining material flow controlled. Mostly, however, the best identification method consists of labels affixed to products, containers or packaging which fulfil additional control functions. As in flow diagrams, a document is symbolized by a rectangle with a wavy base, i. e. the document symbol (Fig. 2.34c).
3. *Lists* give an overview over the documents pertaining to a given business process. They may also indicate sequences, priorities and/or other business process classifications. Typical examples are production plans, lists of operating sequences for individual production processes, picking lists, shipping lists, loading lists. These are symbolized by rectangles with double left and right borders (Fig. 2.34d). Like documents, lists may be distributed in the form of printouts, or accessed on a *display unit* (Fig. 2.34h). Often, feedback data as to times, quality data or other production-generated information need to be entered manually.

4. With the aid of said data, documents and lists, the information flow influences and controls the production processes. These either precisely conform to the documented specifications or independently effect changes and/or adjustments in accordance with the prevailing circumstances. In cases of malfunction-related deviations from the production plan, this so-called *'go-see' scheduling* carried out on the actual shop floor allows decentralized, unscheduled adjustments of the production plan. Likewise, unscheduled changes in customer demand or urgent orders may be triggered by production control, thus overriding the self-provided plan. The symbol for these spontaneous interventions directly on-site is a pair of glasses (Fig. 2.34e), reflecting the flexibility – and unforeseeable nature – of productions in general.

   Often, two or more production areas get together for regular *coordinating meetings* to discuss capacity adjustments in line with the actual demand or to arrange for the speedy channelling through of urgent orders. At the same time, specificities of the current production plan may be clarified; queries resulting from defective data, documents or lists may be handled, special cases unable to be dealt with by predefined forms can be settled. The information flows between the participating production and business processes flow both ways, the respective information flow arrows are therefore equipped with arrowheads at both ends. Such coordinating processes are best depicted by the symbol of a table, i.e. a three-legged oval (Fig. 2.34f).

Several documents, data and/or lists flowing in the same direction between two processes share one information flow arrow. The traditional use of arrows with zigzag-shaped bolts of lightening symbolizing electronic data communication generally leads to confusion, in particular in the case of – often inevitable – crossing information flows and more often than not causes allocation problems and also requires greater graphical effort.

With the aid of above described information symbols we can now depict the information flow of a manufacturing company (Fig. 2.35). In the example diagram below, order processing is divided into *five typical business processes* of partially hierarchic relation to each other. The mapping of the information flow during the second execution of our value stream analysis begins at the top on the right hand side, the customer end, and continues all the way to the supplier. The top level shows three business processes: 'order entry', 'production planning' and 'material procurement and disposition'. In our example these are all supported by the same ERP system. We then go down one level to add the two business processes of 'production control' and 'delivery preparation' for dispatch. In this case, the production control is carried out by an MES system, however, this could also be done by a control centre.

Other subsystems may be connected by way of interfaces like for instance the Inventory Management System (IMS) or constitute isolated applications, such as the 'cargo' dispatch system. Not included in our example is the business process of 'invoicing' which generally has very little influence on the production procedure. Likewise, independently operating superordinate fundamental business processes

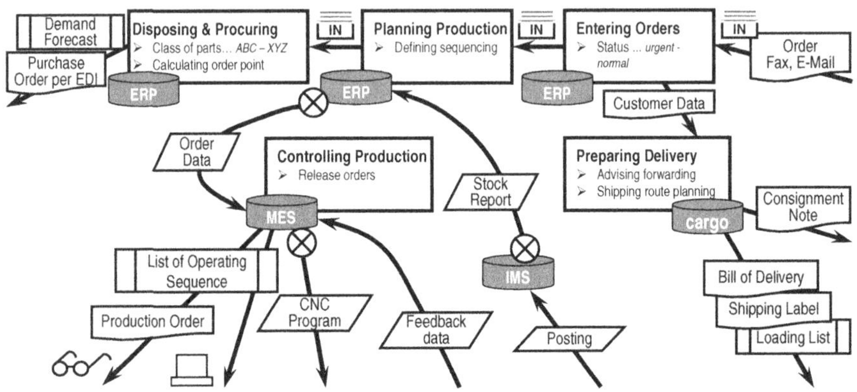

Fig. 2.35 Detailed depiction of a comprehensive information flow

such as 'long-term production planning', 'recruitment' or 'work preparation' incl. work plan development etc. are normally not included in a value stream depiction, they are more important for its design than for its actual implementation.

Between business processes, the respective information flows are recorded together with their pertaining information symbols. Customer orders are received via fax or email and wait in a queue to be classified, for instance by urgency. Material purchase orders are submitted to suppliers via EDI. In production, CNC programs are fed into a production process. The production is controlled by manufacturing orders, the sequence of which is determined in the operating sequence list, which may be adjusted to the respective circumstances as the 'go-see' glasses indicate.

Logistic Linkage

The production planning and control method applied has a strong impact on the material flow; therefore the material flow symbols are adapted to match the respective control logic. Together, the material and the information flow form a unit, because without information flow there cannot be a material flow – rather like a stack of boxes toppling over. In classic production planning with ERP systems, this *logistic linkage* has three main elements: the preliminary production planning, the action-triggering transmission of the production plan to the production processes and the resulting material flow. With this type of logistic linkage, the material flow is the result of the production being conducted more or less according to plan. The production processes 'push' their processed unfinished goods in the direction of their respective subsequent processes. As each subsequent process follows its own production plan, the unfinished goods are first stockpiled; then retrieved by the subsequent process from this planned stock.

This approach is usually referred to as 'push production'; however, this term is rather inaccurate and at the same time misleading. The FIFO linkage also pushes unfinished goods towards the following process, though there is no storage in between (Sect. 3.3.1). Besides, we also know a type of consumption-driven disposition working with planned stocks, which should rather be called, pull production', though. Thus we simply refer to both cases as *planning-oriented* production.

Demand-controlled disposition. As a rule, planning-oriented production works *on the basis of forecast*, in particular with a view to dependent requirements. This method, also called program-controlled production, leads to a *demand-controlled* disposition of unfinished and/or finished goods. In particular, this disposition lends itself to an application in connection with serial productions; however, with an increasing number of variants, the risk of false prognoses also grows. In consumption-driven productions, unfinished goods are produced according to this method in order to shorten delivery periods on the one hand, but also to be able to profit from economies of scale resulting from lot size optimization.

The planning-oriented logistic linkage is symbolized as follows (Fig. 2.36): The business process of 'production planning' generates production plans for all production processes based on prognoses or firm customer orders. These plans comprise a list containing all production orders to be completed within a certain *planning horizon* (PH). The planning horizon of a production plan equals the frequency with which it is created, i.e. the *planning frequency* – in our example once a week. In general, the plans thus triggered are not planned on a rolling basis. The sequences stated in the production plan are not compulsory, but limited by priorities. In addition, urgent orders and plan alterations need to be realized as well.

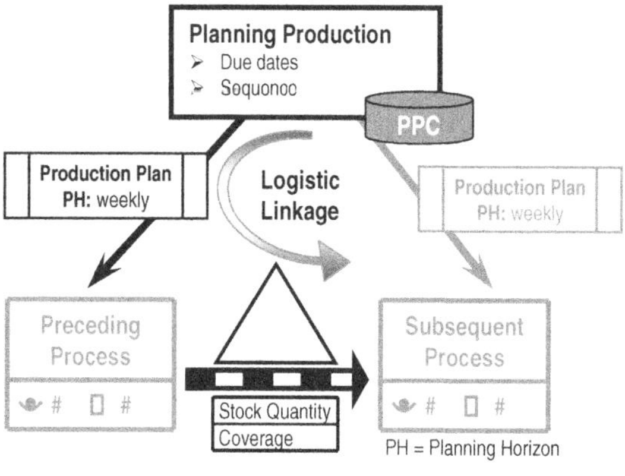

Fig. 2.36 Prognosis-based logistic linkage

The production process controlled by the production plan now manufactures to stock, which is symbolized by a triangle. The pertaining material flow is depicted as a *striped arrow* in order to emphasize the prognosis-related nature of the logistic linkage in the symbol as well (Fig. 2.36). One of the key characteristics of planning-oriented production is the fact that any subsequent process is in turn controlled by an individual plan of its own, perfectly adjusted to the specific requirements of the prior process by the planning department. In order to facilitate said adjustment, the systems allow for a so-called *transition time* between the individual production processes. This time buffer materializes in the form of stocks which are counted during the course of the value stream analysis and entered as inventory below the storage symbol.

Consumption-driven disposition. The second fundamental approach of planning-oriented production is the so-called *reorder level method* with its various forms of appearance. This method creates *consumption-driven* production plans, i.e. production orders are automatically created by stock withdrawals and released by the scheduler. Under normal circumstances the inventory log will conform to the sawtooth graph typical for this type of control (Fig. 2.37, left). Over a certain period of time, average goods withdrawals will gradually decrease the inventory in the respective storage. As soon as the replenishment parts arrive, the inventory zooms up by the delivered lot size. Repetition of this procedure generates the typical 'saw tooth'.

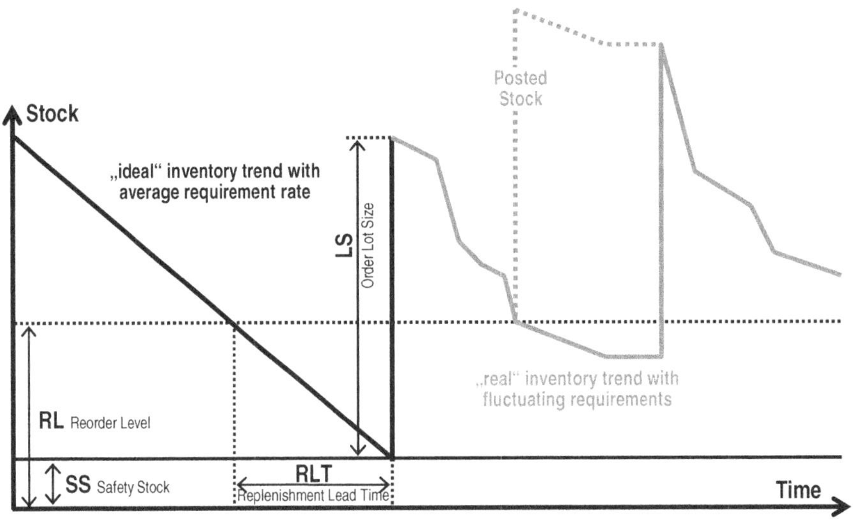

Fig. 2.37 Inventory log in reorder level method

Automatic orders are triggered when material withdrawals decrease the inventory to reorder level. The remaining stocks must then last until the ordered lot is received, i.e. it must match the average daily requirement rates for the entire replenishment lead time. To make sure even short-term increases in demand as well as possible malfunction-related delays in delivery are covered, a so-called safety stock level must be observed. The calculation base for the determination of the reorder level is thus as follows:

$$RL = RLT \times DD + SS \tag{2.15}$$

with: RL reorder level [pcs.]
RLT replenishment lead time [d]
DD daily demand [pcs./d]
SS safety stock [pcs.]

The actual use of material though will never be totally consistent. In cases of large material outflow after order triggering, even the safety stock may be drawn on, whereas in cases of small withdrawals like those shown in our example it will remain unaffected (Fig. 2.37, right). It is therefore vital that the order quantity is posted as an open stock receipt, because only then can an order lot size smaller than the reorder level trigger another replenishment production when the virtual stock falls under the reorder level again. Without said open posting, however, the stocks could possibly run dry completely, should the reorder level not be exceeded through the replenishment delivery of the first lot.

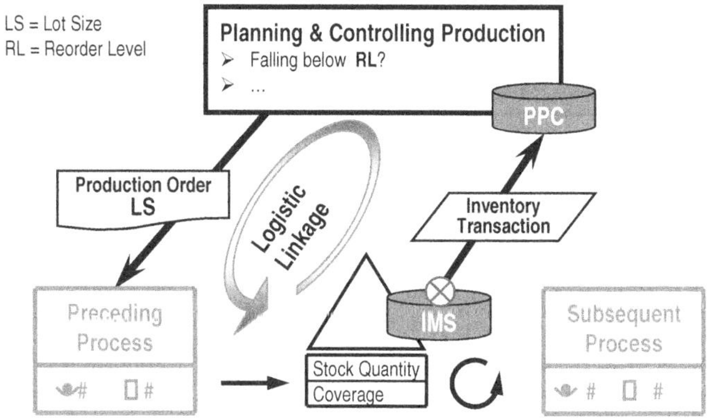

Fig. 2.38 Consumption-driven logistic linkage

In cases of consumption-driven disposition, the logistic linkage presents a closed circuit (Fig. 2.38). The inventory management system transmits the current inventories to the PPS system, the respective planning logic identifies cases of stock level falling beneath reorder level and generates order suggestions for set lot sizes. After release, the respective production order is fed into the prior process. The finished parts are then added to the stock and posted in the inventory management system. Individual procedures may differ from our example, maybe with respect to system-oriented equipment or possibly with a view to the planning logic of the business process described above. The very similar order frequency method for instance uses fixed order intervals and variable lot sizes.

Over and above that, the *reservation* of stocks allows consumption-driven approaches better management of future developments. In order to guarantee the future completion of a delivery confirmation, the required parts and material are irrevocably allocated to the respective orders before the commencement of production, thus ensuring availability as and when the demand actually arises. As the physical stocks are not available for other purposes, replenishment production is triggered earlier, which adds a prognosis-related factor to the process.

Information flow

The value stream analysis depicts *order processing* tasks divided into *business processes*. In particular, the planning and control logic is described in the respective rectangular symbols.

The exchange of data and documents is effected as indicated by the *information flow arrows* linking business processes, customers, suppliers and production processes. A value stream analysis thus includes all *documents* created and dealt with in connection with order processing.

Once the planning and control logic has been defined and recorded with the aid of business processes and information flow, the material flow in its symbolic depiction may be adjusted accordingly. Together, all three components constitute the *logistic linkage* between any two production processes.

When all business processes, all information flow arrows and their respective information carriers have been depicted, the value stream map is complete (Fig. 2.39). The information network connects customers, suppliers and production processes and thus makes the value stream flow – or sometimes causes it to stall. All processes involved in the production procedure, all documents and fabricated unfinished goods and all characteristic key values of the process which describe the value stream in its current state have now been duly summarized. The bottom half of our sheet now reflects the shop floor with the production processes and their connecting – mainly intraplant – material flow, flowing from the left to the right. On the right and on the left, the vertically flowing external transport is depicted together with the connected customers and suppliers. Within this frame we can see the network-like information flow with its starting point at the customer end, spreading out to the suppliers on the left and the processes below and flowing back to the customers on the right via the shipping end.

Case study At *Liquipur,* the current-state mapping of the value stream differentiates two business processes: 'production planning and material disposition' and 'delivery preparation' (Fig. 2.39). According to the demand forecast submitted by the customer, a weekly production is developed. In addition, milling and pre-assembly are equipped with order documents. The material flow resulting from prognosis-oriented planning is always depicted by a striped arrow. Solely the transfer between milling and washing features interim storage and is thus depicted as a conveyor.

On a daily basis, the shipping department's 'ProLog' system is supplied with the relevant dispatch information via SAP interface. Depending on the customers' specifications, carriers must be selected and notified of the impending dispatches incl. volumes. The 'shipping' process receives the delivery documents and a picking list to retrieve the respective oil filters from the finished goods storage.

The customers forward non-binding demand forecasts to the SAP system, reflecting their respective demands for the coming week, the current month and the current calendar quarter – with decreasing accuracy as the time spans grow larger. Call-off orders are triggered daily with agreed same-day-delivery. Based on the demand forecast, SAP generates a weekly delivery schedule which is forwarded to the supplier via EDI. Call-off orders for raw parts are released two days before the intended processing following proof and visual check verification via fax.

For the production, a weekly production plan is developed in the form of an excel spreadsheet. This plan lists all manufacturing orders for milling processes as well as all assembly orders for pre-assembly and assembly in their processing sequence. In daily scheduling meetings, the production plan is adapted to current circumstances by changes in volumes and sequences. Assembly orders are fed into pre-assembly and transferred to assembly together with the respective material.

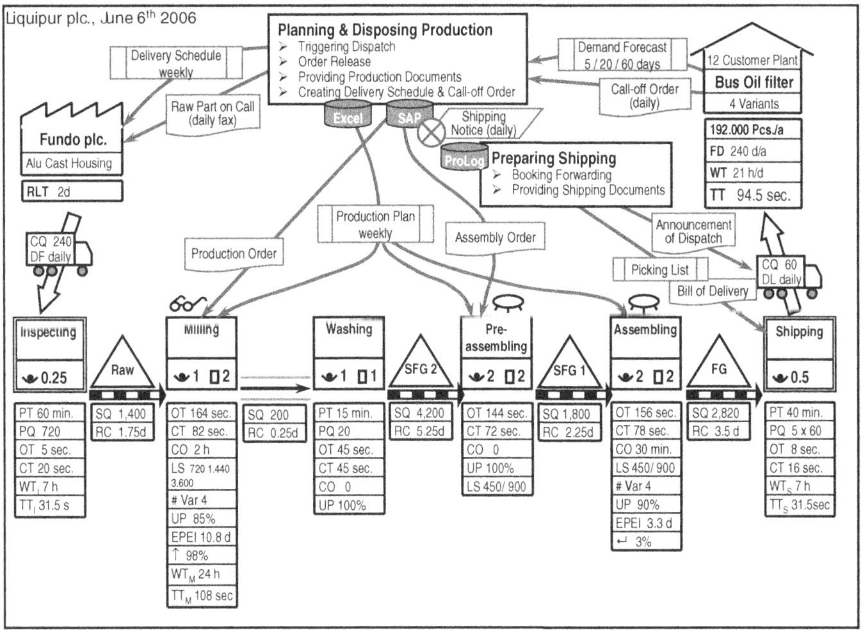

Fig. 2.39 Value stream mapping at *Liquipur* (4): Business processes, information flow arrows and information carriers

2.4 Potential for Improvement

With the completion of the value stream mapping, however, the value stream analysis is far from finished. Rather, we are now in a position to use the value stream depiction to indicate the *potential for improvement* of the production process:

1. The fundamental basis for this is the timeline, which with its comparison of lead times and operation times enables a first evaluation of the production system as such (Sect. 2.4.1). The ratio of these two values, the *degree of flow*, indicates the sluggishness of a production. Major points of production hold-up are to be found in places with large inventories.
2. Besides, a comparison of the efficiency of individual production processes indicates the quality of the machine layout, both with a view to each other and to the customer demand (Sect. 2.4.2). Based on their comparative depiction together with their respective *degrees of capacity utilization* we can work out how to achieve coherent production process capacity with a view to customer takt times.
3. In addition, selective immediate improvement measures may be triggered in cases of easily recognizable *weaknesses* of the production procedure without initiating major conceptional overall changes. The value stream depiction also reveals the relevant complexity and *transparency* levels of the entire production procedure. The more confusing the current-state depiction seems, the more important is a fundamental reorganization according to the design guidelines of value stream design (Chap. 3).

The value stream analysis enables and generates 'management by walking around' by the production manager. Repeated value stream mapping allows an evaluation of the effects of the changes triggered by improvement projects. Besides, possible negative changes with respect to the production procedure resulting from changed requirements and/or bad habits can be comprehensively and accurately pinpointed. The value stream analysis thus serves as a tool for the initiation and evaluation of changes effected to the production process.

Production audit

The value stream analysis is a highly suited approach for the efficient and target-oriented depiction of the current state of a production as well as for the identification of its improvement potential, in particular for *external observers*.

Its degree of abstraction together with the analogue depiction of even highly diverse productions facilitates the comparison of – and application of experiences to – even different industries. Likewise, key figure-based *benchmarks* between companies of the same trade or several factories of the same mother company may be linked through the depiction of their production procedures.

The generally unprejudiced view of external observers paired with their partial ignorance of so-called technical constraints thus offers the companies under analysis solutions outside the tunnel vision of established know-how of traditional production methods. This is less significant in the actual value stream mapping, which internal staff may be trained for just as well, but mainly for the subsequent identification of improvement potential. Mostly, though, the latter can only be realized if we challenge habitual procedures. We must give up seemingly worthwhile and well-proven methods and recognize 'technical constraints' with their potential for sculptability, or changeability, respectively.

2.4.1 Production Lead Time

The term ‚value stream' applies the metaphor of flowing to production, which will initially remind us of the flowing of water, such as in a river or mountain stream. A production, however, hardly ever develops naturally, therefore the picture of a canal would be more suitable. As a production controls both the direction and the volume of the flow, we shall thus be using a pipe, or rather a pipeline, as a model here.

The pipeline model of production shows the products as balls being pressed through a pipe. The production lead time can be pictured as the period of time required for one product-ball to travel from the goods receipt entrance of our pipe to the goods issue exit (Fig. 2.40). Accordingly, we can mark one particular raw part and measure the time until the marked part has reached the shipping end. On its way, the marked part has to pass through all respective process steps with all pertaining operation and processing times as well as all interim storages. While in storage, our part is not permitted to overtake other parts; it will not be retrieved until all other earlier stored materials have been used up.

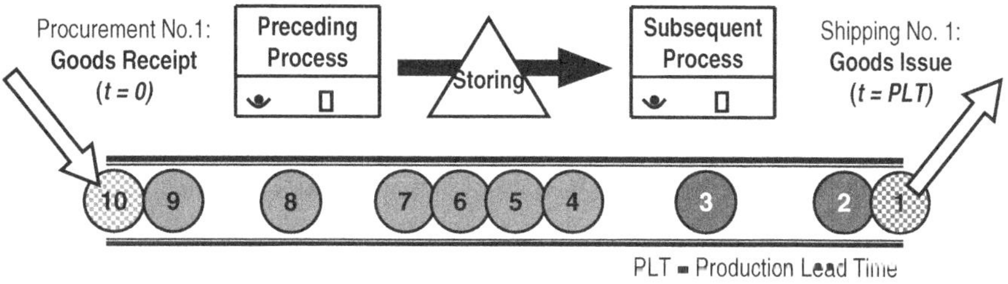

Fig. 2.40 The pipeline model of production

In accordance with our model, the total inventory of a factory tells us the lead time of a product from the first storing of the raw material until its readiness for dispatch. The lead time of a value stream thus equals the total of all ranges of coverage with a view to the slowest branch of the value stream under observation (Eq. 2.13). The impact of operation times and processing times on the lead time is automatically reflected by the work in process (WIP) – which includes even parts being mounted for processing at the time of observation. In value stream mapping, we therefore add the respective work in progress of the following processes to the inventories between production processes and thus determine the lead time as the result of all ranges of coverage:

$$PLT = \sum_{i=1}^{Process\ n} \frac{WIP_i \times \uparrow}{DD \times \# P} + \sum_{i=1}^{Stock\ m} RC_i = \sum_{i=1}^{m} \frac{(SQ_i + WIP_i) \times \uparrow}{DD \times \# P} = \sum_i RC_i \tag{2.16}$$

with:
- PLT production lead time [d]
- RC range of coverage [d]
- WIP work in process [pcs.]
- SQ stock quantity [pcs.]
- DD daily demand [pcs./d]
- $\uparrow$ first pass yield [%]
- # P number of identical parts per finished product

The range of coverage may be expressed as stock quantity or as customer takt time. This interrelation is also referred to as 'Little's Law' – named after the interrelation formulated by John Little in connection with the Theory of Queues, according to which the inventory within a production equals the lead time multiplied with the production rate. If we now substitute the production rate by its reciprocal value defined with the aid of the customer takt time, the simple transformation of the equation results in the relation formulated in Eq. 2.16.

The value stream details are entered on the so-called *timeline* underneath the value stream depiction (Fig. 2.41). First of all, we draw a timeline with two levels – the upper level reflects the material flow with the pertaining stocks and buffers, the lower line is allocated to the production processes. On this line, we record the respective coverages and/or operation times. As a rule, the processing times considerably differ from the operation times, they include simultaneously processed lots but do not include manhours; they are therefore best noted separately underneath the timeline as shown in our example, which in effect results in a three-levelled timeline.

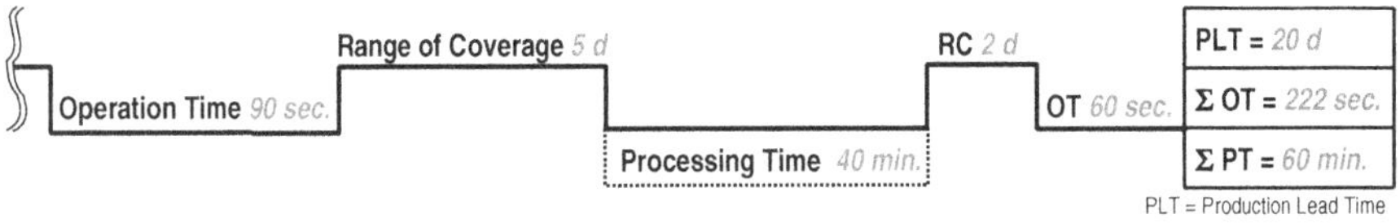

Fig. 2.41 (Partial) timeline and box with totals

The *production lead time* (PLT), i.e. the total of all ranges of coverage, is entered in a box at the right hand end of the timeline. For better comparability, the total of all operation times (Σ OT) and the total of all processing times (Σ PT) are entered there as well. As a result, the almost always highly astonishing differences between operation times (minutes), processing times (hours) and lead times (days/weeks) are visible at a glance. The degree of flow (DF) as an indicator for the dynamics of a production is calculated as the quotient of the operation time, or processing time, respectively, and the lead time:

$$DF = \frac{\sum_{Process}(OT + PT)}{\sum_{Store} RC \times WT} \times 100 \tag{2.17}$$

with: DF degree of flow [%]
OT operation time [time unit]
PT processing time [time unit]
RC range of coverage [d]
WT working time per day [time unit/d]

At a single glance, the timeline indicates parts of the value stream with particularly high coverages. High inventories in production, however, suggest blockages in the overall procedure; material is not passed on fast enough, thus creating distinct hold-ups in the production process. This is most likely just before capacity-related bottlenecks, which the orders literally have to squeeze through (Fig. 2.42). However, high ranges of coverage may also occur without bottlenecks, for instance if the ERP system is set to high transition times, thus creating intended order backlogs. Also, production processes without logistic linkages must feature coverages at least similar to the EPEI values of the respective previous production processes, as otherwise parts would frequently be missing. As long as the production control logic remains unchanged, the range of coverage can only be adjusted through lot size reduction.

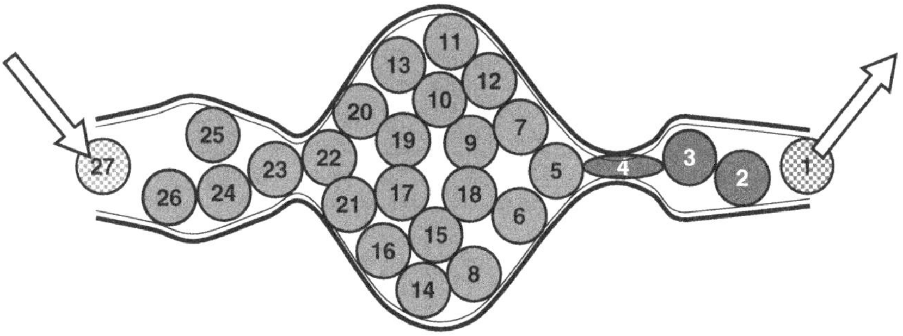

Fig. 2.42 Hold-ups in the production flow indicate bottlenecks

The question: 'Why does my production take so many days to process my product in only a few minutes?' outlines the extent of potential for improvement of our value stream – not only with respect to time. The total operation and processing times show how fast a factory could actually become; the lead time reflects the difference to this theoretical minimal value. The higher this difference, the higher is also the degree of congestion in the production, as it shows possibly disproportionate amounts of partially processed material simply lying about in the factory. As a slight exception, only a high raw materials inventory is less problematic – apart from the related inventory costs – because it will neither clog up the production space nor hinder the production. Ideally, though, the raw material inventory should match the delivery frequencies. Experience has taught us that production hold-ups interfere with all logistic targets, i.e. delivery times, delivery reliability and delivery capacity – problems which time-efficient factories can handle much better (Chap. 3).

With the aid of the timeline, a very simple graphic depiction allows us to follow the interrelation between lead times resulting from inventories and operation times resulting from technical and logistic processing times over the entire course of the value stream. The timeline is entered below the value stream, the box comprising the different overall totals is put on the right hand end below the customer (Fig. 2.43).

Case study Once the timeline has been added, the *Liquipur* value stream depiction is complete. The respective pencilled draft is included here by way of an example (Fig. 2.43). The evaluation of the time characteristics results in the following:

Lead time The total lead time is 13 days. With not quite two days, the raw material coverage is rather low. The work in process amounts to 7.75 days, i.e. a little under 60 % of the entire stocks, also a rather high figure proportionally.

Operation time The total operation time is 522 seconds, i.e. almost nine minutes. Considering the small number of variants and the relatively small container quantities (60 pcs.), the inventory for this production should be easily reducible.

Processing times The processing times of 115 minutes consist mainly of logistics times in goods receipt and shipping, which hardly influence the production flow. The only relevant value would be washing – and this takes only a relatively short 15 minutes at that.

Conclusion A single hypothetical oil filter produced in an otherwise order-free factory would be completed in a little less than 23 minutes (464 seconds for the entire processing plus 15 minutes for washing). Based on the work in progress of 7.75 days, the resulting degree of flow would be a mere 2.3 per mill. In the case of overlapping production of one transport container, the order would be processed in 1.7 hrs. (calculated as the total of 60 times milling, parallel on both machines, plus one washing process, plus pre assembling and assembling). If the entire lot was kept together, the respective lead time would increase to 4.6 hrs. Based on the total inventory of 13 days, the resulting degree of flow would be 1.7 %. This defines the lead time potential.

Accordingly, the savings potential for raw material stocks was apparently already taken advantage of in earlier optimization measures. A noticeably disadvantageous feature, though, is the separation of the two assembling processes by an interim storage location. Even the information flow regards these two processes as belonging together, similar to milling and washing.

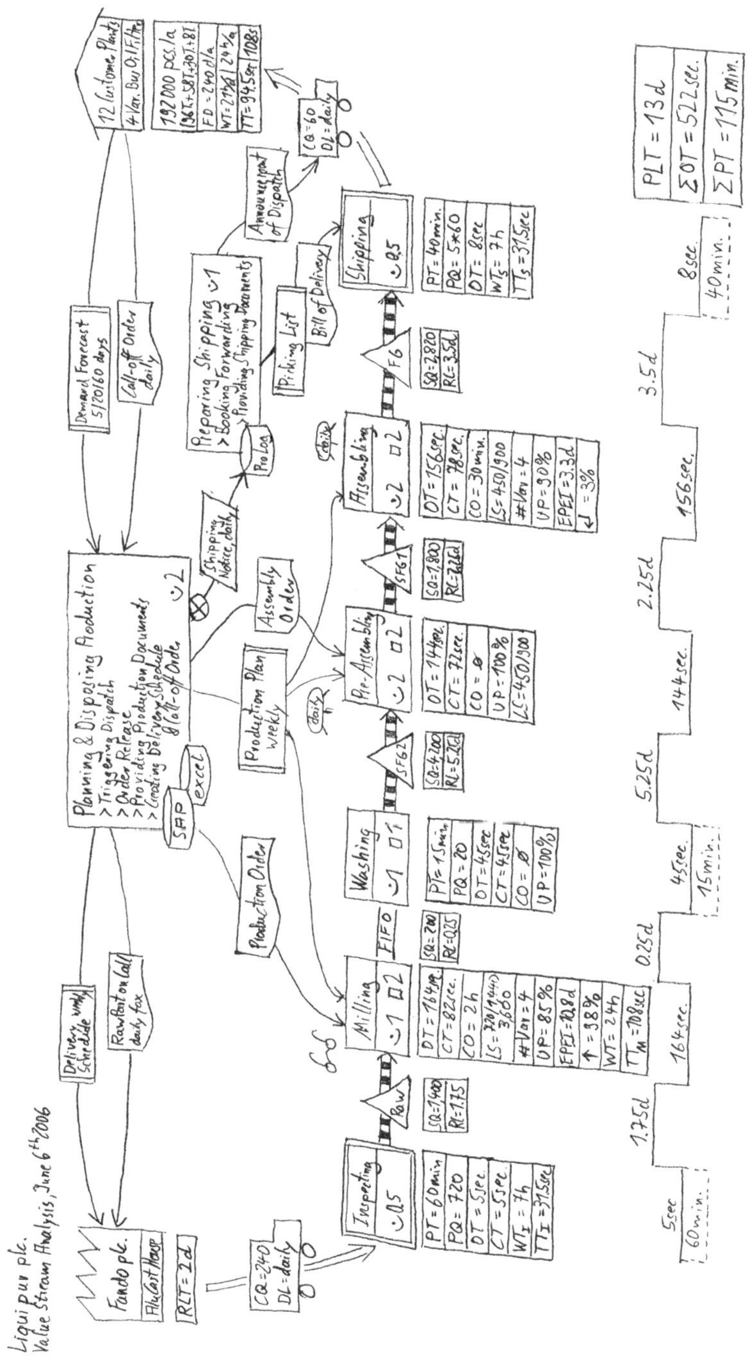

Fig. 2.43 The complete *Liquipur* value stream, drawn on 06 June 2006

Degree of flow

The degree of flow *measures the dynamics* of the entire value stream. Its optimal value, though, strongly depends on the respective product – primarily on the work content, and secondly on its possible serial character. Increasing work content per product facilitates an increase in the degree of flow, while growing variance rather complicates the reduction of stocks and thus lead times.

Accordingly, even a highly dynamic serial production of small variance and work content of between a few minutes and an hour needs a lead time of at least two days, since otherwise lot formation is hardly possible and the maximum degree of flow will still be only a single-digit percentage. In the case of much higher work content, though, for instance 50 to 100 hours in machine assembling, the flow degree will be over 50 %. Reversely, a single-digit percentaged degree of flow would here result in lead times of one to several years.

2.4.2 Line Balancing

The second factor to be regarded in the qualitative evaluation of the value stream is the balance of the processes with respect to each other. For this purpose, the *operator balance chart* (OBC) is an ideal and easy tool. Line balancing aims to evenly allocate work contents to a succession of workplaces. The graphic representation of the allocated work content in a bar diagram helps evaluate the line balancing quality between the individual workplaces (Fig. 2.44). This tool is mainly used in assembly and/or manufacturing lines. In the value stream analysis it is applied to the entire value stream to enable a comparison of all production processes.

In our numerical example below, each operator is allocated one workplace, which in an assembly line may well include several stations (Fig. 2.44). We now record the work content for each individual workplace – in our case between 44 seconds at workplace no. 3 and 92 seconds at workplace no. 4. Accordingly, we actually compare the various operators' degrees of capacity utilization. The highest bar indicates that particular line's *bottleneck*, this determines the maximum total output. In an ideal case, i. e. without malfunctions or additional planned times, the output at the bottleneck's takt time of 92 seconds equals 39 pcs/hr. All labour saving at other, faster workplaces is effectively lost, both with a view to output and in the form of operators' waiting times incurred at their respective workplaces. In particular, the operator at workplace no. 3 incurs large amounts of waiting time in our numerical example, amounting to 52 % of the highest work content, which is to be found at workplace no. 4. These *line balancing losses* are very clearly shown in the operator balance chart.

The customer takt time may also be included in the bar diagram. In order to meet the customer demand under normal operating conditions, each individual workplace must be faster than the customer takt time, i.e. achieve lower takt times. The takt time is integrated in the bar diagram in the form of a vertical line (Fig. 2.44). All workplaces with work content bars that end below this *customer takt timeline* possess sufficient capacity. In the numerical example, 40 pcs/hr are needed to meet the customer demand, equalling a customer takt time of 90 seconds. Workplace no. 4 is thus not only the slowest workplace; with a cycle time of 92 seconds its output is also too small in relation to the customer demand. We can safely presume this workplace to require regular overtime or regular temporary additional help from another employee.

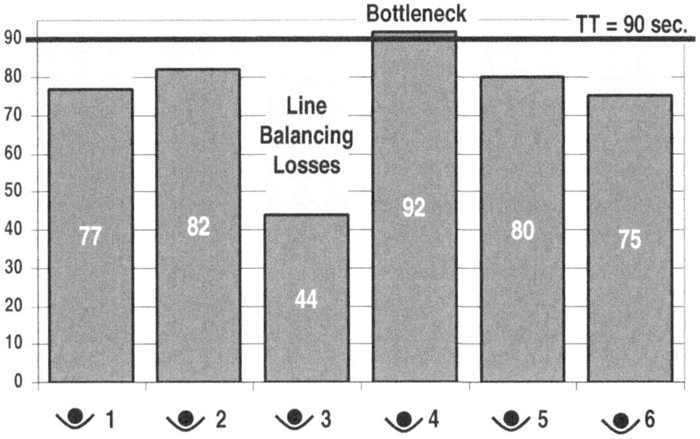

Fig. 2.44 Operator balance chart (numerical example)

In a value stream analysis, the operator balance chart is used to compare the outputs of all observed production processes. The cycle time of a process indicates the time requirement per piece and thus the capacity utilization level of the process in question. The shorter the cycle time, the higher is the related capacity. By recording the cycle times, the entire value stream and its available capacity can be depicted in a bar diagram. The development of the available capacity over the entire value stream describes the *capacity profile* of the value stream. However, this does not refer to the time-related capacity development, but the spatial development, across the various production processes of the value stream under observation. The capacity profile as depicted in the operator balance chart is an important analytical tool in connection with a value stream analysis and as such indicates:

- if the required customer takt time can be met under normal working conditions, without overtime or sporadic help from other employees and without alternative resources, i.e. if a cycle time lower than the takt time can be achieved;
- the exact location of a capacitive bottleneck – the process with the longest cycle time;
- the extent of possible excess capacities, or capacity reserves, respectively, in the value stream – these equal the differences between cycle times and customer takt times.

With the aid of the operator balance chart, the value stream analysis with its fundamental underlying idea of always assuming the customer's point of view clearly illustrates how well (or badly) an individual production processes matches the customer demand. In this context, the customer takt time serves as a measure for the evaluation of a production's capacitive balance. The great improvement potential of a value stream lies in the elimination of thus identified bottlenecks and excess capacities. Said potential for improvement may be assessed by calculating the mean capacity utilization of the production processes of a value stream as follows:

$$DU = \frac{1}{n} \times \sum_{i=1}^{Process\ n} \frac{CT_i}{TT_i} \times 100 \tag{2.18}$$

with: n number of production processes within value stream

DU degree of capacity utilisation [%]

CT cycle time [time unit]

TT customer takt time (process specific) [time unit]

Case study The *Liquipur* operator balance chart shows the two assembling processes to be rather well balanced (Fig. 2.45). The milling process works closest to maximum capacity and lies 13 % below the customer takt time of 94.5 seconds (Eq. 2.2). The washing machine is utilized at almost 50 %. The two logistical processes of 'dispatch' and 'inspection of goods received' work single shifts; their related takt times are thus 31.5 seconds (1/3 of 94.5 seconds). Disregarding availabilities, changeover times and quality problems, all production processes are appropriately dimensioned. The value stream's degree of capacity utilization of 73 % – without the two logistical processes – is determined as follows:

$$DU = \frac{100}{4} \times \frac{(82\,sec. + 45\,sec. + 72\,sec. + 78\,sec.)}{94{,}5\,sec.} = \frac{87\% + 48\% + 76\% + 83\%}{4} = 73\% \tag{2.19}$$

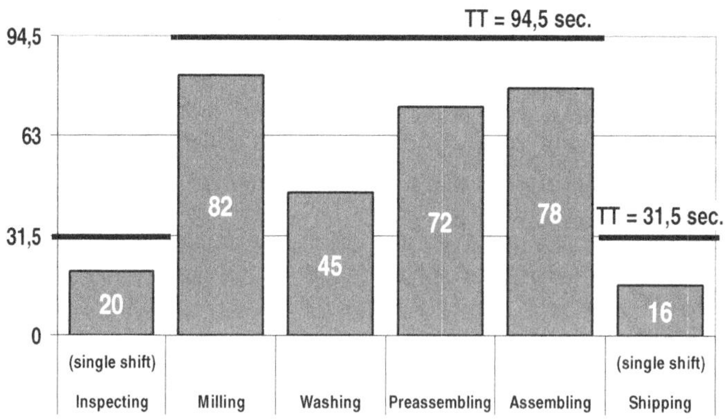

Fig. 2.45 The *Liquipur* current-state operator balance chart

The operator balance chart is also very well suited to take into consideration the decelerating effects of changeover times, uptime restrictions and quality problems. For this purpose, percentaged mark-ups must be added to the so far depicted *net cycle times*, achievable only under ideal conditions, to determine the actually achieved *gross cycle times*. In our operator balance chart, the cycle time bars are thus increased accordingly by the three different types of time loss, thus reducing any possibly remaining time buffers to customer takt times (Fig. 2.46).

The extra times required for changeovers in the case of uniform lot sizes and only one machine are easily determined by the quotient of changeover time and lot size. In the case of variant-dependent lot sizes and/or changeover times, though, the changeover-related mark-up is calculated by way of the annual demand broken up by lot sizes and multiplied by the net cycle time:

$$CT_{CO} = CT_{net} \times COF = CT_{net} \times \sum_{i=1}^{\#Var} \frac{CO_i}{LS_i \times OT_i} \times \frac{Pcs_i}{Pcs} = CT_{net} \times \frac{CO}{LS_{\varnothing} \times OT} \tag{2.20}$$

with: # Var number of variants
CT_{net} = CT; cycle time according to eq. 2.3 [time unit]
CT_{CO} time losses through changeover [time unit]
COF factor for losses through changeover [%]
Pcs. annual piece number [pcs./a]
OT operation time [time unit]
LS lot size [pcs.]
$LS_{\varnothing}$ average lot size [pcs.]
CO changeover [time unit]

The time loss resulting from machine breakdowns is calculated by converting the restricted uptime into a percentage of customer takt time. As uptime refers to the entire working time, not just productivity, time losses resulting from breakdowns cannot be calculated with the aid of cycle times. Thus, the following applies:

$$CT_{UP} = TT \times (1 - UP) \tag{2.21}$$

with: CT_{UP} time losses through breakdown [time unit]
TT customer takt time [time unit/pcs.]
UP uptime [%]

The additional time needed on the grounds of quality problems refers to rework, rejections and scrap inherited from subsequent processes; it is determined as shown in Eq. 2.7. For the determination of the respective gross cycle time, all three types of time losses are then added to the net cycle time accordingly:

$$CT_{gross} = CT_{net} + CT_{CO} + CT_{UP} + CT_Q \tag{2.22}$$

with: CT_{gross} gross cycle time [time unit]
CT_{net} = CT; cycle time according to eq. 2.3 [time unit]

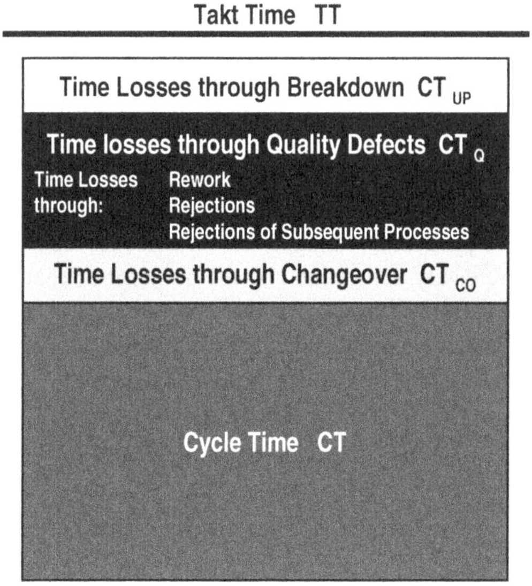

Fig. 2.46 Schematic depiction of time loss caused by changeovers, rework, rejections and uptime restrictions in an operator balance chart

Case study Taking into consideration changeover times, uptime restrictions and quality problems, the *Liquipur* cycle times for assembling, milling and inspection of goods received need to be adjusted. In assembly, time losses occur due to changeovers, malfunctions and rework, resulting in a gross cycle time as follows:

$$CT_{CO} = 78\,\text{sec.} \times \frac{\left(\frac{96.000}{900} + \frac{58.000}{900} + \frac{30.000}{450} + \frac{8.000}{450}\right) \times 1\,\text{h}}{192.000 \times 156\,\text{sec.}} = 78\,\text{sec.} \times 3\% = 2{,}34\,\text{sec.}$$

$$CT_{UP} = 94{,}5\,\text{sec.} \times (1 - 90\%) = 9{,}45\,\text{sec.} \tag{2.23}$$

$$CT_Q = 78\,\text{sec.} \times 3\% = 2{,}34\,\text{sec.}$$

$$CT_{gross} = 78\,\text{sec.} + 2{,}34\,\text{sec.} + 9{,}45\,\text{sec.} + 2{,}34\,\text{sec.} = 92{,}13\,\text{sec.}$$

Milling incurs time losses through changeovers, malfunctions and rejections, thus the gross cycle time is as follows:

$$CT_{CO} = 82\,\text{sec.} \times \frac{\left(\frac{96.000}{3.600} + \frac{58.000}{3.600} + \frac{30.000}{1.440} + \frac{8.000}{720}\right) \times 4\,\text{h}}{192.000 \times 164\,\text{sec.}} = 82\,\text{sec.} \times 3{,}4\% = 2{,}8\,\text{sec.}$$

$$CT_{UP} = 108\,\text{sec.} \times (1 - 85\%) = 16{,}2\,\text{sec.} \tag{2.24}$$

$$CT_Q = 82\,\text{sec.} \times (1 - 98\%) = 1{,}64\,\text{sec.}$$

$$CT_{gross} = 82\,\text{sec.} + 2{,}8\,\text{sec.} + 16{,}2\,\text{sec.} + 1{,}6\,\text{sec.} = 102{,}6\,\text{sec.}$$

For the process of inspection of goods received we need to take into consideration the 2 % of rejections passed down from milling. At a cycle time of 20 seconds, Eq. 2.7 results in a time mark-up of 0.4 seconds, which in view of the high precision in time recording is virtually negligible. The production processes' time losses are entered in the operator balance chart on top of the cycle times in the form of light grey bars (Fig. 2.47).

The cycle time for milling clearly exceeds the customer takt time of 94.5 seconds; this explains why the milling machines have to be kept running during break times. In this case, a takt time of 108 seconds applies to 24-hour-operation (Eq. 2.12). Both milling and assembly feature cycle times only minimally smaller than the respective takt times. The production is thus able to meet the mean customer demand, but is highly inflexible capacity-wise. Both processes could easily become bottlenecks. The existing high inventories are therefore needed to assure constant delivery capability.

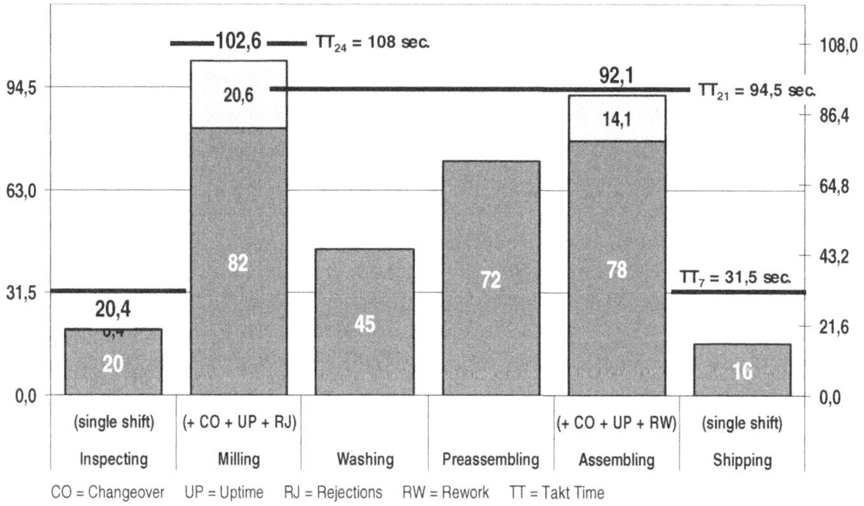

Fig. 2.47 *Liquipur* operator balance chart with gross cycle times

With the identification of the potential for improvement concerning flow degree and line balancing, our value stream analysis has been completed. We may now focus on the original task, for the success of which a sound prior analysis is a fundamental requirement: The reorganisation of our production in order to reach optimal values in line with the respective targets of our production (Sect. 1.2). The first step towards an improved value stream is a value stream-oriented segmentation of the entire production followed by appropriate dimensioning of their respective allocated capacities (Sect. 3.1). The second step aims at making the production flow with the aid of improved production procedures as well as technical improvements of the production processes, i.e. our target is to lower the production lead time. This is done with the aid of the value stream design guidelines introduced in the following chapter (Chap. 3).

Analysis of potential for improvement

Goal is the identification of the value stream's potential for improvement both with a view to production lead times and to line balancing. The *timeline* indicates bottlenecks in the production flow causing material hold-ups and at the same time illustrates by how much the respective production lead time could be shortened – the latter is also identified by the *degree of flow*. The *operator balance chart* reflects how well-balanced the production processes' performances are – with respect to each other as well as relating to the customer demand, the latter is also expressed by the *degree of capacity utilization*.

In particular, the *value stream depiction* shows at which locations and for what reasons some of the following typical problems occur:

1. The lead time is overlong, which results in high inventories and a shop floor cluttered with material.
2. There are numerous urgent orders, which are generally due to overlong lead times.
3. The separation of production processes by stores or buffers disturbs the production flow and increases the logistical effort.
4. The structure of the production with respect to the allocation of products and resources is not clearly defined.
5. The cycle time of a process is slightly higher than the related customer takt time, which often causes production bottlenecks.
6. The employees interrupt their value-added tasks in order to complete logistic, quality assuring or other ancillary tasks unrelated to the part to be processed.
7. The operators incur waiting times caused by missing parts and/or poorly balanced production schedules.
8. Some resources are clearly too efficient for the respective customer takt time or possess excessive automation levels. In addition, waiting times occur for operators at automatically working machines.
9. An excessive EPEI does not permit the prevention of high inventories, causing the production to be highly inflexible in reacting to fluctuations in customer demand.
10. The uptime of a resource is too low for customer takt times to be met.
11. Poor suppliers' reliability and/or quality of parts and excessive delivery times for (costly) parts.

Chapter 3
Value Stream Design

The creation of a value stream-optimized factory is the goal of the value stream method. Our initial value stream analysis was based on the current state of the production (Chap. 2). However, in order to redesign the production with the aid of *value stream design* we have to venture into the open space of a target space yet to be developed. For this purpose, a guideline to help us stay on the right track and not get lost is extremely helpful in this context, which is exactly what the value stream design approach with its clearly defined and structured set of design guidelines provides to be skilfully applied for the redesign and improvement of a production.

As also stated in the first design guideline, the capacitive dimensioning of resources in consistent conformation with the respective customer takt times is a vital prerequisite for the design of any production. Other design guidelines apply to the material flow, controlling and planning. They provide simple and proven production procedure solution components. Thanks to its graphic depiction and symbols, the value stream design clearly illustrates the effects of said solution components on the overall process of a production and by way of the *design guidelines* explained further on enables a systematic approach for the conception of customer-oriented and efficient production.

However, the consistent implementation of the solution components defined in the design guidelines below sometimes turns out to be problematic in spite of their simplicity – or possibly even on account of their inherent generality. Their implementation depends on a relatively strong conceptional adaptation of the components to the specific production-related circumstances. The following comments on the design guidelines shall serve to facilitate this approach. The major part of these solution components was developed in the automobile industry and is herein supplemented and adjusted to the specific requirements of other piece-producing industries. In particular, this requires conceptual expansions with a view to order processing as well as production planning and control methods for customer order-related production and/or the prevention of bottlenecks.

Avoidance of Waste

The fundamental underlying idea of value stream design is the *avoidance of waste* in a production process, directing the main focus of production optimization on cost reduction, i.e. economic efficiency. The term 'waste' as such, however, already specifies that costs may not be cut at the expense of performance, but that

K. Erlach: Value Stream Design: The Way Towards a Lean Factory, LNL, pp. 97–229.
springerlink.com

only those costs may be eliminated which are not related to any immediate benefits to the customer, which the customer has paid for. We must therefore aim to eliminate all non-value-adding tasks from the production process and concentrate on the value-adding ones. Since all technical restrictions, such as changeover times and process quantities (whenever these require deviations from the current customer demand) are considered as waste, the avoidance of waste constitutes an optimization task similar to the one presented by the dilemma of production scheduling (Sect. 1.2). However, beyond the mere definition of the problem, this viewpoint already offers the respective solution approach.

This solution approach is based on the assumption that the main source of waste lies in an *uneven production*. This irregular strain on the production has two implications. On the one hand, it leads to the temporary overburdening of machines and personnel – affecting reliability, quality and work safety accordingly. On the other hand, it causes temporary underload, entailing idle times which will rarely be filled with value-adding activities. This is where the actual waste in the narrower sense of the word – *muda* in Japanese – originates from. The superordinate target of the value stream design is therefore a homogenization of the production – in spite of fluctuating customer demand – and thus the avoidance of waste.

Hitoshi Takeda differentiates seven types of waste occurring in production (Takeda 2006), which we shall be taking a closer look at in the following. In addition, though, there is an eighth type of waste, which is waste through unsuitable and laborious business processes and IT tools in order processing. Four types of waste are immediately attributable to *waste in the production process*; they are caused by rejections, inappropriate motion sequences on the part of the operators, suboptimal machining processes and waiting times.

Overproduction, stockpiling and transport are the three types of waste relating to *waste in the production procedure*, the first two of which are the gravest types of waste we know apart from the production of rejections. These are particularly clearly visible in the value stream analysis, the potential of which explicitly points out these two types of waste: The operator balance chart identifies waste caused by badly balanced production processes and their usually entailing overproduction. The timeline reveals waste resulting from long lead times and their resulting excess inventories. The value stream depiction helps localize the respective causes within the production. In the following, the *eight types of waste* are going to be discussed in detail.

1 Overproduction. Waste through *overproduction* generally describes production in excess of the relevant market demand. In a value stream, however, overproduction means production that is conducted earlier or faster than required by the subsequent production process, i.e. a production process either processes larger amounts than necessary, or before the due date, or with greater yield than required. This may be caused by selective overcapacities and/or production processes poorly balanced with a view to capacities, or cycle times, respectively. In addition, even sophisticated production plans very often have no shop floor rules for cases of short-term deviations from the production plan due to changed requirements and/or machine breakdowns to reduce the overproduction otherwise resulting in other areas. In the value stream design, central importance is attached not only to the decision of when to produce, but also, when *not* to produce. The value stream design is responsible for the designing of a processing logic in which each process produces only what the subsequent process needs and not until it is needed.

2 Stockpiling. Waste through *stockpiling* materializes in inventories of finished goods, semi-finished goods, raw materials and bought-in parts. Quite possibly, stocks are accumulated so often – and defended so persistently – because they offer security against risks concerning material supply and plant availability. Related capital commitment costs and loss of liquidity are readily accepted, almost seem a small price to pay. According to experience, savings in interest will hardly ever pay for technical changes in the short run. Costs resulting from storage technology, higher space requirements, additional personnel effort, administration, and not least the risk of overaging are at least twice as high. But even this comprehensive review of storage costs does not come close the real problem behind stockpiling.

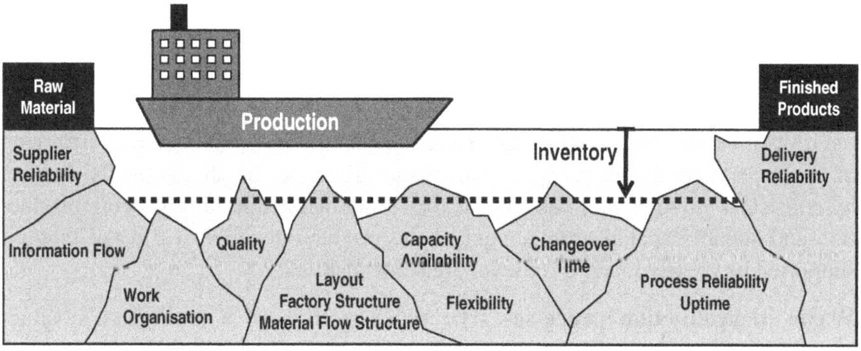

Fig. 3.1 Inventories camouflage production deficiencies

One could well argue that storage costs are a reasonable price to pay for inventories which assure the consistent operation of a production. But a factory production does not always necessarily run like clockwork – rather the contrary. The essential disadvantage of high inventories, though, lies in the fact that they *camouflage* production problems and *prevent their elimination.* Whenever something doesn't go according to plan, there's always a stopgap to make up for it – this is why inventories make us feel so secure. Inventories enable a production to run unevenly and yet pretend that everything is running smoothly. But numerous costly wastes of all kinds are hidden – as shown in the well-known illustration of the ship of production, which on high sees safely sails across the cliffs of shop floor problems and weaknesses (Fig. 3.1). High inventories smooth out various weaknesses. Thanks to pre-production, poor schedule reliability with a view to the production plan for instance will only partially affect delivery reliability. Also, rejections may be compensated for by overproduction and scrapped without problems. All resources may be run with large lot sizes and a high degree of capacity utilization regardless of the actual customer demand – and machine breakdowns take a while to negatively affect the entire value stream.

Only through *continuous reduction of stocks* can the various shop floor problems be made visible. This is awkward, because it puts the ship of production in constant danger of collision with the various problem cliffs. But this is exactly what is needed to finally enforce a reduction of the waste-creating problems. Consistent improvement of the production may thus be pursued by consistent reduction of the inventory levels with the value stream design determining the correct amounts of safety stock in the right places and thus creating the right conditions for consistent, decrease-oriented inventory management.

3 Conveyance. Waste through *conveyance* and other logistics activities such as the transfer of goods to, from and between storage locations as well as sorting and commissioning procedures may occur for two different reasons. One of these reasons is of spatial nature, as inconvenient layout increases the length and frequency of transport routes in comparison to an ideal layout of resources and storage locations and may also result in more frequent transfers to and removals from stock. The second reason is an organizational one: Interruptions of partially processed orders may increase the transport effort. Besides, transport routes and sequences are rarely optimized with a view to distance. Sometimes, parts are provisionally put down randomly because storage places are not defined, stores are overflowing or a transport order is interrupted. Conveyance effort can be almost totally avoided by consistent value stream-oriented factory planning, a smoothly flowing production and suitable spatial arrangement of the production processes in the factory, supported by clearly marked retrieval areas for material.

Waste in production processes. The actual portion of a production's value-adding activities in which parts are processed and finally assembled is comparatively small compared with the overall labour input. A far greater portion of the production process does not constitute any *effective performance* as such, but consists of ancillary activities and obvious waste (Fig. 3.2). Incidental activities such as the grasping of a tool, the clamping of a part into a fixture, the switching on or adjusting of machines as well as general quality control create a *apparent performance* because – though indispensable technology-wise – they do not create any value at all as they do not directly convert the part into a product. These resource-related support processes are of value-decreasing effect, entailing costs without raising the value of the product. Obvious waste, such as searching for parts and/or tools or the repacking and sorting of parts are pure *idle performance*. These activities generate no value increase at all. They occur in connection with organisational and/or logistic activities and can be avoided without any technological changes.

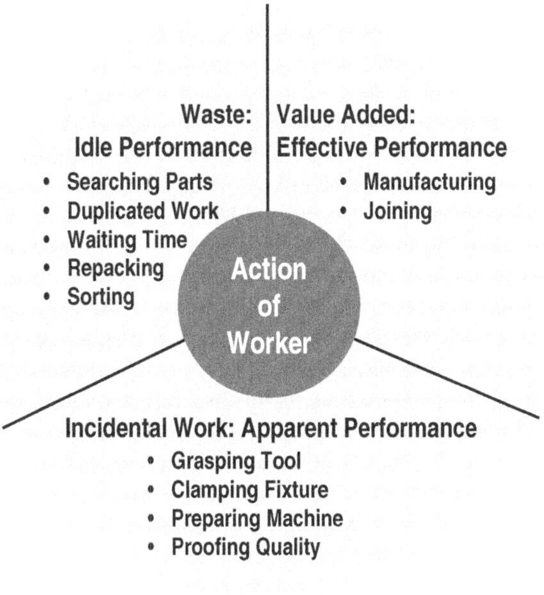

Fig. 3.2 Value creation and waste in workers' activities

4 Rejections. The most critical activity, however, is the *defective performance*, i.e. waste through production of rejections. In the case of manufacturing faults, we are not just looking at partial non-creation of value, what is more, the respective activity loses its value entirely, it even (at least partially) destroys value created in previous value-adding processes. Unless the fault is identified immediately, even subsequent, normally value-adding, activities are turned into waste (in the case of rejections) or require additional production processes in the form of rework. It is now up to the value stream design to reorganize the process in such a manner that only flawless parts are passed on, as any later detection results in decidedly higher fault elimination costs. However, quality controls themselves also constitute a type of waste, so that at the same time the pertaining quality control effort must also be minimized. Preferably, of course, flawless quality should be produced from the beginning rather than achieved through subsequent testing, though this may not always be possible from a technical point of view.

5 Motion. Waste through *inappropriate motion sequences* of operators is caused by ancillary activities and poor ergonomics. Inadequate workplace design with inconvenient layout of parts and tools to be reached for as well as appliances and machinery to be handled lead to ergonomically unfavourable strain for operators, i.e. unnecessarily high effort in motion sequences with long reaching distances and body rotations. Additional loss of efficiency may be caused by one-handed operation, re-gripping and repeated picking up and putting down of parts plus unnecessarily long walking distances due to over-spacious work area design and retrieval of or searching for implements, tools or parts not readily available as well as duplicate work.

6 Processing. Waste through *adverse processing procedures* at certain machines and other technical installations in the narrower sense results from poor machine layout. This inadequate technical quality either refers to oversizing in general or poor design insofar as a machine is not immediately ready for use but needs to be prepared by the operator first. This may mean reference surfaces having to be cleaned repeatedly, cuttings having to be removed by hand, machines or fixtures needing adjustments, machines having to be moved to their operating positions or being subject to excessive feed distances, parts having to be fixed manually during processing procedures, switches needing to be pressed down or maybe machine operations having to be triggered manually. In a broader sense, though, the application of unsuitable technology may well be the major source of all waste, for instance if a level of automation is badly chosen with respect to product variance. In general, batch processes are not really suitable for flow-oriented productions and should, if at all possible, be substituted by continuous flow processes. In particular, technological alterations may be triggered from a value stream perspective and aligned with the production process accordingly.

7 Waiting time. Waste resulting from *waiting times* incurred in the production process may be due to two reasons. Lack of material, for instance, may indicate problems in intraplant logistics or maybe in production control. In this case, it is important to prevent substitute activities bridging over material shortages and make the waste visible rather than covering it up. The second possible reason consists of operators with no specific tasks simply monitoring the correct operation of a plant during automatic processing. The same applies to dead man devices, where for reasons of work safety a switch must be pressed down without interruption. In both cases, waste may only be avoided if one operator is responsible for several workplaces, for instance in multi-machine handling or production lines.

8 Order processing. Other than the previously described seven types of waste, waste in *order processing* is not directly related to the production processes. The depiction of the information flow in a value stream analysis clearly points out possible deficiencies in order processing – from order entry to order receipt. Extra costs incurred as a result of incompatible software in order processing, materials management, and production planning and control are particularly noticeable here, generally accompanied by scattered and partially multiple data management. Very often, routine tasks such as lot size calculation, changeover optimization, sequence scheduling, generation of production documents and data collection are conducted manually in spite of an existing production planning system – either due to its unsuitability or poor implementation. On the other hand, highly efficient planning and control systems sometimes lead to complex control mechanisms and complex order (re)scheduling. An inconvenient division of tasks between the departments involved in distribution, construction, purchasing and production may lead to problems in coordinating and duplicated work.

Approach

The *conduction* of value stream design for the complete redesign of a factory in the manufacturing industry consists of five steps (Fig. 3.3).

1. **Production structuring.** The value stream to be redesigned refers to a certain section of the factory's product range. Similar to the creation of product families in the value stream analysis, the value stream design defines production segments. In connection with this structuring of production, certain resources are firmly allocated to the value stream (Sect. 3.1).
2. **Dimensioning of capacity.** Optimal capacities are defined for the individual production processes which are then redesigned in accordance with the requirements of the overall process and aligned with the customer takt times. An important dimensioning factor is the planning case-related period under observation (Sect. 3.2)
3. **Production control.** The redesigning of a production procedure begins with the production control guidelines which determine the manner in which the individual production processes are to be linked and thus determine the material flow for the entire production process. At the same time this defines the triggered production process as the pacemaker process for the entire value stream (Sect. 3.3).
4. **Production planning.** The design guidelines for production planning determine how production orders are to be planned and released. In particular, differences between the customer demand and the requirements and limitations placed on the respective production processes need to be compensated in this context (Sect. 3.4).
5. **Improvement measures.** The conceptual design concludes with the definition of concrete improvement measures for the realization4 of the future-state of the value stream, supplemented by the pertaining implementation planning (Sect. 3.5).

Only after the production process design has been concluded in all five steps does it make sense to commence with the actual planning of the factory building and the factory layout. Location-specific and factory-related integration must be considered, different options may be explored by way of direct site comparison (Sect. 4.1). The work organization as a vital component of any production system finally enables a value stream-optimized factory operation through strict conformance with the relevant standards and constant assurance of high product quality, fundamental components of any production system (Sect. 4.2). The two issues of spatial and organizational implementation, however, shall only be glanced at herein.

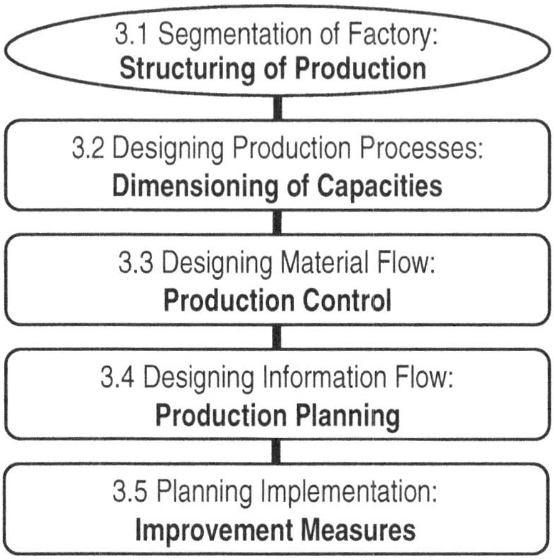

Fig. 3.3 Value stream approach

Value stream design for the redesign of a production

Goal is the complete redesign of a production to achieve an efficient and customer-oriented value stream. Alignment with the customer takt time and consistent assumption of the customer's perspective in our analysis assure customer orientation. The efficiency is further increased by avoiding all eight types of waste. In this context, particular attention is given to waste caused by overproduction, stockpiling and transport in the production process as well as to waste resulting from unnecessarily complex order processing. Waste due to quality problems, adverse processing procedures, waiting times and conveyance between production processes are minimized through consistent improvement.

Approach In a proven set of value stream design guidelines the fundamental components for the ideal design of productions are formulated. The goal-oriented experience-related application of these clear design guidelines systematically leads the current-state value stream towards the future-state, optimized in line with the four goals of production. The value stream perspective helps pinpoint technical and organizational requirements leading perfectly aligned processes. Accordingly, technical innovations are triggered old production management habits must change. Individual measures leading towards our desired target state are defined in an implementation plan, scheduled, and allocated to the responsible parties.

The *result* is a transparent and well-arranged depiction of the targeted future-state with all production processes and their pertaining logistic linkages. The application of the design guidelines under consideration of the entire value stream leads towards overall optimization of the entire production process – not just partial suboptimization of individual production processes.

3.1 Production Structuring

Within a factory, different technologies are usually applied for the production of different products. Similar to the value stream analysis, the value stream design refers to one particular value stream, i.e. a certain part of a factory. The definition of this part is a task similar to the task of creating a *production structure* in factory planning, which segments the entire production into areas, or segments, respectively. As far as the structuring of productions is concerned, there are two basic approaches with a view to production procedures, namely horizontal segmentation and vertical segmentation. The first relates to the characteristics of the required materials, the second to the production procedures and characteristics of the respective products.

Resource-oriented Segmentation

Competencies. The horizontal subdivision of a production; also across production procedures, includes different types of *resource-oriented segmentation*. Once production sequences lose significance as criteria for suitable factory structure, the structuring of said factory by bundling of competencies becomes expedient. This may be done by *types of equipment* or by *qualification* of workers (Fig. 3.4). A typical result of resource-oriented segmentation is the job-shop production, where areas are defined by the respective job-shop organization pertaining to their operators and/or machines. The individual job-shops are called grinding or milling shops. Workers with similar qualifications are pooled, such as turners and millers. However, the greatest strength of this approach at the same time constitutes its greatest weakness. The concentration of technological competencies and technical skills in a confined area specifically furthers specialization and top performance, but at the same time, though, distances are created between areas needed for the same products.

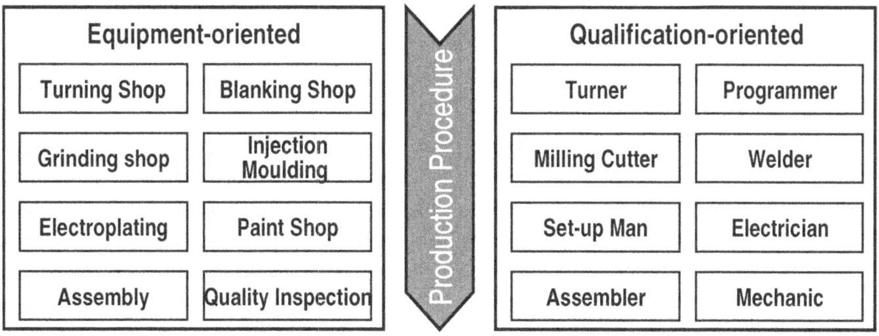

Fig. 3.4 Resource-oriented segmentation

Monuments. The structuring across production procedures, however, contradicts the value stream idea, as it impedes the continuous observation from the shipping end to the receiving dock, thus handicapping the flow principle rather than furthering it. We shall therefore not delve into this structuring principle any further herein. Many factories, though, possess structures developed in accordance with said principle – the value stream analysis then identifies the problems resulting from the respective interfaces. In large *centralized plants* in particular, resource-oriented segmentation seems reasonable. Due to economies of scale, these technical monuments often show favourable costs per piece, in particular if production processes place specific requirements on infrastructure and media supply, such as soundproofing, dust-free rooms, test media or special processing plants. These economies of scale, however, are always accompanied by higher caused by complications in the production process, namely when it comes to channelling all parts produced in the various previous processes through this one monumental chokepoint.

Product Family-oriented Segmentation

Product families. Vertical segmentation breaks down a factory in accordance with one particular product family's production procedure. All production processes required for this particular product family are thus grouped together in a segment. Other than with horizontal segmentation, the same technologies will still be required by different segments. Similar to the value stream analysis (Sect. 2.1.2), the product families are determined with a view to production procedures as well as raw materials, parts and products (Fig. 3.5). The product family-oriented, value stream-appropriate structuring of a production will always result in *production procedure-oriented* and *product characteristic-related* segmentation. Accordingly, in value stream design, the first step in the structuring of a production is always the determination of product families.

The structuring of productions by product family-related logic based on standardized production process sequences is a challenging task, in particular in connection with the job-shop production of parts and subassemblies. The successful stringent determination of part-related families will turn an intransparent job-shop organization with complex control mechanisms into a flow-oriented parts production. In this context, a revision of existing work schedules with a view to process sequences and machine allocations might be a good idea. In some cases, process sequences may be swapped around by changing certain fixtures or allocating machines differently. Special process steps for certain parts may possibly be conducted parallel to the primary operating time and thus do not need to be considered specifically, even though their sequence order may vary for different parts.

Business types. For the purpose of conducting a value stream analysis, the segmentation into product families is sufficient. For the redesign of a production, however, these product families must be reviewed with respect to their suitability for the creation of an appropriate production structure. Accordingly, a further sub-division of our product families according to criteria like piece numbers, demand development, market trends and competitive environment may be expedient from a *market perspective*. The second step of the product family-oriented segmentation thus identifies business types to help certain production segments to selectively satisfy specific market segments by way of one particular product family (Sect. 3.1.1).

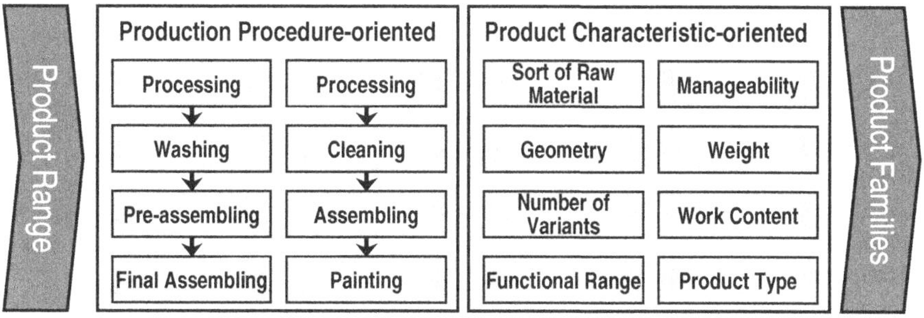

Fig. 3.5 Creation of product families – first step of product family-oriented segmentation

Resources. At times, though, the accumulated capacity requirements of one product family are not high enough to occupy *entire resources*. In this case, two product families partially sharing resources should be grouped together into one production segment. Only with the allocation of resources in step three of our product family-oriented segmentation the practicability of a production structure can be validated (Sect. 3.1.2). Accordingly, value stream design is conducted by production segments, not product families.

Production structuring targets

A Goals requiring product family-oriented segmentation for optimal realization:

- Disentanglement of material flows and transparent shop floor structure.
- Clear allocation of product responsibilities as to targets concerning flexibility, quality, time and profitability.
- Minimization of fault prone interfaces.

B Significant goals in product family-oriented segmentation which may equally be achieved through resource-oriented production structure:

- Higher segment competencies with respect to segment-specific success-related decisions.
- Each segment focused on market-specific customer requirements
- Staff involvement in changes as to production sequences and processes for enhanced motivation and performance.
- Increased staff flexibility through multiple qualification and job rotation.
- Allocation of indirect areas such as logistics and quality assurance to individual segments for the improvement of performance profiles und cost transparency.
- Visualization of target achievements on segment or team level by way of production control boards.

C Goals which in product family-oriented segmentation may only be realized to a limited extent:

- Concentration of technological competencies and technical skills.

3.1.1 Business Types

The criteria used for the determination of product families are of purely technological (production process) and technical (product design) nature. The customer perspective, a factor of central importance in the value stream analysis, is not given any particular attention just yet. In an analysis, a value stream may refer to highly heterogeneous customer requirements and markets. However, certain customer expectations concerning delivery times, packaging, mode of dispatch, order quantities, order frequency and/or distribution channels often require a different type of production control. These different types of production control may be realized within one segment or individual production segments may be created for certain customers or market segments.

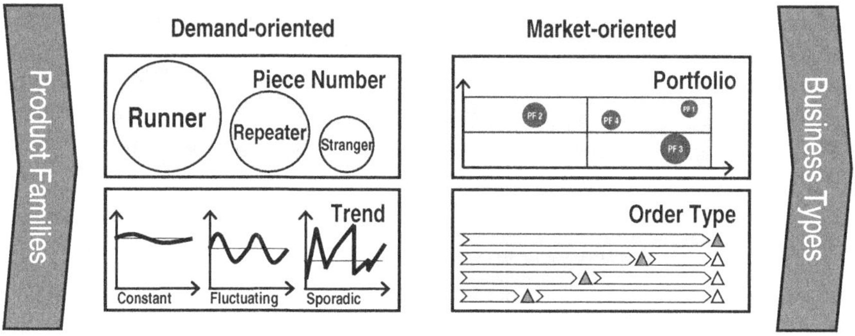

Fig. 3.6 Definition of business types – second step of product family-oriented segmentation

The market-specific sub-division of product families in the second step of our segmentation may be conducted in two different ways: as a data analysis of demand quantities and development on the one hand, or, on the other hand, in the form of a strategic classification of the products as to their competitive positioning in the respective markets (Fig. 3.6). These observations lead to the differentiation of different *business types* with a view to the production procedure within a product family.

Demand-oriented Segmentation

The so-called ABC analysis is particularly suited for a *data-oriented* analysis of the customer demand with respect to the different variant quantities within a product family. Owing to their major significance for the segmentation and the conception of production control it is going to be introduced in greater detail in the following.

The *ABC analysis* enables the sorting of large numbers of items according to relevance and to classify the same as 'A', 'B', or 'C' items with the aid of threshold values. Besides, the results can be graphically illustrated in a sorted cumulative curve. The classification is based on the assumption that the evaluation parameters of the analysed items significantly differ in relevance; accordingly, a small percentage of items will constitute a significant part of the accumulated evaluation parameters. Though comprising a relatively small number of elements, the A-items form a rather high portion of the overall volume, while the relatively large number of C-items only accounts for a very small contingent. With this method we may thus pick out items of particular importance and allocate them to classes, choosing different evaluation parameters. The evaluation by article numbers suggested below emphasizes the logistic aspect of the production.

Class of Article	Variant of Classification: regular	Variant of Classification: inverse
A	70% of Quantity	10 % of Articles
B	+20% of Quantity	20 % of Articles
C	+10% of Quantity	70 % of Articles

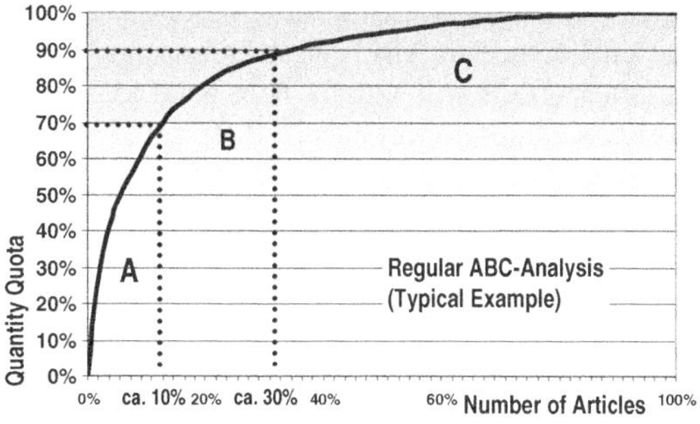

Fig. 3.7 Regular and inverse ABC analysis

Selected variants or item numbers, respectively, are grouped together in a product family, observed from the production perspective and evaluated with respect to their annual piece numbers. The articles are sorted by numbers of items, i.e. article no. 1 will be the one with the highest number of items, while the article with the highest ordinal number will comprise the smallest number of items. This results in a typical cumulative curve illustrating the accumulated annual demands of several articles (Fig. 3.7, right). Equal distribution would result in a linear relation; the respective graphic illustration would be a diagonal. Depending on the respective spread as to article numbers, the curve will bulge upwards to a greater or lesser extent. The example below shows a rather flat curve, well within the typical range according to experience.

ABC classes may be determined in two different ways. The *regular ABC analysis* is based on piece number thresholds, where articles allocated to class A comprise 70 per cent of the cumulative piece numbers. The following 20 per cent consist of class B articles, while class C articles account for the rest (Fig. 3.7, left). The suggested thresholds of 70/20/10 were determined for pragmatic reasons, but they could easily be determined differently – the relevant literature includes numerous other examples, defining A class articles as those with piece numbers of anything between 60 and 80 per cent, which according to experience equals 5 – 15 per cent of the overall volume. The *inverse ABC analysis*, on the other hand, starts from the percentaged proportion of all articles, with class A for instance being defined as 10 per cent of all articles, and from there works backwards to determine the respective piece numbers. According to experience, the classification thresholds for an inverse ABC analysis are best set at 10/20/70 (Fig. 3.7, left). In the classification by piece numbers, both the regular and the inverse ABC analyses provide typical illustrations of the accumulated annual demand, though with slightly shifted classification limits.

With the terms '*runner*', '*repeater*' and '*stranger*', the classes determined by piece numbers are given descriptive denominations. Depending on the respective product characteristics, a runner's requirements on production facilities and production process will normally differ from those of a stranger. While the one requires a higher degree of automation, the other will be hindered by higher changeover effort. While runners achieve a high stock turnover, even in make-to-stock production, strangers remain in stock far too long or are never available when needed. Accordingly, a difference in ABC classes may suggest the formation of an additional segment, otherwise it will usually be a good idea to implement different planning and control procedures for articles of different classes within one segment (Sect. 3.3.3). Depending on the respective A, B or C class, we may be looking at different business types, which will either be allocated segments of their own or be processes in individual production procedures of their own within the same segment.

Another form of mass data analysis is the so-called *XYZ analysis*. This mass data analysis is based on the variation coefficient and classifies articles according to whether their time-related customer demand remains stable, shows high fluctuations or maybe occurs only sporadically. As this approach is only of minor importance with respect to segmentation, it shall not be discussed any further here.

Market-oriented Segmentation

For the selected targeting of certain market segments or major customers it may be a good idea to create specific segments. Also, some customers like to audit the production of their products – the provision of customers with different requirements from highly diverse industries may also give rise to customer-oriented segmentation. Different distribution channels may place different requirements on packaging, modes of dispatch, delivery times, delivery quantities and container quantities. One and the same product may for instance be shipped to individual customers via parcels service, large amounts on pallets may be transported to a marketing subsidiary via own truck or be sent to retailers on mixed pallets via forwarding agent. Depending on the respective vertical range of manufacture, these distribution-related aspects may also influence segmentation. All such *market-oriented* product classifications are of strategic nature and – contrary to mass data analyses – conducted by qualitative methods.

The *portfolio analysis* is particularly suited for market-oriented product classification. Two parameters define the two axes of a portfolio, which as a rule may assume two values, i.e. high and low, though depending on the parameters chosen, other values would also be possible. The result comprises four fields illustrating the four categories of a two-dimensional matrix. Each of these categories is characterized by their different targets which must be strictly observed in order to maintain said categorization. Each field of the portfolio may thus be allocated specific targets and options for action, which then become binding for all elements of the respective field. Two particularly useful portfolios with respect to segmentation, the market opportunities and market attractiveness portfolio and the success factor portfolio are going to be introduced in the following. Besides, business types may be defined with the aid of the customer decoupling point.

Market opportunities and attractiveness portfolio. For the purpose of classification in accordance with the market opportunities and attractiveness portfolio developed by the Boston Consulting Group, the already defined product families are further categorized by their respective current *market shares* and expected *market growth* according to the criteria of 'high' or 'low'(Fig. 3.8). They are then entered into said portfolio as circles, the size of which reflects their current *turnover*. The 'star' field contains product families of sound competitive standing in fast growing markets. This field aims at increasing or at least sustaining an already high market share. The 'cash cows' generate the highest turnover. However, as they are not part of the specific competencies of the company in question, a gradual downsizing of this particular market share may be sensible for strategic reasons. As far as the 'question marks' are concerned, strategic reasons call for great caution and intensive market investigation with possible selective upsizing or downsizing, respectively. Product families with small market shares in markets with little opportunity for growth ('poor dogs') should be eliminated. As a superordinate goal for the two fields with low market shares, low turnover should be aimed at.

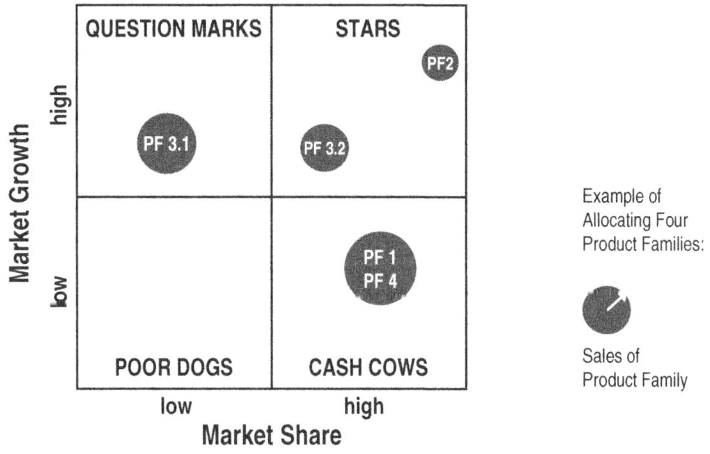

Fig. 3.8 Market opportunities and attractiveness portfolio

In the portfolio, however, the *product families* segmented by production process and product characteristics in our first step, will not always be able to be *allocated* without ambiguity. Product families with products from two different fields will have to be broken up into two business types accordingly (cf. product family no. 3 in PF 3.1 und PF 3.2 in our example below, Fig. 3.8). Product families with identical market criteria may be depicted together in the matrix (product families 1 and 4 in our example, Fig. 3.8) though the existing separation into product families remains unaffected. However, differences may also be reflected within one field (in our example, product families PF 2 and PF 3.2 are 'stars' of different specificities, Fig. 3.8).

The strategic decisions derived from the portfolio have a certain impact on the allocation of *resources* and on investment. 'Cash cows', for instance, need reliable machines; their high volumes and very infrequent changes enable make-to-stock production for an already saturated market which expects short delivery periods for such standardized products. For the 'stars', though, capacity utilization plays a minor role, not least because potential growth must still be possible. 'Question marks' require high flexibility to enable possible alterations in resource application in reaction to changing market developments. Apart from that, order-related production should be aspired rather than make-to-stock production, which would entail the additional risk of goods becoming outdated. 'Poor dogs' are rather not invested in instead they will be allocated older machines or machines which are less suited for other product families. Accordingly, the achievement of logistic goals is of lower priority here than for other products.

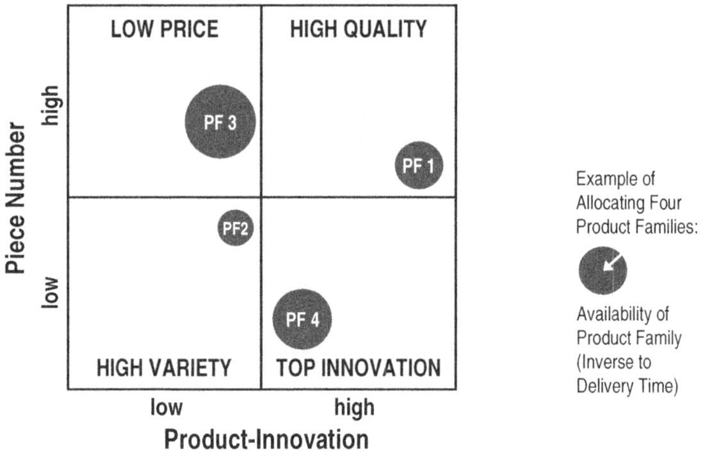

Fig. 3.9 Success factors portfolio

Success factors portfolio. The success factors portfolio classifies product families by *piece number* and *level of innovation* (Fig. 3.9). Depending on the value – high or low – different goals may be pursued to succeed in the market. In order to succeed with a low degree of innovation, market-related, favourable prices will enable high piece numbers in order to assume cost leadership with the respective product family ('low price'). Inversely, highly innovative products can only be placed on the market in small numbers ('top innovation'), as their lack of practical experience entails functional risks and a small customer base. Large piece numbers of innovative products may only be achieved once high quality can reliably be assured and the respective product family has become leading in the market with respect to competence ('high quality'). Low piece numbers of non-innovative standard products should be avoided unless a position as a niche supplier can be assured and a large variety of customer desires can be satisfied by way of conventional technology ('high variety'). Easy and fast product availability, i.e. the *delivery time,* is a success factor which figures in all four fields with varying degrees of priority and preferably should be as high as possible. Accordingly, product families with short delivery periods are entered in the portfolio as large circles. This portfolio takes up the production targets formulated in the logical target square (Sect. 1.2.2).

Depending on the respective success factors, the required resources as well as the implemented production procedures will have to meet different requirements. Innovative products cannot very well be produced in aged low-cost factories, unless potential customers are to be scared away rather than shown around the factory – in particular if the product innovation is mainly due to product design. In line with a high-tech factory, low-priced products could be produced in a separate segment with highly automated machines. Delivery periods should be minimal for such standard products, as numerous competitors are to be expected in this field. However, automated production of low-cost products is only profitable at locations with high wage levels. The production of such product families at low-wage locations would be better carried out with cheaper, manually operated equipment. Depending on the location, one and the same product family may thus place highly differing requirements on the production and therefore also on the segmentation. The explicit observance of the success factors in the structuring of production thus enables *segmentation across locations*. However, numerous other factors need to be taken into consideration in this context, from supply logistics to distribution logistics, but shall not be discussed in further detail here.

Customer decoupling point. Business types derived from the various customer specificities usually differ with a view to the position of the *customer decoupling point*. This is the point where materials, parts or products are allocated specific customer orders, before this point the pre-production is aimed at unspecified, anonymous customers. Depending on the respective levels of variance or customer specificity, the customer decoupling point is to be found at different points of the process sequence. In cases where the types of required resources depend on the range of variants or customer specificity, the creation of different product families is recommended in accordance with the criteria of product characteristics (Sect. 2.1.2). On the other hand, though, and regardless of the aforesaid, different business types may well be determined within one product family depending on variance and customer specificity. This is expedient whenever in spite of similar resource requirements considerable differences in order processing are needed. Accordingly, customer-unrelated make-to-stock production is only feasible if the products in question do not feature any customer specific properties. Likewise, it is not recommended to have all variants of a make-to-order production on stock, in spite of possible technical feasibility.

Business types

Depending on demand behaviour, market strategies, success factors and the position of the customer decoupling point, different technological and logistical requirements may be placed on a production. Identical requirements are described by the same business type, so that besides clearly defined one-on-one allocations, product families may sometimes include several business types. Likewise, several product families may be allocated to the same business type, which needs to be taken into consideration when segmenting.

Case study. At *Liquipur*, five business types may be differentiated (Fig. 3.10). The first business type is a customer-unrelated make-to-stock production, with only part of the allocated products being painted. The fifth business type, starting with the unfinished housing, conducts the entire production procedure in an entirely customer-specific manner, allowing for customer-specific boreholes in the milling process for instance. Products manufactured in accordance with business type no. 4 may be painted according to the customers' wishes, but are made from processed standard parts. Business types no. 2 and 3 both allow for a high degree of variance in assembling.

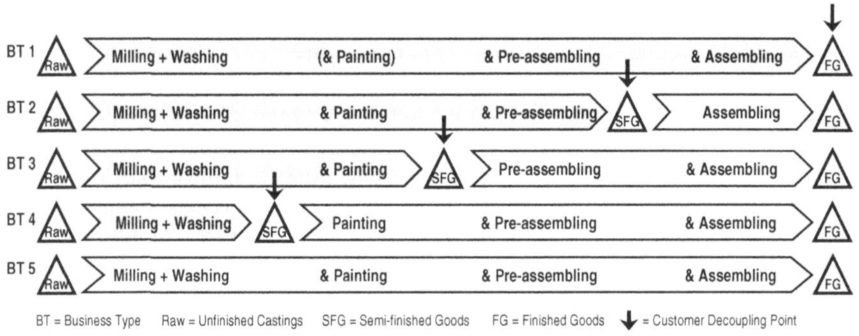

Fig. 3.10 *Liquipur* business types

3.1.2 Allocation of Resources

Now that product families have been determined and business types have been defined in accordance with production procedure-related, product characteristic-specific, demand-oriented and market-relevant criteria, each product family/business type combination needs to be *allocated* the required *resources* in the final segmentation step. The numerous relevant criteria for the structuring of a production generally lead to a large number of different product families and business types. However, the resource perspective soon reveals that there aren't enough different resources to cover all product families and business types individually, – initially a desirable effect, as a relatively detailed subdivision of the product range seems to be a prerequisite for a successful restructuring of the production detached from its current state.

In our third segmentation step, the product families already broken down into business types are grouped together in such a way that mutually utilizes resources are used in one segment only. It is crucial that the entire value streams of a *product family/business type combination* are comprised in one segment, a value stream may not flow across segments. The production procedure schemes are thus partially grouped together in accordance with relevant resource-related constraints (Fig. 3.11). As a result, one segment may include one or more product families with one or more business types; the allocated resources determine the respective

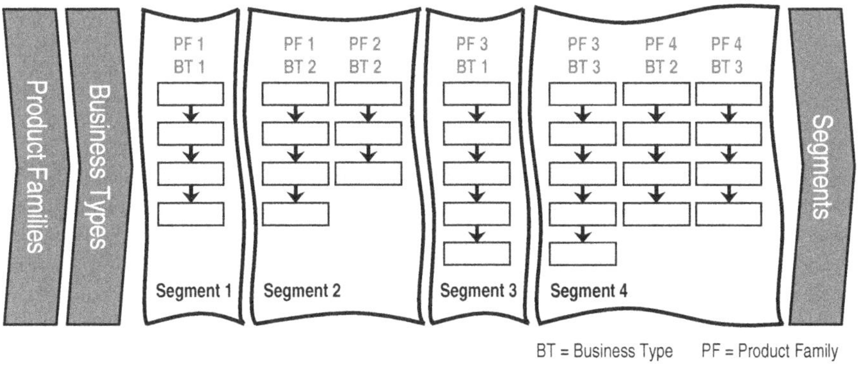

Fig. 3.11 Allocation of resources – third step in product family-oriented segmentation

segments. The illustration above shows various possibilities – one segment including two product families, one or – by way of exception – two business types, or one to three combinations grouped together.

Each segment must only comprise similar product families and/or business types in accordance with the following three *similarity criteria*:

- Similar *production procedures*, i.e. one of the product families will typically skip a production process or pass through an alternative process while all other processes are identical.
- Similar or identical *equipment* required for the overall process: None of the respective product families feature any product characteristic-related requirements that are not shared by all other product families of the same segment, i.e. requirements are identical with a view to degree of automation, media used, installation space of resources, means of transport, etc.
- Identical or similar *business types*, e.g. product families grouped together having the same customer decoupling point and/or use identical parts, but needing different resources.

Sometimes, though, certain groups will be formed in spite of otherwise significant differences, for instance in cases of identical resources being required by several product families within one production process. This typically happens with resources of higher capacity than others, often purchased because of their lower costs per piece. This type of equipment is highly suited for resource-oriented segmentation and is thus somewhat at odds with the value stream observation intended herein. This type of product family grouping, however, will always constitute a compromise rather than a first choice, as most value streams of different product families overlap at some point – sometimes even get entangled in the process. The seeming flexibility of equipment able to produce all manner of different parts soon turns into an impediment, though, an obstacle to flexibility, as parts will frequently have to wait to be processed in large lots at such costly bottlenecks.

However, these unfavourable exceptional cases usually lead to the creation of a *central resource* frequented by (almost) all respective value streams. Strictly speaking, product family-oriented segmentation should be impossible in this context, i.e. one such technical monument would virtually undo the advantages of segmentation. In such cases, however, such large indivisible equipment may be accepted as an intersegmental exception. The monument's available capacities are temporarily allocated to the different individual segments, a production planning task realized with the aid of an intersegmental levelling box (Sect. 3.4.2) which works more or less like the widely used capacity reservation, but is more flexible, as the segment-related capacity allocation may vary according to the respective demand mix. However, capacitive overloading of the individual segments should be duly avoided.

Case study. At *Liquipur*, seven product families are grouped together in four segments (Fig. 3.12). Segments no. 2 and 3 conform to the product families with identical numbers and aluminium cast housings. Their business types, however, differ: Segment 2, just like segment 1, is subject to make-to-stock production, segments 3 and 5 together constitute special customer-specific customizations, starting as early as the milling process (Fig. 3.11). Segment no. 4 comprises two serial product families, which are assembled in different lines, but share a jointly used automatic paint shop. Segment 1 includes various products of high variance and low piece numbers. The production procedures of this segment differ with a view to vertical manufacturing integration, but use similar resources in the form of stand-alone assemblies and manual paint stations.

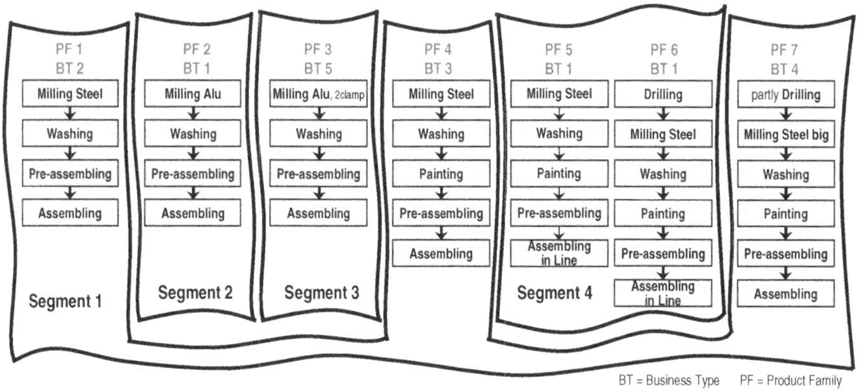

Fig. 3.12 Segmentation at *Liquipur*

With the completion of the segmentation, the individual production processes are allocated to their respective resources. These are usually identical to the current state, or, in cases of an impending capacity extension, supplemented as 'clones' of existing resources. This allocation defines the various segments and should be considered a first rough preliminary design, a merely interpolation from the current state. It constitutes the basis for the subsequent dimensioning and redesign of the production processes (Sect. 3.2). However, customer takt time-oriented improvement of processes and resources may significantly influence the segment-specific capacity and resource requirements, which may in turn retroactively effect the segmentation, i.e. the respective segments may need to be iteratively adjusted accordingly.

Product family-oriented segmentation

The essential *goal* of product family-oriented segmentation is a strict disentanglement of the material flows to create a transparently structured shop flow with a view to products and production processes, thus implementing the customer perspective in the factory structure.

Approach In the first step, the range of products is sub-divided into product families with a view to production procedure and product characteristics. In the second step, these product families are allocated suitable business types depending on demand behaviour, market strategies, success factors and customer decoupling points. Finally, in the third step, the resulting product family/business type combinations are allocated their required resources.

Result The segments of the factory have now been determined and may subsequently be further developed technologically and organizationally.

3.2 Production Process Design

The segmentation of a factory (Sect. 3.1) is followed by technological production process design through dimensioning and redesign. The *dimensioning* starts with the capacitive layout of the resources required for the respective production process. If the relevant operation times result directly from the applied technology, the number of required identical resources must be determined (Sect. 3.2.1). In addition, however, the respective resource's performance profiles and/or technology applied may be reassessed. However, the resulting changes to the resources may accordingly also change the number of required alternative resources per process step. Capacity dimensioning is of particularly high significance in manufacturing processes, as poor dimensioning leads to insufficient capacity utilization or obstructive bottlenecks.

In the subsequent redesign of the production processes through maximum comprehensive *integration* in continuous flow processes, the previously dimensioned resources are grouped together accordingly, possibly across various production steps (Sect. 3.2.2). The sometimes drastic changes effected to individual production processes compared to the current state may also strongly affect the utilized resources and alter the resource requirements accordingly. Process integration is particularly suited for assembling processes, as experience has proved this to be where the highest efficiency increase may be achieved. Similarly, processing machines may be coupled to form lines, and/or individual manufacturing processes may be integrated into assembly lines.

3.2.1 Capacitive Dimensioning of Resources

In value stream design, the customer takt time determines the target for the capacitive dimensioning of the resources. The operator balance chart reflects the level of success with a view to the *capacitive dimensioning* of a production in relation to the takt time. In general, cycle times should be as uniform as possible and should never exceed the customer takt time. Compared to the current state, the individual production processes' cycle times must now be better aligned and appropriately dimensioned with respect to the customer takt time through elimination of bottle necks and excess capacities. The customer takt time as a standard of evaluation provides the first design guideline for an optimal value stream:

> **Design Guideline 1:** *Adjustment to takt time*
> The capacity availability of a production has to be adjusted to the customer takt time throughout in order to reach a balanced capacity profile of the value stream.

Layout of the available capacity to match the customer demand, however, does not automatically require each and every production process to be equipped with resources of identical cycle times. Based on the respective planning and control logic, however, additional requirements may be placed on the capacitive dimensioning of the value stream. As far as flexibility and delivery reliability are concerned, excess capacities, i.e. decidedly shorter cycle times, may result in distinct technological and organizational advantages for some parts of the production flow; the related effects on costs may differ considerably depending on the prevailing circumstances. Therefore, the correct balancing of the capacity profile (Sect. 3.4.3) is a design tasks of vital importance. The dimensioning of resources may also need to be *iteratively* revised during the course of the value stream.

A segment's *resource requirement* for each production process is indicated by the number of its parallel resources (always a whole number). Should the operation time exceed the customer takt time, at least one additional resource will be required for the respective production process in order to make the cycle time – the quotient of operation time and number of resources – shorter than the operation time. If we use the customer takt time to match the available capacity to the customer demand, the minimum number of required resources is determined through the division of operation time and customer takt time. The result may have to be rounded up, as resources are of course indivisible. This relation is true under ideal conditions, such as those found for instance in flexible, low-maintenance assembly lines, though, changeovers, maintenance and malfunctions as well as quality problems will usually reduce the available capacity of a resource (cf. Sect. 2.3.1). As this often decidedly lowers the actual capacity of the relevant resources, the number of the respective resources must be increased accordingly. The number of required resources is thus calculated as follows:

$$\#\,\text{Res} = \text{ROUNDING UP}\left[\frac{OT}{TT \times (1 - COF) \times UP \times (1 - QL)}\right] \tag{3.1}$$

with: # Res resource requirements
OT operation time [time unit]
TT takt time [time unit]
COF factor for losses through changeover [%]
UP uptime [%]
QL rate of quality losses [%]

In make-to-order productions with differing operation times for different variants, however, the capacitive layout requires greater calculation effort. In this case, the consistent operation time in equation 3.1 is substituted by the mean operation time weighted by piece numbers.

As a result, the production process capacity should be dimensioned in such a way as to keep the cycle time slightly shorter than the customer takt time. As the resources' available capacities are scaled in a *step-fixed* manner, rounding up leads to excess capacities. This obligation to round up to whole resources has a particularly negative effect on the capital expenditure in cases where mercantile rounding would have resulted in rounding down, i.e. where the resource requirement to be rounded is less than half a machine. Alternatively, the resource requirement may be decreased to achieve the lower value, using one of the following possible approaches:

- Increased productivity of the respective resource through reduction of the operation time per part.
- Increased uptimes through higher maintenance efficiency, i.e. conducted in a faster and better organized manner to better prevent malfunctions.
- Considerable cuts in changeover times through technical and organizational measures such as changeovers parallel to the primary operating times, changeover-optimized equipment and/or larger tool magazines.
- Increased operation times of resources compared to the original calculation base, for instance through introduction of weekend shifts or by working through breaks.
- Re-allocation of product variants to different product families in different segments, thus shifting problematic resource requirements to another segment with excess capacities.

Another possible way to prevent excess capacities would be to round down the calculative capacity requirement and allocate the remainder to a differently dimensioned resource. This could possibly be an existing older machine with a lower degree of automation or a lower-priced machine of lower efficiency could be purchased. This would of course raise the unit labour costs – but only proportionally and at decreased unit production costs, too. In this case, the available capacity would consist of combined *heterogeneous* resources, thus enabling better vernier adjustment.

Case study. The *Liquipur* dimensioning is illustrated by the number of milling machining centres (Fig. 3.13). In oil filter product family no. 2, an annual piece 192.000 pieces of an operation time of 164 seconds need to be processed. In order to achieve a balanced value stream, the dimensioning is based on a regular three-shift model. To fulfil the relevant customer takt time of 94.5 seconds, however, a minimum of 1.7 machines, rounded up to 2 machines, is required. Taking into consideration a changeover loss factor of 3.4 per cent, an uptime of 85 per cent and a 2 per cent scrap quota (cf. Eq. 2.24), the resource requirement rises to 2.2 per cent. As a result, three costly, highly automated machines with a joint excess capacity of 36 per cent would be required. In the current state, though, only two machines are required. These keep up unmanned operation during break times, thus raising the customer takt time to 108 seconds (Eq. 2.12) and pushing the resource requirement down to 1.9. The future-state conception will now aim to get by with only two machines through decreased changeover effort, increased uptimes and minimized rejections. The respective result is shown in the operator balance chart (cf. Fig. 3.65).

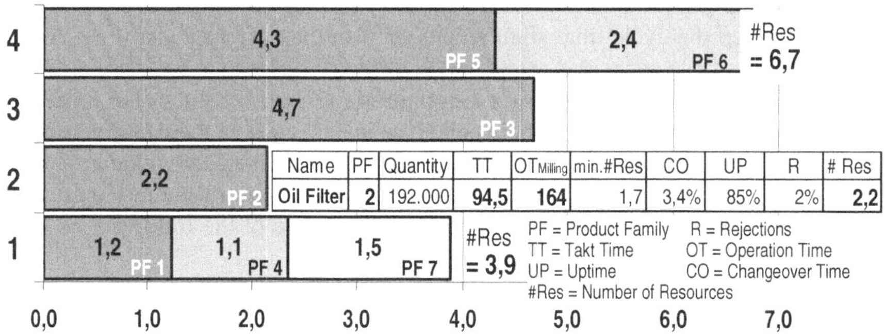

Fig. 3.13 Dimensioning at *Liquipur*

Regardless of the aforesaid, though, *excess capacities* may well be created on purpose – even beyond rounding up, as this is the only way to flexibly react to short-term demand fluctuations with a view to overall volume or variance at short delivery periods – especially in cases of strongly differing processing times per variant. With a low degree of automation, such capacity reserves may be highly cost-effective and are thus particularly suited in assembly, which is located further downstream, closer to the customer and therefore require much higher flexibility in any case.

Over longer observation periods, however, the targeted planning of excess capacities may also be significant with respect to dimensioning. As a rule, value stream design is applied in the *planning case* of continuous improvement of existing production, aiming at an immediate realization of individual measures, with the future-state conception indicating the direction for the individual steps and the current sales rates being used to define the customer takt time.

On the other hand, though, value stream design is equally suited for medium- or long-term rescheduling and/or complete re-planning. In both cases, the customer takt time must be determined in due consideration of the intended increase (or possibly decrease) in piece numbers. In capacity dimensioning concerning re-scheduling and re-planning, the customer takt time is thus prognostically forecast, which results in excess capacities compared to the current state. As a tool for long-term production planning the operator balance chart is thus an inherent part of the value stream management (cf. Sect. 3.6.2).

Capacity dimensioning

Goal For the capacitive dimensioning of resources, the cycle times of the individual production processes must be balanced and aligned with the customer takt time.

Approach As a capacity dimensioning tool, the operator balance chart illustrates the available capacity in relation to the capacity requirement. Since all resources must be provided in whole numbers, the respective available capacities are scaled in a step-fixed manner and therefore not easily aligned with the capacity requirement. The design of resources and production processes is thus faced with the task of precise matching of the customer takt time through increased production performance or minimization of changeover times in order to avoid rounding up in the determination of the resource requirements.

3.2.2 Process Integration and Continuous Flow Production

With completed definition of type and number of the required resources, the production processes are roughly dimensioned. Usually, existing resources are drawn on or later versions of similar resources are purchased as replacements or supplements. The next step reviews the distribution of the value creation with respect to the production processes and their allocated and dimensioned resources. The previous current state of the production, so far only structured by product families, is now to be further developed towards the newly configured future state of the relevant production processes. On the one hand, applied technologies are examined as to their suitability in a value stream–oriented production, on the other hand, machine-related, organizational boundaries of the individual production processes must be redesigned.

An *ideal production* combines industrial division of labour with manual piece production. It realizes productivity gains achieved through structuring of the work tasks and specialization of working materials and workforce. However, it is also perfectly in tune with the customers' wishes because each required product may be manufactured individually. Providing possible technological obstacles can be eliminated, this ideal production can be realized with the aid of the flow principle.

The flow principle spatially arranges individual workplaces by work task sequence (DIN 33415). The division of labour is implemented by distribution of the work tasks among the workplaces where each work place must be manned by precisely one operator processing exactly one part; the result is called *continuous flow production*. After completion of each process step the part in process is immediately passed on to the following process step. This is very appropriately also referred to as 'one-piece-flow'. Starting from the raw material, ideal production thus manufactures a product which was only just ordered in an uninterrupted chain of process steps used to optimal capacity.

Basically, a continuous flow production is a single production process subdivided into a sequence of process steps. The ideal value stream may thus be depicted as a process with inflowing raw material, and in which an end product is manufactured without the slightest interruption of the production process. This target state is achieved through *integration* of formerly separated production processes. The first design task in the future-state mapping of a production is therefore the conception of such 'one-process-factory' through consolidation of as many process steps as possible in an integrated production process or continuous flow production, the respective capacities of which are aligned to the customer takt time. Accordingly, the following design guideline leads towards an ideally laid out production:

Design Guideline 2: *Process integration*
Production processes have to be consolidated as continuous flow production or integrated production as far as possible to minimize waste.

We differentiate two basic approaches of process integration – a technology-related one in manufacturing and a production procedure-oriented one in assembly. In the first of the two, manual deburring or drilling, parallel to the primary operating time, already constitutes a simple form of process integration. Real technological integration, however, is achieved through consolidation of several machining operations, such as the three metal-cutting procedures of drilling, milling, and – to a limited extent – turning in an automated multi-axle machine centre. Accordingly, the sequential steps of turning, milling in two clamping processes and drilling, normally carried out by several machine tools, may be substituted by *complete machining* on one machine only. This approach is improved through further machine-related technological development such as the application of rotating tables in turning, integrated grinding spindles, or integrated lasers for surface treatment (hardening).

In the second case, several assembly steps and component assemblies are combined in a continuous *flow assembly*. This approach is focused on increased work performance through significant improvement of primarily manual work processes. The integration of individual semi-automatic machine tools converts a continuous flow assembly into a flow production. Accordingly, both approaches may be combined to fit the respective requirements.

In order to consistently implement the flow principle in a factory, however, batch processes must converted into *continuous flow processes*. In some cases, washing machines may achieve on-piece-flow-cleaning of parts, possibly even immediately following the upstream metal-cutting process. In surface and thermal treatment (painting, galvanizing, hardening), though, the plant-specific batch-processing monuments apparently seem indispensable to the respective manufacturers.

Over and above the basic characteristics and advantages of process integration, the *conceptional approach* of a flow production shall be outlined in the following sections. Fundamental components of this concept are the determination of the suitable automation level and the line balancing approaches and methods which enable the allocation of work content to the various operators. Differently weighted work distribution makes the flow production even more flexible.

Characteristics of Continuous Flow Production

A comparison of the continuous flow production with traditional discontinuous batch production makes the advantages of this production principle particularly clear (Fig. 3.14). Most striking are the drastically cut production lead times. In addition, the elimination of interim storage results in major improvements with a view to the cost effectiveness of the production. Also, make-to-order production enables variant mixes perfectly aligned with the customer demand. Another, not immediately obvious positive effect of the flow production lies in the system-related improvement of the production quality. These four advantages, which incidentally conform to the four goal dimensions (Sect. 1.2.2), shall be discussed in greater detail in the following.

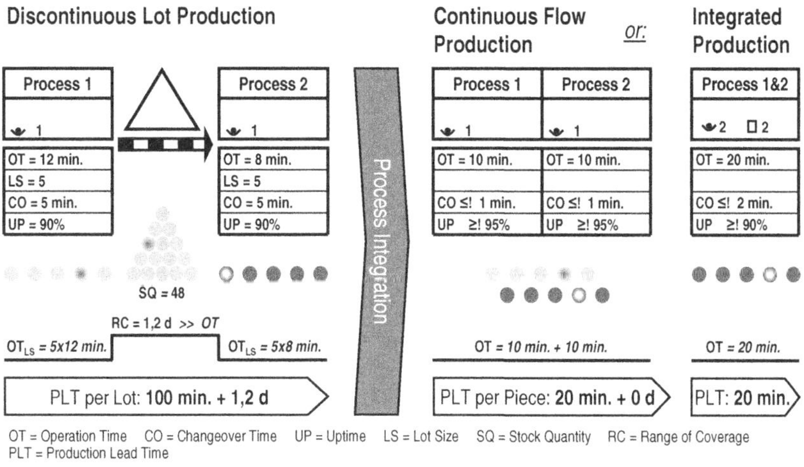

Fig. 3.14 From discontinuous batch production to continuous flow production or integrated production through process integration

Production lead time. The drastic *shortening of production lead times* in batch production amounts to at least the total of all operation times multiplied by the lot size. In general, though, process interruptions account for additional, decidedly longer interim storage or buffer times. In our illustrated numeric example of two processes of one minute's operation time each and a fairly small lot size of 5 pieces, the resulting production lead time is 10 minutes (Fig. 3.14). The interim storage depends on the respective planning and control logic, but will typically take hours (for example in cases of storage areas close to the workplace) or even days (in cases of interim storage until order release for further processing). In a flow production, this time portion, which is often hard to calculate, is eliminated completely. Operators only ever process one part at a time which they then immediately pass on to the subsequent workplace – or, depending on the respective work organization, personally take with them to the next workplace. In our numeric example, the production lead time is thus reduced to the operation time required for one piece, i.e. two minutes.

Cost effectiveness. Thanks to the elimination of interim storage, less capital is tied up and all activities related to transfer to and removal from stock, transport to and from storage locations as well an inventory management become obsolete, thus resulting in significant *productivity gain* and eliminating all related storage space requirement. Process integration also enforces a space-saving workplace layout in order to enable the immediate passing on of the products at all. In assembly, in particular, this leads to a significant increase in *space productivity* as opposed to the traditional production of small batches. In addition, this densification of space cuts out long walking distances and thus also contributes to the productivity increase. However, this applies only in cases where close proximity of the respective resources is actually practicable, i.e. where there are no impeding constructional and/or technical obstacles.

Variant mix. As the flow production is in fact a piece production, any user-defined variant mix may be produced in perfect conformance with the customer demand; likewise, customer-specific production may easily be realized. The implementation of such *maximum variability*, however, is subject to two conditions – minimal changeover times and variant-independent operation times.

Piece production of flexible variant mixes is only possible as long as no changeover times or other technical restrictions apply to the respective sequence. The 30 seconds of changeover time in our numeric example equal 10 per cent of the operation time of the entire lot (Fig. 3.14). Unamended, this changeover time would equal 50 per cent of the operating time, which would make no sense at all in a serial production of high variance. The almost total elimination of changeover times is therefore a vital prerequisite for the implementation of flow production. Minimized changeover times, however, may remain – though of course these constitute waste. A reduction of the changeover time to six seconds in our numeric example wouldn't really effect any changes, therefore the changeover time should be brought down to less than six seconds. In cases where the changeover times cannot be sufficiently reduced, a flow production with *overlapping lots* may alternatively be realized. This takes away the sequencing freedom, but all other flow principle advantages still prevail.

In addition, though, free choice of variant sequence is only possible as long as the operation times per process step are unrelated to the type of variant – as shown in our example. A variant-specific *spread* of operation times may – depending on the respective implementation details – lead to hold-ups or waiting times at the individual workstations, i.e. when fast variants catch up with slow ones and vice versa. Up to a certain extent, this may be avoided through work organization-specific measures and/or minor decoupling of process steps (Sect. 3.3.1); as a result, though, the flow production output will pulsate accordingly. The coordination of a direct succession of variants with very high and very low work contents is rather problematic. According to experience, the maximum spread of operation times in a flow production should not exceed 30 seconds. Products outside this flexibility range should rather be allocated to a different product family. Apart from variant-oriented numbers of operators allocated to the respective processes, a viable alternative would be the definition of *sequencing restrictions* of varying degrees with respect to the succession of the different variants and their subsequent strict observation in the production process.

Even bigger problems are faced, however, with operation times of individual process steps fluctuating in varying degrees depending on the respective variants. In this case, line balancing requires a variant-dependent allocation of operators to the various work stations, which in turn needs to be monitored accordingly. If this is the case for only one work station, that particular station will probably not be integrated in the flow production at all, but be otherwise connected to the remaining flow production (Sect. 3.3).

Fault detection. The flow principle also affects the quality of a production. In batch production, a fault incurred in a process and not be detected until the following process step, will go unnoticed for a long time. In the example, defective parts are marked by differently coloured circular product symbols (Fig. 3.14). The defective second piece of the first process is initially stored and recognized as a rejection as the fifth piece of the second process, by which time the cause of the defect will not necessarily be traceable anymore. Not so in a flow production: The bad part is passed on to the subsequent process straight away and can thus trigger an immediate investigation of causes, which enables significantly improved quality management. In cases of systematic faults, failure costs may be reduced by way of process-related *early fault detection*, as no complete batch of rejections must be produced before the fault is detected in the following process (or interim quality inspection).

Uptime. Similar to immediately detected bad parts, *malfunctions* of resources directly influence the production procedure – i.e. completely disrupt the same. This leads to additional requirements on the flow production design. Batch production, however, works the other way round: If any of the machines in a separated production process fails, all other production processes can simply carry on processing their respective lots. Downstream processes will be able to draw on interim storages – and upstream ones will replenish the stocks before the failed process. Unconnected resources allow for significantly higher uptimes for the entire production and, accordingly, higher output, too.

In our numeric example, both processes use resources of an uptime of 90 per cent each, also resulting in a total batch production output of 90 per cent compared with the theoretical maximum volume (Fig. 3.14), providing a generously dimensioned buffer layout covers the longest period of malfunction. In a flow production, however, the uptimes multiply, resulting in a total uptime of only 81 per cent in our numeric example – and the longer the process chain is, the more drastic becomes the resulting reduction. In order to achieve the same output in our example, the respective uptimes would need to be raised to 95 per cent each. Accordingly, excellent uptimes of the individual components are a vital prerequisite for the smooth functioning of a flow production. This is why flow productions are mainly to be found in assembly and less in mechanical manufacturing.

Readiness to assume risks in lean production

In a flow production, quality problems and malfunctions promptly become noticeable – and accordingly necessitate speedy reactions. Manufacturing seems to be riskier, as malfunctions cannot be cushioned by inventories. On the other hand, though, there are no inventories containing large amounts of initially undetected rejections, which in turn are not available for further processing when needed. In any case, the decisive factor is that the flow production makes both problem areas visible, thus exacting process improvements. In the opposite case, waste could only too easily be hidden by inventories.

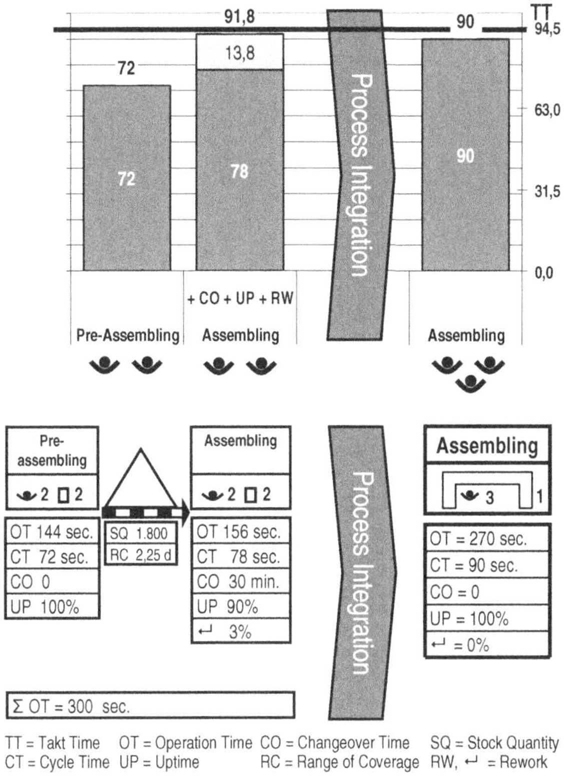

Fig. 3.15 Continuous flow assembly at *Liquipur*

Case study. The *Liquipur* future state integrates the two processes of pre-assembling and assembling into one flow production laid out in a stylized U-shape. Flow production processes are characterized by a symbol in the shape of a stylized U (Fig. 3.15, right). The newly designed assembly line is going to be improved as follows compared with the current state: eliminated changeover times, uptimes increased to 100 per cent, eliminated rework. The cycle time will be slightly below customer takt time – 5 per cent have proved well worth, already allowing for additional planned times. The intended cycle time for the assembly line is thus going to be 90 seconds. With three operators, an operation time of 270 seconds should be feasible. The total operation time at the old assembly workplaces amounts to 300 seconds. A 10 per cent decrease would lower the respective total personnel requirement from four to three operators, resulting in a total *productivity gain* of 25 per cent.

In addition, process integration will get rid of inventories of 1,800 pieces stored between the previously separated assembling processes. Thus, the *production lead time* will be reduced by 2.25 days. The future-state value stream of oil filters will thus consist of five processes, namely shipping, assembling, washing, milling and unloading (Fig. 3.16). The relevant parameters like number of staff, operation times, changeover times, etc. for the flow assembly have already been determined. The process parameters for the other production processes are yet to be determined depending on their respective logistic linkages. This will be discussed in greater detail in the following sections.

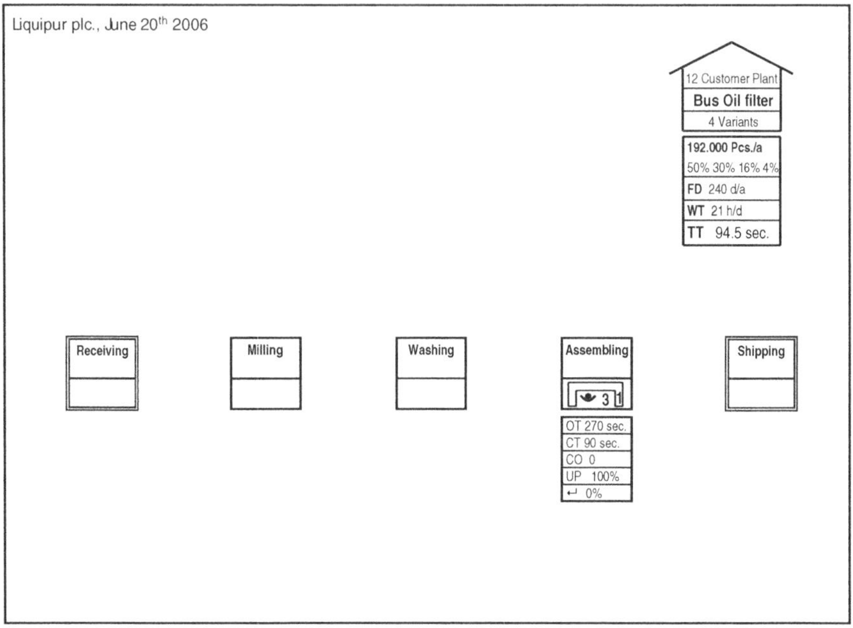

Fig. 3.16 Value stream design at *Liquipur* (1): Production processes and customer demand

Determining the Degree of Automation

The designing of a continuous flow production begins with the determination of the various work content and its pertaining time requirement and the definition of its suitable degree of automation and respective machine capacity. To begin with, the entire work process is mapped and subdivided into individual *work elements*. Video analysis can be very helpful, especially when it comes to the determination of the time requirements of individual work elements. Each work element is then classified as to its value-adding, ancillary, or purely wasteful nature. Finally, the work contents reduced by their *share of waste* are illustrated in the operator balance chart. The results of this so-called 'paper-kaizen' indicate the time parameters for the redesign of the work process to be achieved through better ergonomical layout, allocation of parts and equipment design.

In a flow production, the correct choice of the appropriate resources is a significant prerequisite for low-waste operation. A minimum requirement, for instance, as opposed to a totally bare workbench, would be the application of holding devices for the parts to be processed or assembled, respectively. Very often, the left hand is misused as a flexible and seemingly cost-effective holding-device, which, however, prevents two-handed, value-adding work. Assembly workplaces in particular need ergonomically beneficial retrieval areas for parts with small, assembly-appropriate containers arranged within easy reach. For mechanical process steps, the best possible *degree of automation* must be determined. In this context, four different levels of automation may be differentiated besides the purely manual design of a flow production (Fig. 3.17). As a general rule, an increasing degree of automation entails rising capital costs and decreasing unit labour costs, though at times quality requirements may necessitate a certain minimum of automation. Besides, the type of automation applied also significantly influences flexibility and changeability.

By trend, automated systems may be flexibly applied within their preset limits. Alterations exceeding pre-planned applications, though, would require extensive and costly changes, thus rendering these systems unchangeable. The first degree of automation exclusively refers to either cam-controlled or CNC-controlled processing procedures, including the substitution of inspection tasks of purely 'observing' operators through self-monitoring functions (Fig. 3.17, column 2). In the case of disruption of the normal production process or general deviations of verified parameters from their respective nominal values, the process comes to a halt and the machine emits an alarm. Only in the case of a disruption does the employee need to turn his attention to the respective machine. This principle of so-called *autonomous automation* ('jidoka' in Japanese) was first applied by Toyota for automatic looms, which with each recurring thread breakage automatically stopped and raised a respective alarm ('andon' in Japanese), thus enabling one operator to monitor several looms. This principle may suitably be applied to all automated production procedures, including job-shop production machines. Thanks to the separation of human labour and machine running times through *multi-machine handling*, the relative work portions may be decreased by the respective factor.

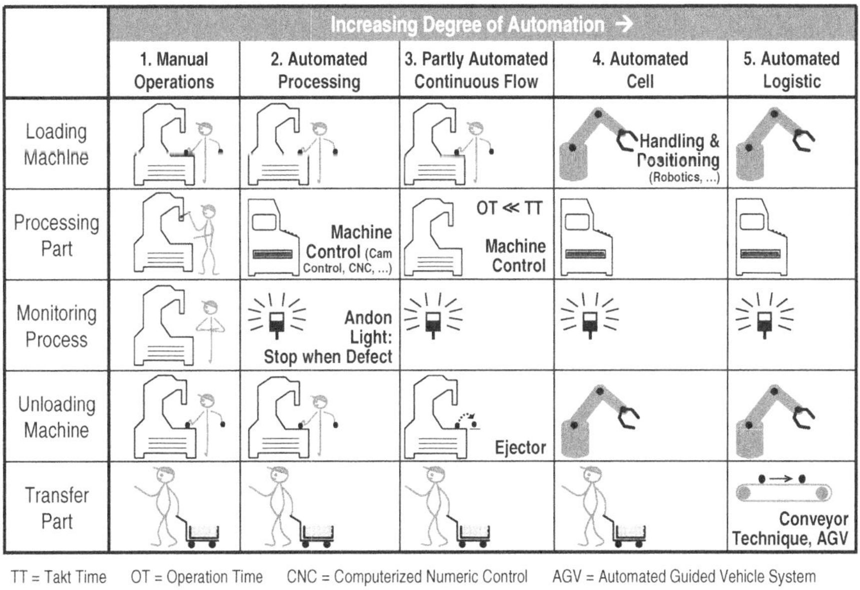

Fig. 3.17 The five levels of increasing degree of automation

In a semi-automated flow production, an additional element is introduced: the *automated ejector*, which after completion of each machine cycle automatically ejects the part from the holding device. The machine is thus prepared for reloading by the operator without any need for extra preparation of the holding device (Fig. 3.17, column 3). As a rule, the automation of the ejection process is very inexpensive, as the ejecting does not require any precision. On the other hand, it saves immense working time, as one entire handling process is eliminated. Another important principle for the realization of this cost-effective type of automation – also referred to as 'low-cost automation' – is a generous *overdimensioning* of machines, a labour-cost-optimized approach: No human needs to wait for a machine – the machines wait for the humans who during machine running times may attend to other tasks at other workstations. In addition, the overdimensioning of machines in a flow production helps increase capacity flexibility.

Decidedly more challenging than automatic ejecting, however, is the implementation of automatic work piece infeed, which requires highly precise positioning and securing (Fig. 3.17, column 4) and may thus hardly be achieved without the aid of *robotics* or similar features suitable for the selective handling of pieces. This comparatively high technical effort, though, is accompanied by a wide range of applications. Other than in semi-automated production, highly complex motion sequences may be automated. Previously purely manual activities may be expanded to handle greater weight and conduct operations in hazardous or even harmful environments. Both would also be possible without automation, i.e. manually controlled, but only with the aid of appropriate machinery and handling aids.

Very tempting is often the automated transport of the parts to the respective subsequent process step by way of *conveyor technique*s such as conveyor belts, chain conveyors or automated guided vehicle systems (Fig. 3.17, column 5). This would provide maximum automation, even a factory 'devoid of people' would seem possible – at least with respect to activities directly involved in the production process. Like all automatisms, this holds a particular fascination – but equally detracts from the inherent grave disadvantages, namely the high costs involved: for the initial installation of the respective conveyor technique – with in some cases may even 'pay off' – but also for later alterations, should a component need to be changed or supplemented due to changed procedures. This could even constitute a risky impediment to innovation, should conversion costs for conveyor technique impede possible improvements to the production process.

Degree of automation

The *appropriate* degree of automation must be chosen in a suitable way to achieve a sufficiently flexible, top-quality, cost effective production of speedy flow for each individual case. Generally, multi-machine operation should be aimed for, which requires the implementation of fault-reporting signalling devices in line with the principle of autonomous automation. In a semi-automated flow production only the most basic partial operations are automated, thus consistently preventing operator waiting times – providing machines are used to capacity.

The clearly more expensive and costly automation of mechanical handling enables a completely different design of the production process and should thus not simply imitate human movements. Automatic transfer of the work pieces, however, may negatively influence the changeability of a production.

Line Balancing in Continuous Flow Production

Process integration in flow production requires *line balancing* between the integrated process steps. While isolated production processes may adjust their capacity to the customer demand with little or no problems through overtime and/or addition or elimination of individual shifts, flow production requires all processes to observe identical customer takt times to avoid costly waiting times of operators.

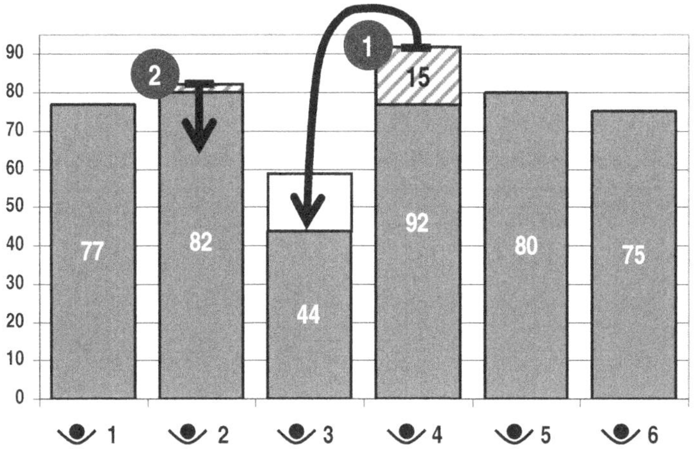

Fig. 3.18 Line balancing approaches

The allocation of work contents to individual operators may be clearly illustrated by way of an operator balance chart. It is easily possible for one operator to handle several neighbouring workstations. Basically, there are two possible approaches to achieve suitable line balancing between individual operators – the *shifting* of work elements to different workplaces and the *reduction* of work content (Fig. 3.18). The shifting of individual work elements to another workplace, or, as an easier solution, the allocation of workstations to different operators, allows a more balanced distribution of work content. In the numerical example, a work process with a work content of 15 seconds may be shifted from operator no. 4 to operator no. 3, resulting in these operators' effort to change to 77, or 59 seconds, respectively.

Another possibility would be a targeted reduction of the work content of the operators with the highest work load. Through application of handling aids, for instance, or automation, individual work elements may be sped up or become obsolete. However, it should be noted that the *amortization* of the additional resources must be calculated differently. If the work content of operator no. 2 can be lowered by two seconds, this will affect all six operators, thus reducing the labour cost-related time effort per product by 12 seconds. The same measure applied to operator no. 1 would not trigger any saving of labour costs, unless the reduction was combined with an additional shifting of work elements.

Line balancing serves the distribution of work tasks for a balanced capacity utilization of all workplaces in the work system (cf. DIN 33415). Many may now be tempted to distribute the work content *evenly* (Fig. 3.19, left). This form of line balancing, however, leads to waste – either in the flow process itself, or in the superordinate production process. When all operators produce in line with the customer demand, they run the risk of not utilizing free periods of time in a productive manner. This creates waste through the respective waiting times or as a result of workers 'looking for' additional tasks, which as a rule results in pseudo performance. The alternative, i.e. an increase in output up to maximum operator utilization – though optimizing local employee productivity – leads to overproduction with a view to the customer demand. Line balancing effecting a balanced work content distribution furthers increased output to prevent productivity loss. Accordingly, *utilization pressure* at evenly distribution work content only seemingly improves the production.

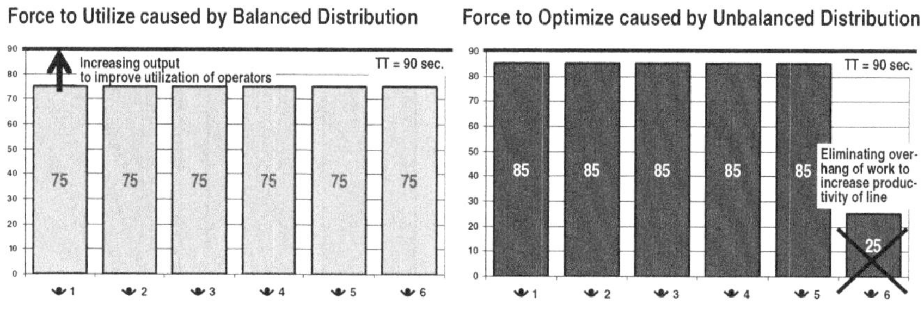

Fig. 3.19 Line balancing methods in flow production

The value stream-oriented solution approach, on the other hand, strictly conforms to the customer takt time and thus leads to an *unbalanced* distribution of the work content (Fig. 3.19, right). As a result, the work loads of all operators except for one are balanced with the customer takt time, the flow production's excess capacity is concentrated on one operator. This is done to make the proportion of waste in the system particularly clearly visible, thus creating an *optimization pressure* to eliminate the respective work backlog and achieve additional productivity gain by cutting out the now redundant operator. This finally results in optimal line balancing perfectly aligned with the customer takt time. To this end, the previously described tools for a targeted reduction and shifting of work content may be repeatedly applied; the unbalanced distribution also serves as an incentive for the consistent improvement of the work system.

Contrary to the preliminary layout of resource capacities formulated in equation 3.1, the targets as to the layout of a flow production are achieved by rounding down:

$$\#\mathrm{OP} = \text{ROUNDING DOWN}\left[\frac{\sum OT_i}{TT \times 95\% \times \#\mathrm{Res}}\right] \tag{3.2}$$

with: # OP number of operators in continuous flow production
OT_i operation time at station i [time unit]
TT takt time [time unit]
Res number of parallel lines

In many cases it may be advisable to leave a little extra room for the customer takt time to be utilized as *additional planned time* – for instance to cushion possible uptime and/or changeover losses. Accordingly, up to 95 per cent of the customer takt time are filled with work content. In our numerical example, an intended cycle time of 85 per cent was chosen, equalling 94.5 per cent of the customer takt time of 90 seconds.

> **Line balancing**
> The work content of a flow production adjusted by obvious waste is distributed in such a way as to achieve approx. 95 per cent of the customer takt time per operator. Taking into consideration additional planned time, each operator is thus used to capacity in accordance with the customer demand. Any possibly remaining work content is successively eliminated through reduction and shifting until optimal line balancing of the flow production directly adjusted to the customer takt time has been achieved.

Work Distribution in Continuous Flow Production

Once the line balancing has been completed, each operator is allocated a workplace with the newly balanced work content. In a manner of speaking, this is the basic solution for the allocation of work to operators, which with a line-shaped spatial layout is particularly suited for *production lines* (Fig. 3.20, case 1). Each operator carries out his firmly allocated tasks at one particular work station and immediately

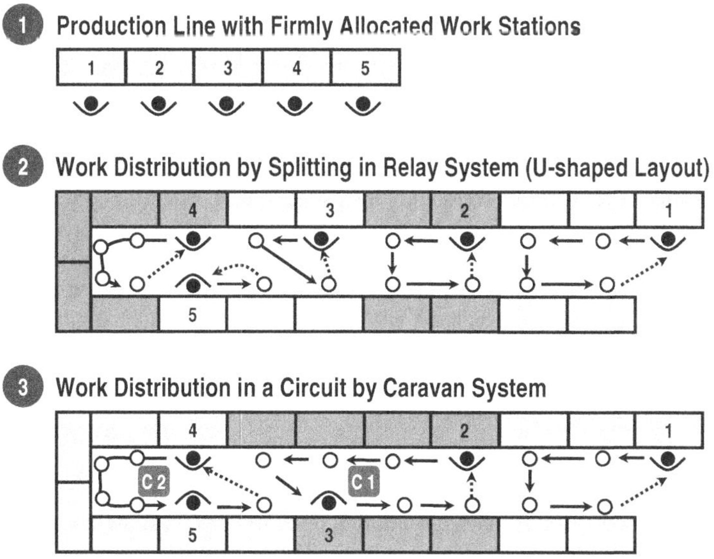

Fig. 3.20 Distribution of work to operators in a flow production

hands over each part to the subsequent station. This layout, however, requires perfect alignment of each balanced work content and its allocated workplace.

In the case of many process steps of low work content requiring different appliances or machines, the number of workplaces will exceed the number of operators needed to meet the customer takt time. As a result, each operator must be allocated various workplaces in a way to utilize all operators roughly to the same capacity (Fig. 3.20, case 2). The major advantage of several workplaces being allocated to each operator lies in much operator-friendlier ergonomics as opposed to sedentary or standing activities at fixed workplaces. As the operators pass the processed parts on to their respective co-workers, this system is also referred to as *relay system*. Thanks to the higher number of workplaces, optimal line balancing may be achieved with greater ease than in the first case. The shifting of work content does not require any alterations of workplaces; a simple change in operator allocation is all that is needed.

The different line balancing possibilities also depend on the spatial layout of the workplaces. If these are organized in a line, only neighbouring workplaces may be integrated; if organized in a narrow *U-shaped array*, opposite workplaces may also be combined. In the U-shape, however, the first and the last workplace should always be handled by the same operator, as this facilitates conformation with the respective work cycle. At the numerous handover points of the relay, the operators may occasionally allocate the respective work content differently. A faster, more experienced operator could, with every other cycle, for instance, take over the work content of the subsequent station manned by a new employee yet to be worked in. Likewise, similarly qualified operators may mutually change their work rhythms.

A uniform work rhythm may be achieved by way of the *caravan system* (Fig. 3.20, case 3). One operator carries out all work processes one after the other, walking from work station to work station in the direction of the material flow. Another operator follows several working steps behind. This type of work organization achieves the greatest standardization effect with a view to work organization. However, both operators need to work in total synchrony. As a distinct advantage, the line balancing here concerns only two operators and is thus much easier optimized. In a flow production consisting of two operators, there will be no line balancing losses whatsoever. In larger systems, identical percentaged takt time deviations are halved. No work station may be allocated disproportionately high work content, as this would result in operator hold-ups. This type of *circular walk*, though, requires experienced staff with similar levels of qualification and work pace. Experience has shown this system to be most suitable for no more than two operators – three workers may already incur problems interfering with each other and each other's work rhythms.

Distribution of work

The distribution of Work content among several work stations per workplace arranged in a U-shaped a relay system facilitates the balanced distribution of work content to the respective operators. The combination of work stations for two operators in a caravan system additionally facilitates line balancing.

Capacity Flexibility in Continuous Flow Production

Several traditional methods may be drawn on to alter the output of a production line, such as overtime before or after a shift, in clever two-shift systems or between shifts, additional shifts at night or on weekends, depending on the prevailing shift system. The capacity of a flow production may be additionally varied by utilizing a different number of operators, which causes the flow production to change into a different *mode*. By changing the mode of a flow production, we may alter its output in a controlled and clearly defined manner. This approach is particularly suited for medium-term capacity adjustments and is thus an ideal tool to balance seasonal fluctuations in customer demand.

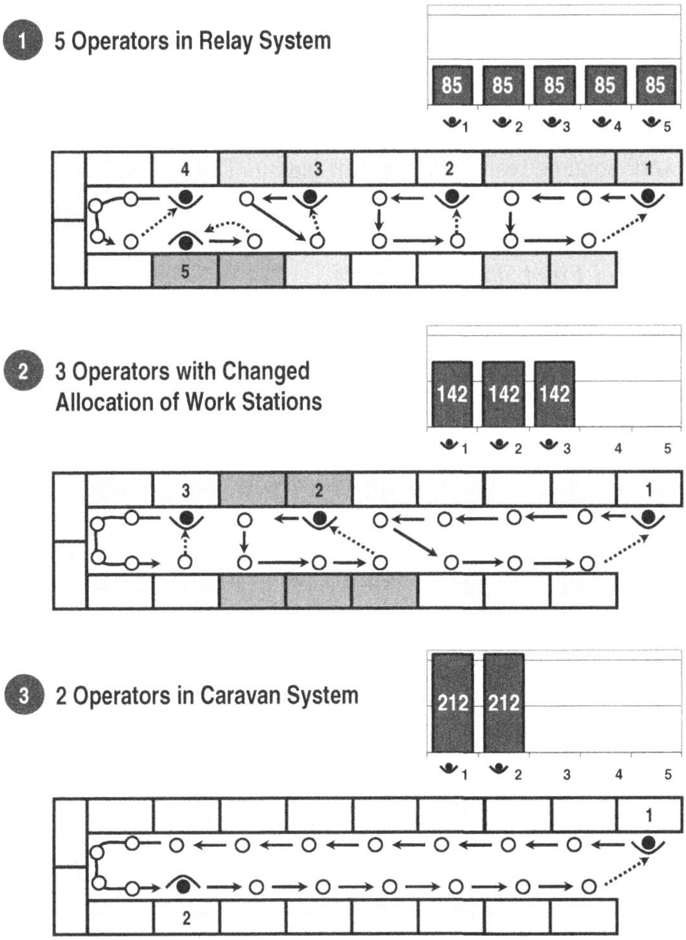

Fig. 3.21 Different modes of output in flow production

A different flow production mode, however, also requires a different allocation of the work elements, so that for a flow production of *flexible capacity*, different ways of line balancing must be developed. In our numerical example, five operators in a relay system generate an output of 40 pieces per hour (Fig. 3.21, case 1). Following a reduction to three operators, the work stations are reallocated; the output drops to 24 pieces per hour at a cycle time of now 124 seconds (Fig. 3.21, case 2). A further reduction to two operators allows for the application of the caravan system, so that no more line balancing is required. The calculational output decreases to 16 pieces (Fig. 3.21, case 3); as there are no line balancing losses, though this value should be slightly higher in reality. On the other hand, at least one of the other two values should be lower in practice, as the realization of two such perfectly balanced lines in a work system is highly improbable.

Capacity flexibility

Flow productions may easily be set to different modes of performance. Each mode with its different number of operators and accordingly balanced distribution of work content, results in different output. This enables extremely high capacity flexibility without resource-related changes.

Flow Principle in Fixed Station Assembly

However, the successful implementation of a flow production will generally exclude products unable to be moved without excessive effort as is typically the case in plant construction. Often, a plant cannot be transported at all in its assembled state – and accordingly needs to be disassembled into modules after bringing into service and/or start-up before it can be transported to the customer. The application of the flow principle to a *fixed station assembly* therefore constitutes a special design challenge. However, the solution approach at hand simply substitutes the physical flow of a flow production, in which each assembly progress is firmly connected with a change of workplace, by a simple assembly progress with clearly defined work steps.

An assembly sequence in plant construction may generally be roughly broken up into mechanical assembly, electrical installation, bringing into service, start-up, and finally the disassembly of the respective modules. A possible further subclassification, for instance into basic assembly and surface mounting, wiring and installation, mechanical adjustment and electrical bringing into service, start-up and customer approval, etc. then depends on the respective plant (Fig. 3.22). The times required for the various work steps should be roughly identical. As soon as two or more operators working on the plant jointly share in a work step, the work content may of course be raised by the relative factor as opposed to work steps carried out by individual workers. The numerical example includes seven work steps with a work content of between 5 and 16.2 hours each. Putting these operation times into relation to the respective numbers of operators per work step results in cycle times of between 5 and 8.1 hours at a customer takt time of 7.6 hours, i.e. the customer takt is slightly exceeded in two cases. This inevitable line balance deficiency, however, may be made up for through local overtime or with the temporary aid of other employees – such as operators from work steps with cycle times below customer takt times.

The layout of the seven plants under construction as per Fig. 3.22 suggests a material flow, which of course does not exist – nor does it need to. These seven assembly places may be arbitrarily located anywhere in the assembly shop(s). Once a work step has been completed – in our numerical example this would take exactly one shift, i.e. one day in regular one-shift operation – the operators move on to the plant on which the respective previous work step has just been completed. The illustration thus shows the logical position of each respective plant in the value stream – and would accordingly be mapped in the value stream analysis. The production lead time of the assembly thus results from the assembly in progress of seven plants with seven days of single shifts in our example.

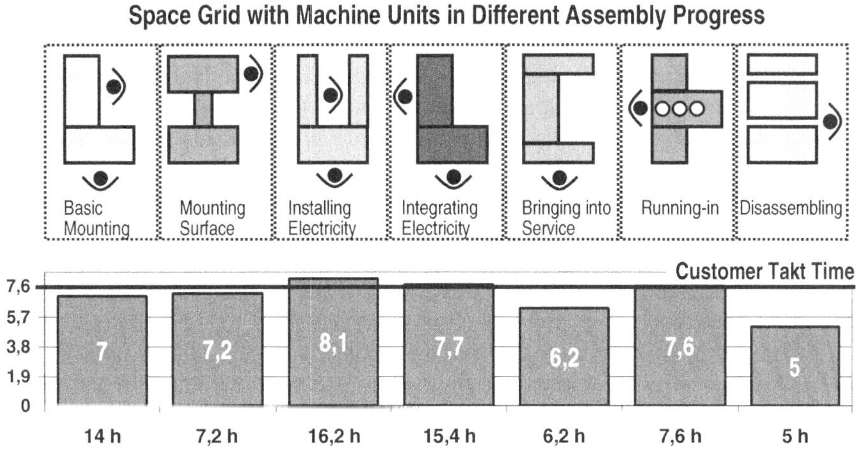

Fig. 3.22 Flow principle in fixed station assembly

In the graphic illustration, the plants move one step forward at a time, so to speak, but without physically moving at all. In the ideal case depicted here, the plants to be assembled are of roughly identical size, which allows the application of a uniform *space grid*. Other cases would require additional spatial planning with time-related interlacing of the plants in question.

Obstacles and Constraints

At times though, the desirable consistent implementation of a flow production is impeded by *technological obstacles*, the elimination of which may on principle be possible but would not be economically practical. Providing a uniform work sequence has been realized for all products of a segment, the following constraints must still be taken in to consideration:

- Some resources just cannot be *spatially* arranged in a suitable manner. This may be due to high alteration costs or constructional restrictions; in which case replanning from scratch would have a distinct advantage over the existing location. In some cases, this may be due to the sheer size of the resources, for instance in cases where several product families merge in one central resource and later separate again. Besides, certain processes may place contradictory requirements on their structural environments (dust evolving processes vs. cleanroom requirement, thermal processes vs. cooling processes, fire protection), so that physical proximity would lead to significantly increased effort.

- Some resources possess immense, irreducible *changeover times*. In these cases, flow production may be implemented, but only in the form of overlapping batch production. In case of considerably differing changeover times, the longest one determines the interruption periods for all integrated resources. This may necessitate the installation of capacities exceeding those of a regular batch production.
- The resources do not possess sufficient *reliability*. This obstacle is generally due to the age-related quality of the respective resources and thus constitutes a simple investment factor; the risk of being left behind by competitors is accordingly high. This does not apply in cases of production processes with a high proportion of stochastic *rejections*, the reduction of which would not be feasible at that point in time for technological reasons.
- The *operation times of the different resources* vary considerably, making line balancing virtually impossible. Accordingly, the stamping process of a simple sheet metal piece and the processing of a cast part and their respective subsequent assembly processes cannot be adjusted to the same takt time. Though the investment in appropriately constructed machines would be possible, the advantages achieved by the ensuing flow production would never be able to make up for the costs. Instead, the faster resource will likely be utilized for the parallel flow production of several product families.
- In general, *operation times* are strongly *variant-dependent* – and to a different extent with each work step at that. It would of course be possible to adjust everything to the slowest process with the most time-consuming variant, but this would in turn result in an insufficient mean capacity utilization of the resources. In this case, economic reasons may necessitate the decoupling of certain processes.
- A flow production often requires the acquisition of highly product-specific resources. In cases of short *product life cycles* and follow-up products with significantly changed technological and capacitive production requirements, however, continued utilization of the respective resources may be doubtful. Due to their very specific design, they would need to be highly versatile, which for technical and economical reasons is only possible to a limited extent. However, the problem of insufficient changeability does not only affect flow productions, but to a certain degree also occurs in solitary plants – accordingly, the respective decisions depend on the individual circumstances.

Most of the above described obstacles arise from *factory structures grown over time* and the *habitual construction* of resources including appliances, much of which could be designed as a flow production from the start in case of an entirely new construction. On principle, no general obstacles prevent the consistent implementation of flow production as such, most counter-arguments arise from specific situations and need to be evaluated for each individual case before other possible ways of designing a production are even considered.

With the conclusion of this conceptual phase, part of the production processes of a value stream have been integrated in one or more flow productions. These flow productions and the remaining production processes are then linked. These logistic linkages will finally create the value stream. The intraplant logistics consisting of the material flow and pertaining control are going to be designed in the following (Sect. 3.3). Subsequently, suitable production planning to match an ideal production control may be/will be developed (Sect. 3.4).

Conceptual approach to flow production

The *ideal* way of producing will be realized in a continuous flow production. Production lead times are minimized, maximum production quality is enabled and required, waste in the production process is reduced and variant mix may be aligned with the customer demand.

The *approach* consist of five central steps:

1. Selection of the product family and determination of the pertaining customer takt time.
2. Determination of the required work elements and their related work content adjusted by waste.
3. Design of the required resources and related suitable degrees of automation.
4. Layout of the work stations on a line or as a U-shaped cell.
5. Distribution of work among the respective operators (line balancing) and development of different capacity models.

3.3 Production Control

Once a value stream's technological production process design has been completed by maximum implementation of a flow production (Sect. 3.2), its conceptual design is approached. With the aid of the *control principles* introduced in the following, the material flow is designed to flow smoothly all the way from the goods receipt to the shipping end. This requires reliable production processes, as efficient value stream design seems to be best realized if the value stream flows as evenly – and thus as predictably – as possible.

To this end, three design guidelines must be applied to the production control to determine fixed processing sequences (Sect. 3.3.1), low inventories to assure minimum production lead times (Sect. 3.3.2) and order release information triggered at precisely one particular production process (Sect. 3.3.3). The strict prohibition of sequence deviations on the shop floor and the low production lead times result in high *controllability* of the production. The triggering of order release information at only one point of the value stream prevents poorly adjusted or even contradictory control pulses and aids the intended smooth production flow.

Order release constitutes the interface to production planning. Once the control rules have decided how released orders are to be controlled to form a smooth production flow, the pertaining production planning may be defined (Sect. 3.4). Axiomatic control principles prohibit the prognosis-oriented planning of production processes where individual processes are planned according to forecasts and then independently activated. Among other things, the design guidelines for production control introduced here aim to protect the production from the results of the *prognostic dilemma of production planning* and are therefore not compatible with prognosis-based control approaches. Before the individual control principles are discussed in greater detail, though, let us have a closer look at this prognostic dilemma and its impact on production control design.

Turbulence Profile

Prognostic dilemma. Production planning faces exactly the same prognostic dilemma as all planning procedures: As it is based on a never completely foreseeable future, it can never be 100 per cent exact. This applies even more with longer planning periods and growing complexity of individual processes. At best, the respective permissible deviations may be planned. The interrelation between customers, production processes and suppliers in a factory must surely rank among the most complex processes man is trying to dominate by (sheer) planning. To describe this interrelation, the term *turbulence* has asserted itself. Derived from Latin, it stands for swirls and in physics describes eddy currents. Its characteristic feature is the fact that the movement of parts in turbulent currents cannot be mathematically described – which conforms to the inability to plan the disposition of orders and products in our context.

The prognostic dilemma is due to three different turbulence-creating contributors. First of all, numerous disturbances usually prevent the production from running according to plan. Machines break down, undergo unplanned service, operators become sick, scrap rates rise or fall. Secondly, customers usually don't ever buy what the forecasts would like them to. The customer demand not just fluctuates, what is more, it never fluctuates according to plan. And finally, suppliers aren't always reliable. Despite clear disposition, required material is sometimes not available in sufficient quantity or reasonable quality. As a result, delivery dates cannot be met in spite of precise planning. Production planners thus aim to minimize all doubts as to the observance of delivery promises through efficient planning. All this, though, gives rise to the following question: If this is almost sure to regularly fail as a result of order planning turbulences – why should we plan at all?

Apart from the unforeseeable behaviour of our contributors, *transition times* are a major critical factor. Transition times form the time buffers allocated to orders between two production processes to allow for necessary move times as well as waiting times before the next process, which may still be occupied by another order – there may even be a waiting line. Warned by previous momentous disturbances and accordingly expecting the worst, we plan transition times with large safety buffers, thus triggering a continuous material flow even in cases of lengthy downtimes of upstream machines. Long transition times in failure-free processes, though, result in all orders being completed early and queuing up before their respective subsequent processes, which will still be working on other orders according to plan. Inventories grow with lengthening transition times and make the production increasingly sluggish. Also, as a result, the observance of standard delivery dates ceases to seem urgent, as everybody knows how much 'leeway' is always included in the planning. Really important orders are simply classified as express orders, overtake others and slow them down even further. In this manner, planning and control actually create their own turbulences.

Turbulence germs. On planning level, turbulences may be triggered by quantity *fluctuations*, *variations* of time periods and *modification frequencies* of technical characteristics. On control level, turbulence germs occur in the form of unexpected *deviations* of quantities, due dates and qualities (Wiendahl 2007). Each of these four types of triggers may be caused by each of the three initiators described above – customer, production, and supplier. With the aid of the following schedule, the search for turbulence germs in a company will reveal the following:

- Fluctuations occur if different quantities are required over different periods of time. Fluctuations as to demand and variant mix are caused by the customers. The production may trigger fluctuations in capacity availability, for instance through planned maintenance, start-up of new resources, employees' leave times. Suppliers may influence previously identified demand through changed prices as a result of price scaling and/or mark-ups for small purchase quantities. In unlucky correlation with the production, also in connection with lot sizing, this may cause larger and larger swings in demand and thus result in the so-called bullwhip effect.
- Time-related deviations occur in the form of different product or order-related lengths of time within a certain period. In plant construction in particular, delivery times vary with each order. Batch formation or priority-related sequencing in a production changes production lead times. Replenishment lead times can vary considerably with complex order-related bought-in parts, such as large customized cast iron parts, especially if quality defects in the cast part (Lunker bubbles) necessitate rework.
- Frequent modification of product characteristics, product technologies, materials and bought-in parts also causes turbulences and the respective modification management for such implementations and phase outs must be included in the production planning accordingly.
- In cases of unexpected deviations after order release, stochastic events require controlling intervention, which will entail further unforeseen effects and require additional control. Possible triggers include erratic customers changing dates, volumes or specifications after order release as well as sales or management retrospectively changing order priorities. Other disturbances may arise from constructional alterations conducted to rectify constructional defects. Unreliable production with machine breakdowns, insufficient staff and increased scrap rates constitute another trigger. Finally, deliveries by unreliable suppliers may cause deviations as to volumes, delivery dates or materials.

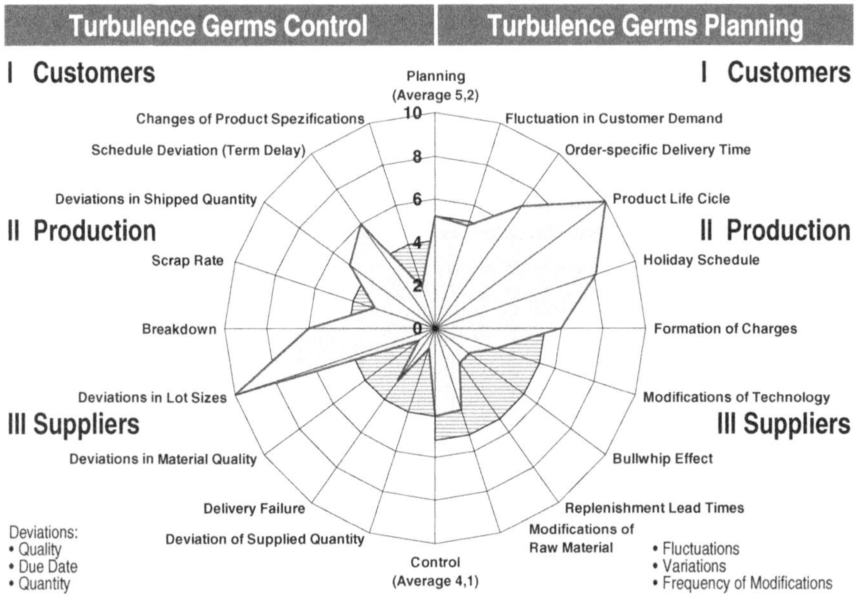

Fig. 3.23 Turbulence profile with numerical example (based on Wiendahl 2007)

The value stream-relevant turbulence germs are descriptively illustrated in a *turbulence profile* (Fig. 3.23). The relevance of each turbulence germ can be evaluated by way of points; the resulting overview indicates the problems to be solved in the future state conception. The respective mean planning and control values show whether the current state suffers from planning problems or if the disturbances are caused by deviating current values. In order to be able to depict the turbulence profile in its entirety, the relevance of each type of turbulence germ is evaluated by points in the numerical example. In practical application, a selection of ten or twelve highly significant turbulence germs will most likely be entered. The respective mean planning and control values are indicated by hachure in the example.

Turbulent and Laminar Order Flow

Production control knows two basic design approaches to handle turbulences. The turbulence-oriented approach tries to dominate the turbulences by way of differentiated planning and control, while the value stream-oriented approach aims at the prevention of turbulences to enable uncomplicated planning based on mean values. The first case requires complex control systems, while the second case combines simple control principles with an appropriately reorganized shop floor.

The *turbulence-oriented* approach considers turbulences to be inevitable and thus focuses on dominating the same. To this end, a sufficiently complex production control system is applied with numerous opportunities to intervene in the production. A turbulent order flow may be visualized by the model of an irregular bundle of pipes subject to cross flow (Fig. 3.24). From the customer's point of view, the whirls lie behind the turbulence germs of the production. From the material flow's point of view, however, it is the other way round, the swirl near the turbulence germ of machine breakdown is now located just before the respective machine in the form of an order backlog. This is exactly where the control elements need to intervene and modify order priorities, or reallocate orders to other, failure-free resources. In turbulence-oriented order processing, order waiting loops are thus controlled and influenced accordingly.

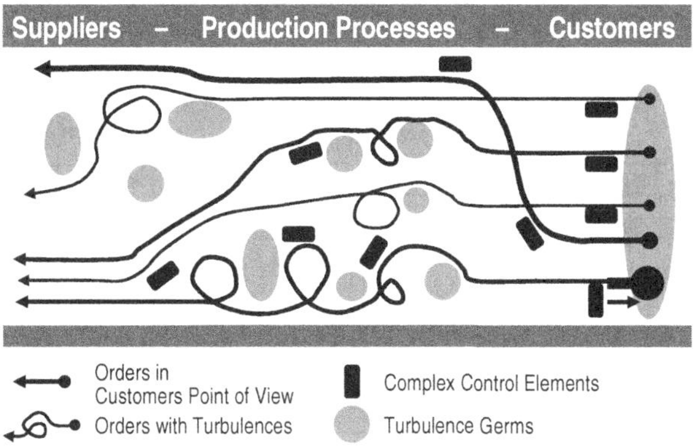

Fig. 3.24 Turbulence-oriented order processing

The success of this approach depends on how far the effects of the turbulences can be minimized. Vital tools for this purpose are increased flexibility on the one hand, in particular with a view to resources, and an expansion of the safety buffers with respect to planned times, transition times and/or inventories on the other hand. Each order is treated differently depending on its respective requirements as to delivery times and volumes. Numerous ancillary conditions and restrictions resulting from the highly complex interconnectedness of sometimes only partially reliable production processes may need to be taken into account. Even in the face of highly diverse requirements, individual orders can be handled in a custom-fit manner; the production can make 'everything' possible. This, however, is accompanied by immense *planning and control effort*, as high complexity increases the *intransparency* of the production procedures and to a certain degree also of order processing, and leads to certain *stability risks*, mainly due to imprecise data and poor data feedback. Even highly intelligent planning and control systems will not be able to totally eliminate the prognostic dilemma and the problem of transition time planning.

The value stream-oriented, *turbulence-avoiding* approach, which is going to be pursued in greater detail herein, aims to eliminate or at least considerably reduce the causes of turbulences. For this purpose, the production is initially segmented to obtain uniformly structured partitions customized to certain requirement profiles (Sect. 3.1), which may be modelled as a parallel pipe system (Fig. 3.25). In the specialized segments, very simple control principles then create a *laminar* production flow which due to the simplicity of its behaviour is highly predictable.

Instead of numerous control elements, one order release is placed before each segment like a turbulence filter. Each order is allocated to a specific segment and then released in conformance with the rules applying to that particular segment to assure a laminar flow within the segment.

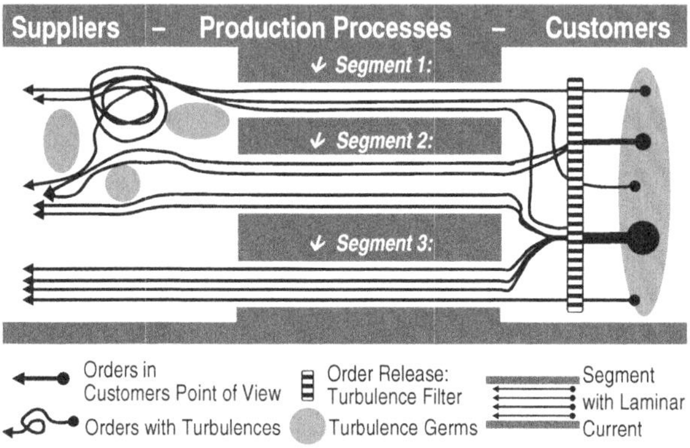

Fig. 3.25 Value stream-oriented, turbulence-avoiding order release

The aim is to carry out as many procedures as possible in an identical way, i.e. to reduce the complexity of different production procedures to few *standardized procedures*, which will greatly improve the internal logistical performance. Decreased order and production lead times in particular can eliminate turbulence germs very early on, because deviations are hardly given any time to develop. In addition, specific *entry barriers*, such as supplier evaluations or prices tied to logistical performance may help unreliable suppliers shape up and at the same time avoid priority orders, customizations and order modifications which are of little or no importance to the customer. Within the production, the implementation of preventive maintenance is highly important for the avoidance of disturbances.

Production control targets

The production control is designed to be as straightforward and standardized as possible. Strict conformance with determined standards on the part of staff and management is of vital importance. Reliable, balanced production processes aligned with the customer takt times are significant prerequisites. The production planning is responsible for the application of this type of control to enable flexible yet balanced reactions to a turbulent factory environment (Sect. 3.4).

3.3.1 Direct Coupling of Production Processes

In cases, where production processes cannot be integrated in a flow production in accordance with design guideline no. 2 (Sect. 3.2.2), individual production processes may be coupled in the form of *line production.* This is done by way of a

FIFO line, at which acronym FIFO is standing for 'first in – first out'. Whichever part reaches the production process first must also be the first to leave – and not let other parts overtake. This strict rule ensures conformance with the defined production order sequence across all production processes. The obligation to adhere to the defined sequence keeps another principle at bay which is very common in daily factory operation – namely the FINO rule 'first in – never out'. As a second characteristic trait, each FIFO line is assigned a finite length, i.e. maximum permissible inventory levels. The following design guideline is to be applied to the coupling of separate production processes:

> **Design Guideline 3:** *FIFO coupling*
> Consecutive production processes, which can't be integrated in continuous flow production for technological or organizational reasons, should be coupled as line production with limited inventory level as far as possible.

Lean management literature refers to the FIFO coupling as 'sequential pull'. This emphasizes the 'pulling' done by the customer as opposed to the prognosis-oriented push systems. Thanks to the *limited inventory level,* orders are not released until a part is retrieved by the customer at the other end. Line production thus allows perfect alignment with the customer takt time and at the same time irrevocably determines the *sequence* of the orders to be processed. It should be noted, though, that seen from the material flow's point of view, the FIFO line constitutes a push production, as each upstream production process pushes a product along the value stream. At best, the order release logic could be referred to as 'pull'. Accordingly, the 'push-pull' distinction may not really suffice to describe control systems in a reasonably differentiated manner, which is why it is largely dispensed with in this book.

The following section will initially present the *function logic* and its respective symbolism for the coupling of production processes by way of a FIFO line. Similar to the various technological obstacles encountered in the implementation of a flow production (Sect. 3.2.2), various different possible *cases of application* for FIFO coupling have the power to overcome several of said obstacles. In this context, particular attention should be given to the different *branching* and merging possibilities of value streams.

Function Logic of FIFO-Coupling

Signals. A FIFO line is symbolized by two parallel lines connecting the coupled production processes. In its centre, an arrow equipped with the FIFO acronym indicates the obligation to adhere to the determined sequence throughout the value stream (Fig. 3.26). When two production processes are coupled by a FIFO line, the downstream process emits a *release signal* whenever it withdraws a product from the FIFO line to ensure that the maximum stock is never exceeded. The respective control pulse is depicted in the form of an arrow – for reasons of clarity we use a dashed line in graphic illustrations. The signal itself is symbolized by a circle, here marked as 'ConWIP' which stands for 'constant work in progress'. Thus, the parts on the shop floor constitute a constant contingent of work in progress; the decisive control value, i.e. the maximum stock, is entered in the circle as 'max. *pcs.*'.

The release signal may be transmitted electronically, or, alternatively, laminated cardboards corresponding to the maximum number of pieces may be provided. These ConWIP cards then accompany the parts on their journey along the FIFO line. Whenever the processing of a part in the subsequent process begins, its card is returned to the previous process, which thus receives the release order to commence production and in turn attaches the card to the now current product. The release information in the form of the ConWIP signal to the upstream process constitutes the actual *production order* which defines the next variant to be produced according to the production planning specifications. Depending on the respective requirements and the pertaining production planning design, production orders may release either individual parts or entire lots (Sect. 3.4.1).

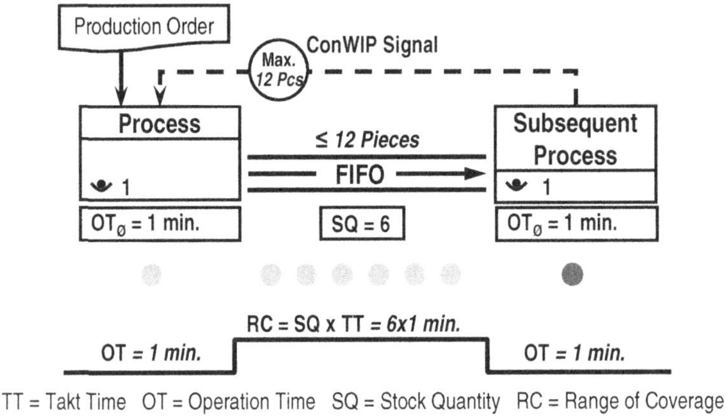

Fig. 3.26 FIFO coupling

Stocks/inventories. In our numeric example, the ConWIP is fixed at a maximum of twelve pieces (Fig. 3.26). For simplification purposes, the maximum stock may be entered immediately above the FIFO symbol instead of the ConWIP signal. Underneath the FIFO line symbol, the respective stock quantity (SQ) as determined in the value stream analysis is quoted. This may lie below – or in the case of control errors also above – the maximum quantity entered in the circle. Since the FIFO line takes over the line balancing and/or other *buffering functions* between the production processes, the actual inventory must be lower than the determined maximum value, as otherwise there would be no room for buffering. With correct layout and no safety stock exceeding the buffers, the mean inventory will equal 50 per cent of the defined maximum stock.

A customer takt time of one minute, for simplification purposes identical to the mean processing time of the two production processes in our example accordingly results in a mean inventory of six parts and thus a buffer coverage of six minutes, i.e. the production lead time for one part amounts to eight minutes. Compared with the two minutes of production lead time in the flow production (Fig. 3.14), the production lead time has risen considerably – but still remains way below the respective value of hours or even days measured in cases of interim storage. The FIFO method regulates inventories on a previously defined level, limits maximum quantities and depending on the respective cases of application keeps it almost constant. This makes the production lead time highly predictable and minimizes the scheduling requirements on the planning and control system.

Quality control. Constant conformation with required quality standards is a vital prerequisite for the smooth functioning of the FIFO coupling. Rejections passed on and not detected until the FIFO line hits the following process require additional production processes and/or rework and thus disarrange the order sequence. As a result, production lead times fluctuate in an unforeseeable manner and the ensuing turbulences on the shop floor impede the correct prediction of completion dates. In the case of inevitable scrap in a production process, *in-process inspections* for instance could make sure that only good parts are passed on to subsequent processes. A clever alternative solution for low-variance productions could equip all follow-up processes with safety stock of all variants, which could be slotted into the process as replacements if necessary. This could indeed be a viable solution; at the same time it would constitute a perfect example for the standardization of waste processes.

FIFO chains. If several production processes are connected in series by way of FIFO lines, one control pulse, transmitted from the last to the first process in the chain, is sufficient (Fig. 3.27, bottom). In this case, the maximum stock to be entered in the ConWIP symbol is mutually determined for all related processes and not allocated to each individual FIFO line. This means, though, that each of the respective FIFO lines must be able to accommodate the entire maximum stock. Accordingly, the maximum stocks before processes no. 2 and 3 in the numeric example add up to a total of fifteen parts, therefore process no. 2 now needs fifteen buffer places instead of five. This will not increase the total work in progress of the entire chain, though, as the two FIFO lines cannot both be filled to capacity. This has the added advantage of temporarily *migrating bottlenecks* merely causing the respective buffers to increase with no need to adjust the respective maximum stocks.

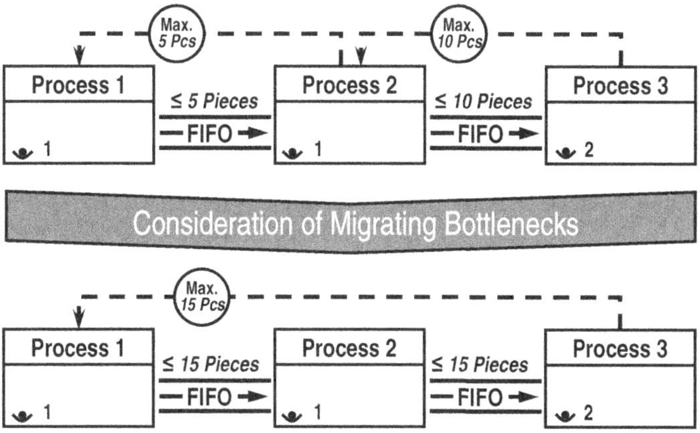

Fig. 3.27 Simplified FIFO coupling of several processes

Application of FIFO-Coupling

In the most straightforward case, FIFO coupling serves to overcome large spatial distances between two production processes. As a buffer, the FIFO system compensates takt time discrepancies, differences in changeover times and disturbance-related downtimes. We can differentiate five different cases of application in a production:

To facilitate the exact *dimensioning* of the ConWIP inventory, simple calculation formulae are stated herein for each case of application. As a rule, a FIFO line must be able to cover several buffer functions at once, for instance in cases of simultaneously occurring takt time deviations and downtimes. In addition, certain spatial distances will generally need to be bridged, as otherwise we would be looking at a flow production. In this case, the different buffer portions are simply added.

As the calculation includes several assumptions as to process behaviour, maximum downtimes, fluctuating production lead times and move times, the results – especially under complex circumstances – may never be more than rough guidance values. Therefore, the simple empirical process of continuous improvement through gradual decrease of initially fairly high ConWIP inventories will be the best possible solution. As soon as the first temporary material hold-ups occur, the production processes involved must be improved or the buffer inventories must be slightly increased.

Bridging of distances. FIFO coupling is suitable for spatially separated production processes which may not be physically located next to each other due to technical reasons or excessive alteration costs. For the actual implementation of the FIFO line, especially in cases of large distances, belts or other conveyor technique may be applied. As already suggested in the graphic illustration, this would even enable single-piece-flow, simply prolonged by the additional move time. In many cases, though, the physical coupling of production processes is not desirable for cost and changeability-related reasons. In these cases, manual *transport* of small parts will be conducted in containers rather than individually. Containers are symbolically depicted by an open top rectangle in which the respective container quantities (Fig. 3.28, case 1) are entered. Each of these containers is equipped with a ConWIP card. The maximum stock per container is calculated from the move time (MT) and the required operation time as follows:

$$ConWIP = \frac{(OT \times CQ) + MT}{TT} = \frac{(60\,\text{sec.} \times 5\,Pcs.) + 20\,\text{min.}}{60\,\text{sec.}} = 25\,Pcs. \tag{3.3}$$

with: ConWIP limited inventory level [pcs.]

OT operation time at preceding process [time unit]

CQ container quantity for conveyance [pcs.]

MT move time [time unit]

TT takt time [time unit]

Two types of packaging units are to be differentiated. Different variants may be arbitrarily bundled in *transport containers* which then shuttle back and forth between the thus coupled processes. This method allows for one-piece-production as required with customer-specific parts and is extremely helpful with large numbers of variants. In serial productions with few variants, homogeneous packaging units may be formed. These *bundles of identical parts* constitute small lot sizes running though the entire production.

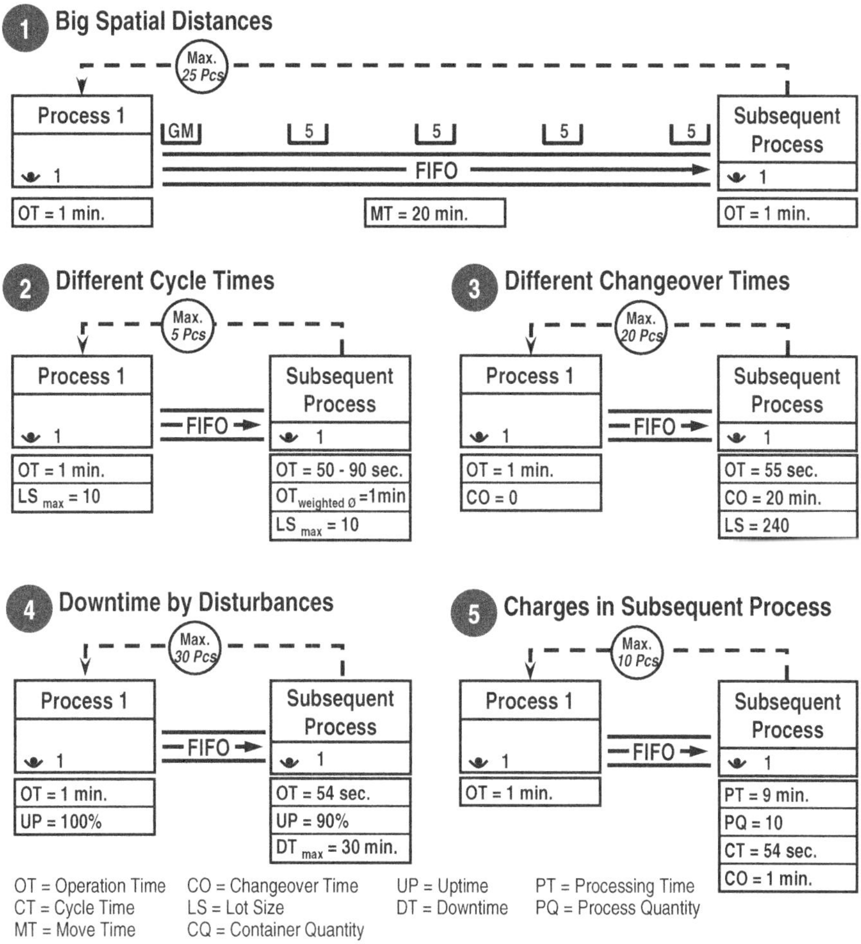

Fig. 3.28 Cases of application in FIFO coupling

Line balancing buffers. The relatively low inventories in a FIFO line enable the decoupling of production processes to the extent where variant-dependent *deviations in takt time* may be balanced, for instance by significantly increasing or decreasing the speed of the subsequent process in comparison to its predecessor (Fig. 3.28, case 2). All demands fluctuating in relation to each other are thus buffered by the FIFO line. Accordingly, a flow assembly could for instance be followed by an inspection process with an operation time slightly higher than the previous flow assembly's cycle time. The capacity of the respective inspection process will be sufficient whenever the mean operation time weighted by variant piece numbers ($OT_{weighted.Ø}$) equals the cycle time of the flow assembly.

The required minimum length ($ConWIP_{min}$) of the FIFO line is calculated from the takt time deviation of the maximum operation time for a specific variant as opposed to the respective operation time of the previous process – multiplied by the maximum number of products of the same variant produced in direct sequence. In the numeric example, the customer takt time is one minute, the maximum operation time of the subsequent process is 90 seconds at a lot size of ten pieces (Fig. 3.28, case 2). In this case, the FIFO buffer will need to hold at least five pieces in order to cushion a temporary overproduction of the previous process. This buffer will automatically shrink during later, shorter inspection cycles, thus creating space for the slower discharge of parts in longer inspection cycles.

$$ConWIP = \frac{(OT_{max} - OT)}{TT} \times LS = \frac{(90\,\text{sec.} - 60\,\text{sec.})}{60\,\text{sec.}} \times 10\,Pcs. = 5\,Pcs. \tag{3.4}$$

with: ConWIP limited inventory level [pcs.]
OT operation time at preceding process [time unit]
OT_{max} operation time at preceding process [time unit]
TT takt time [time unit]
LS lot size [pcs.]

Even if the buffer is chosen slightly bigger as a precaution, the variant mix will be subject to sequencing restrictions depending on takt deviations and buffer size, which need to be taken into account in the production planning (Sect. 3.4.2).

Changeover time buffers. The FIFO buffer may also be utilized to buffer parts produced by the upstream process during changeover of the subsequent process (Fig. 3.28, case 3). This will only be required however, if the downstream process itself is subject to no or minimal changeover times. In the numerical example with an operation time of one minute in the upstream process and a twenty minute changeover time of the subsequent process, the resulting time difference is exactly what needs to be buffered. Accordingly, the maximum stock of twenty pieces is calculated as follows:

$$ConWIP = \frac{CO}{TT} = \frac{20\,\text{min.}}{60\,\text{sec.}} = 20\ Stck.$$

$$LS_{\min} = \frac{CO}{(OT - OT_S)} = \frac{20\,\text{min.}}{(60\,\text{sec.} - 55\,\text{sec.})} = 240$$

with: ConWIP limited inventory level [pcs.] (3.5)

CO	changeover time [time unit]
TT	takt time [time unit]
OT	operation time at preceding process [time unit]
OT_S	operation time at subsequent process [time unit]
LS_{min}	minimal lot size [pcs.]

Five seconds of *production time difference per piece* between the two processes leads to the minimum lot size required by the subsequent process to make up for the time lost in changeover in comparison with the upstream process. The minim lot size of 240 pieces is calculated by division of the changeover time by the difference in production time per piece (Eq. 3.5).

Downtime buffers. Just like changeover times, disturbance-related downtimes of subsequent processes may be buffered. To this end, the expected maximum *downtime* (DT_{max}) to be buffered is determined based on empirical values (Fig. 3.28, case 4). The downtime consists of the time needed until the disturbance is noted and announced to maintenance, the waiting time incurred up to the arrival of an available maintenance person, the time required for diagnosis and repair including spare parts procurement, as well as restart. This list clearly illustrates the extent to which the downtime depends on the quality of the respective maintenance organization and how it may be reduced by optimized procedures. Besides reactivity and qualification of the maintenance personnel, great importance is attached to availability and simple accessibility of spare parts and tools. The FIFO buffer must be designed as large as the maximum downtime to be buffered:

$$ConWIP = \frac{DT_{\max}}{TT} = \frac{30\,\text{min.}}{60\,\text{sec.}} = 30\ Pcs. \approx \frac{DT_{\emptyset} + 3 \times \sigma_{DT}}{TT}$$

with: ConWIP limited inventory level [pcs.] (3.6)

$DT_{max / \emptyset}$	maximal / average downtime [time unit]
σ_{DT}	standard deviation of downtime [time unit]
TT	takt time [time unit]

Based on statistical normal distribution, a buffer size of medium downtime plus three times the standard deviation of downtime (σ_{DT}) will equip the buffer with an uptime of 99.9 per cent. Should, against all odds, no downtimes occur, though, the FIFO buffer will run dry, because in spite of sufficient release signals the upstream process will not be able to keep up with the production, but lose six seconds per piece as shown in the numerical example. In these cases the material flow may break and cause waiting times for the subsequent process.

Batch buffers. The FIFO coupling also enables *batch formation* in the subsequent process, i.e. the accumulating of enough parts before the batch process to make up the required process quantity. The FIFO buffer will of course then need to be big enough to hold at least the number of parts included in the respective batch. The maximum stock in the numerical example corresponds to the process quantity of 10 parts (Fig. 3.28, case 5). However, this only works if the sequence supplied by the upstream process may indeed be grouped together in a batch, different cast parts with customer specific bore holes could for example be run through the washer in a basket together. In cases of variant-specific surface treatment (painting, PVCD coating, galvanizing), though, the parts may never be processed in random order, which needs to be taken into account in the planning of the respective sequences (Sect. 3.4.3).

Supplier Connection

The FIFO coupling may also be implemented in a supply chain. This is necessary whenever variant-specific bought-in parts must be procured instead of raw materials or standard parts supplied by way of a supplier kanban (Sect. 3.3.2). In cases of high variance, it is not desirable or indeed possible to have all of the variants on stock. Instead, *just in sequence* (JIS) delivery of the variant mix is arranged (Fig. 3.29). Whenever the last production process in the value stream is ready to produce the next product, a release signal is sent to the procurement process authorizing the procurement of the next part of the respective variant. The related downstream process functions as a pacemaker.

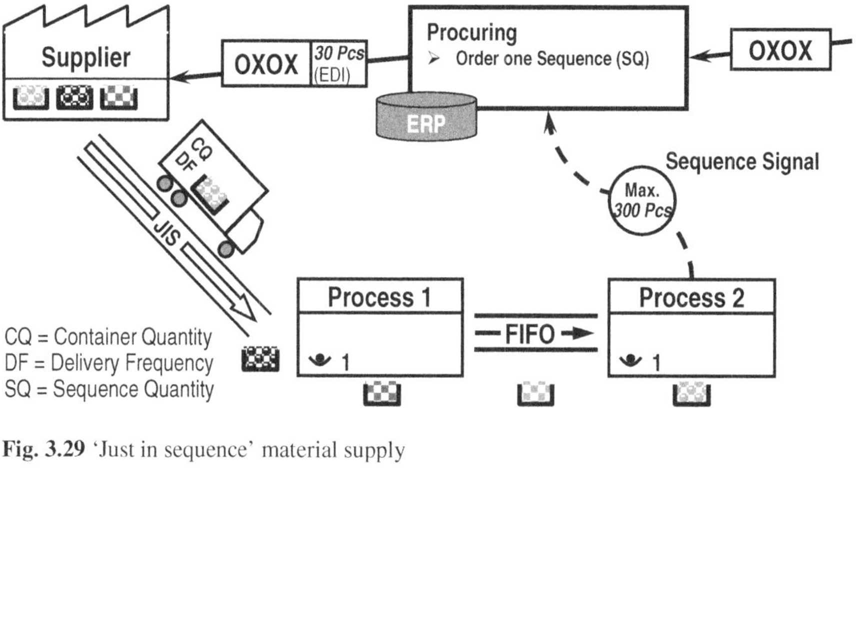

Fig. 3.29 'Just in sequence' material supply

In practice, however, parts will not be ordered individually, among other reasons because delivery is not done in individual parts but in truck transport lots. Several sequence signals are therefore accumulated and a certain *sequence quantity (SQ)* is ordered. The sequence symbol consists of the letters XOXO framed by a rectangle. The respective variant mix is determined by production planning in accordance with certain rules (Sect. 3.4.2). The supplier produces the variants in the desired sequence and sorts them into the containers which are then arranged on the truck accordingly. The symbol for the supplier connection is a *JIS line* framing the broad white arrow of external transport. Customer-specific parts may be acquired in the same manner. In cases of low order frequency, such as in plant construction, the usual purchase process will be kept up with individual orders subject to a framework contract.

Branched Value Streams

The FIFO coupling allows the branching of value streams while keeping up the line principle. In the most straightforward case of branching, several product variants will simply skip a production process (Fig. 3.30, case 1). The branching symbol indicates the relative strength of each branch as a percentaged value. However, note should be taken of the fact that variants with an extra production step will be overtaken by the other variants. In variant mixes, more or less consistent utilization of the lateral branch should be ensured.

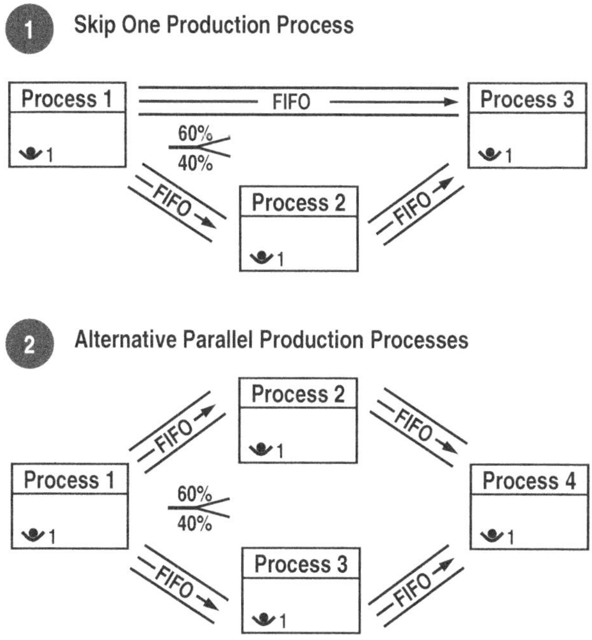

Fig. 3.30 Branched value streams

In value streams of higher complexity with highly differentiated products, partially parallel value streams are also possible (Fig. 3.30, case 2). Before they can be reunited in a mutual production process, though, the respective FIFO lines must merge in one and the same retrieval area. This, however, will disarrange the order release sequence should the processes of the different branches possess different production lead times (buffer length and operation time), therefore production mix planning must aim for a balanced utilization of both branches (Sect. 3.4.2).

Synchronisation. The possible merging of two partial values streams may also be utilized to unite processed parts from different sources in one joining process (assembly, welding). One production process in one of the upstream branches is activated by way of a ConWIP signal and a production order just like in a regular FIFO chain, thus turning it into the main value stream, which should be the one pre-producing the most important part. The lateral value stream(s) will then be triggered by the main value stream by way of a so-called *golf ball signal* (Fig. 3.31). This control logic facilitates the variant-specific *synchronization* of lateral branches with the respective main value stream branch.

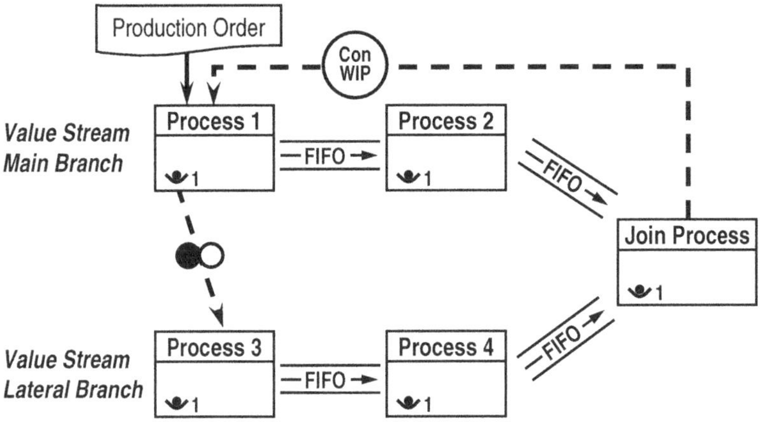

Fig. 3.31 Synchronisation via 'golf ball' signal

The term 'golf ball' refers to a specific implementational feature of this control logic: Depending on the next variant to be produced, the operator in the main branch picks a certain colour-coded golf ball and sends the same to the relevant operator in the lateral branch via tube mail. This enables the operator in the lateral branch to produce the required variants in the correct sequence. The 'golf ball' and pertaining variant sequence are symbolized by a light and a dark circle (Fig. 3.31). This approach ensures strict conformation with the FIFO principle in the connection of assemblies with different pre-production areas (component assembly, pre-production). It is easily applicable and may be implemented using other tools as well – visual ones, for instance, such as little coloured or numbered flags on main parts, electronic ones, or maybe printed cards.

FIFO coupling in line production

Goal In production processes coupled by FIFO logic, orders are pushed through the production process in a preassigned and unchangeable sequence. The observation of maximum stocks by way of release signals results in calculable production lead times and smaller inventories on the shop floor.

Application The buffer function of the FIFO line enables transport between distant production processes incl. supplier connection. Lots of upstream processes may be accumulated for batch production. In addition, deviations in takt times, changeover times and disturbance-related downtimes between the coupled production processes may be buffered within defined limits.

The merging of several value streams is synchronized via golf ball signal, which enables order release in branched value streams by way of one single triggering control signal.

Case study. At *Liquipur,* milling and washing are already connected via FIFO line in the current state. The washer works well with wash baskets containing twenty parts each. Investment in a new washer able to individually wash arbitrary variant mixes on a conveyor belt would allow integration into a flow production. This would not result in any additional flexibility, though, as this is a serial production of low-variance. The current state thus remains unchanged (Fig. 3.32). Instead, exactly ten holding areas for wash baskets are

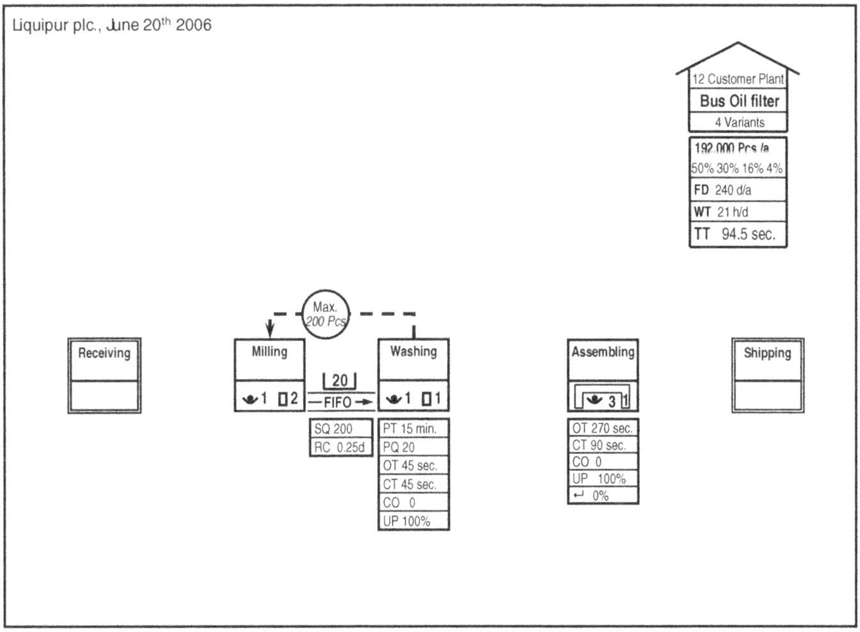

Fig. 3.32 Value stream design at *Liquipur* (2): FIFO coupling

marked before the washer. Whenever there are no vacant holding areas left, milling must be halted. This implements the release signal in accordance with the pertaining shop floor design in a visual manner which does not require any additional information flow. No FIFO coupling with other production processes is intended. This shall be further explained in the following.

3.3.2 Consumption-Oriented Kanban-Regulation

Application. In a line production, the production processes are firmly coupled, leaving the sequence of the different variants unchangeable across the entire value stream. Accordingly, the variant sequences required must be defined at a very early stage. In value streams with many production processes and comparably long production lead times, the production will not be able to react to the customer demand fast enough. This may be cushioned by pre-fabrication, in particular in cases of high-variance productions, a *decoupling* of upstream processes from the downstream processes will be required. Also, extensive line productions with numerous coupled production processes may easily become instable when too many of the almost always occurring small disturbances mutually amplify, in which case a separation of long process chains into two decoupled portions may prove expedient.

In joining processes such as assemblies, several value streams usually merge, each providing internally produced parts to be joined. In addition, bought-in parts must be made available. This requires the *synchronization* of parts which all arrive at different times in different quantities and which need to be allocated to the correct variants – something that would normally not be feasible in FIFO coupling.

Besides, line production is best suited for production processes with no or minimal changeover times, i.e. significantly lower than the pertaining operation times per part. In cases of overlapping lot production, lot sizes are determined to match the process with the longest changeover time. However, all coupled production processes should have more or less identical changeover times (Sect. 3.3.1), otherwise, processes with significantly shorter changeover times would never be able to be used to capacity due to the changeover-related waiting times generally incurred in FIFO coupling. This will necessitate resources of larger dimensions than normally required by the respective operation and changeover times to prevent loss of variant flexibility, as the prevailing frame conditions will define excessive lots. As a rule, productions include production processes of different changeover times, which for reasons of capacity utilization and flexibility leads to the production of multiple usage parts in *varying lot sizes* for different process steps, rendering coupling after the model of a line production virtually impossible.

Kanban logic. The simplest form of *lot production* control is not to rigidly exert governing control, but to control by regulation. The tools for this automatic regulation are *control cards* which are called 'kanban' in Japanese, a regulation system corresponding to the functional principle applied in supermarkets, where the removal of products from supermarket shelves automatically triggers replenishment from a storage room or initiates a respective repeat order with a distribution centre. The ordering is usually done in packaging units, i.e. not every single tube of tooth paste removed is immediately reordered, but a package with ten tubes is ordered upon sale of every tenth tube. This principle and its already mentioned variations are reflected in kanban control.

Kanban control connects two coupled production processes in an internal customer-supplier relationship where the removal of parts from the supermarket shelf triggers the respective re-production of that particular part variant in the same quantity at the supplying process. This ensures re-production of the exact quantity used up. This eliminates planning errors and thus avoids overproduction in lot production, providing the kanban system is correctly specified and maintained. The respective design guideline is summarized as follows:

> **Design Guideline 4:** *Kanban control*
> Production processes for multiple usage parts with high changeover times, low reliability or highly differing cycle times should be connected by lot production with supermarket stores.

Because of its lot formation, kanban is only suited for the production of multiple usage parts. Production processes are easiest connected by kanban if there are few variants with a high rate of consumption and *steady demand*. Each deviation from these ideal conditions necessitates an increase in the range of coverage of the supermarket. Higher fluctuations in demand require higher inventories in order to keep up constant material supply for the customer process. With increasing numbers of variants, more storage space is required. At low consumption, however, the range of coverage increases as a result of lot production.

The kanban system is very similar to the 'usual' demand-oriented (Sect. 2.3.3) disposition. The most notable difference lies in the visualization of kanban control in production. The firmly allocated and labelled storage spaces of a supermarket's inventory enable easy visual monitoring of the inventory. Besides, in a kanban system all containers include information on the parts stored inside as well as their production procedures, while in conventional storage mere part information is supplied until the respective parts are removed from stock and are equipped with a part identification tag, which contrary to the kanban system, though, needs to be separately made out. In the kanban system, the inventory movement is triggered directly by the production, while demand-oriented disposition usually requires a detour via a planning and storage maintenance system.

Over and above these organizational differences, the following fundamental differences apply: The kanban system always uses identical lot sizes in accordance with the respective container quantities, while in demand-oriented disposition lot sizes may be re-calculated and 'thus' optimized from time to time. This, however, leads to irregular strain on the production, which not only reduces transparency, but also significantly complicates production control. In addition, demand-oriented procedures enable the reservation of inventories for future demand, for instance for fix customer orders. This allows earlier reactions than in the kanban system, but makes matters more complex and decisions more difficult with respect to why and when reservations should be permissible and whether or not reserved inventories should be released in urgent cases. Also, there is no pooling of demands in the kanban system – several kanban lots being required simultaneously will still be produced individually and not grouped together in one single order like in conventional planning and control systems.

Overview. All in all, we can differentiate three groups of kanban control. If the main objective is the decoupling of two production processes, kanban control is to be implemented to the effect that the delivery process produces lot sizes exactly conforming to the volumes removed by the customer process. Further information on this standard case of a *production kanban,* supplemented by detailed information on the functioning of materials logistics and dimensioning will be provided in the following.

If the objective is the coupling of two processes with significantly deviating changeover times, however, then kanban control will be implemented to enable changes in lot size. The two types of *signal kanban* introduced above resemble a demand-oriented production control. The determination of the correct lot size is of vital importance in the dimensioning of a kanban system. The changeover portfolio illustrates how to achieve this with the aid of the EPEI value.

In addition, kanban control provides an easy method for the procurement of raw materials and bought-in parts. The bought-in parts portfolio indicates which types of parts may be effortlessly obtained with the aid of the *supplier kanban*. Raw materials place special requirements on the dimensioning of storage spaces and order volumes.

Production Kanban

Function logic. The production kanban works on the basis of a supermarket store between the two coupled production processes. This store always contains all variants processed by the supplier process to guarantee constant supply of the customer process. The *supermarket store* is symbolized by a stylized storage rack with three shelves (Fig. 3.33). The shelves are open towards the left, towards the supplier process, which is responsible for the inventories which ensure said material availability. The connecting material flow is depicted by way of two arrows: Replenishments being 'pushed' into the store after completion by the supplier process are illustrated by a *straight arrow*, the withdrawal by the customer process according to demand is indicated by a 'pulling' *circular bended arrow*. This is also referred to as the 'replenishment pull system'. Triggered by the demand-oriented pulling withdrawal of goods, the inventory is replenished in a consumption-specific manner.

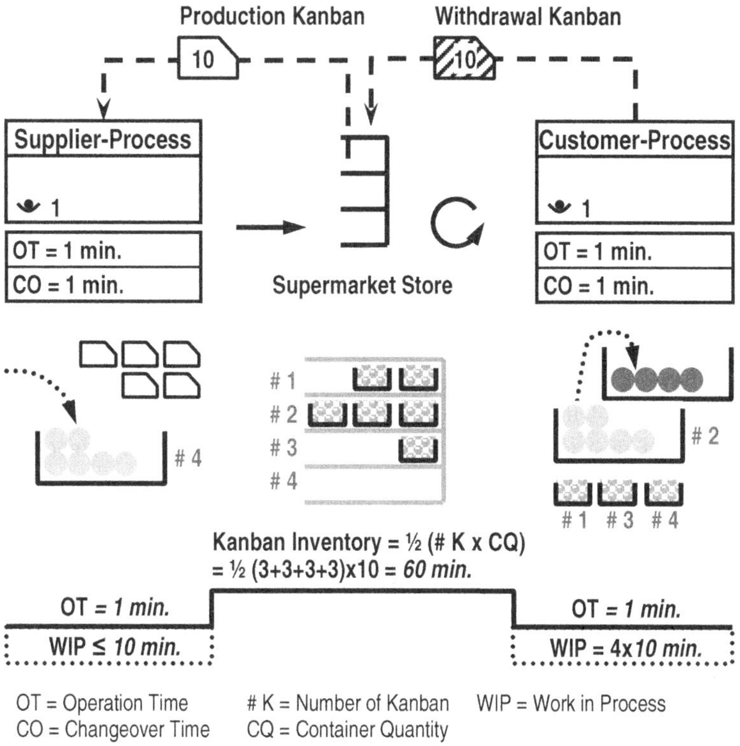

Fig. 3.33 Kanban control with numerical example

The material flow between production processes is regulated by kanban. A kanban is symbolized as a rectangle with the top right corner cut off (Fig. 3.33). Each kanban refers to an order unit to be entered into the kanban symbol. An arrow (herein dashed for better visualization) indicates the flow of the control pulse transmitted by the kanban. As a rule, two kanban types will be required. The *withdrawal kanban* triggers the transport of a container from the supermarket store to the customer process. The *production kanban* activates the supplier process to produce the parts now lacking from the supermarket inventory as a result of the previous withdrawal. For better differentiation, the symbol of the withdrawal kanban is hatched.

The supplier process re-produces the production kanban in the exact sequence corresponding to that of the withdrawal. Thus the *lot size* equals the kanban volume. The kanban yet to be produced queue up waiting in front of the supplier process. The different variants are produced by the supplier process in the sequence of their withdrawal. In order to ensure consistent production lead times, order sequences may not be altered, as may seem expedient for the purpose of lot size increase, for example. All parts are made available to the customer process by the supermarket store in the exact lot size in which they are re-produced and could easily change to piece-production as long as all variants are readily available on location. These may be made accessible in addition to the regular supermarket store up to one kanban quantity or the supermarket store may take over the order function as well.

Each kanban normally corresponds to exactly one *container*. The respective *container quantity* thus equals the kanban quantity. Wherever possible, containers of standard measurements are used, which may then be used with different inserts for many different part variants, but there will only ever be one container type per article code. The containers should be easy to handle for the customer process and enable direct retrieval at the place of usage. The quality of the parts must be protected during transport and storage (cleanliness, corrosion, abrasion). Depending on the geometry of the respective parts, pallets, skeleton boxes or trolleys may also be used. Consistent use of clearly defined part-specific container quantities highly standardizes the production. In addition, container quantities should be coordinated with the respective order release quantities (Sect. 3.4.1).

Stocks In the *data box* for the supermarket store all significant denominations for storage, material and container type are entered (Fig. 3.34). The number of variants (# Var) indicates the number of shelves labelled accordingly in the supermarket store. Kanban logic does not permit chaotic storage, but requires firmly allocated storage places in accordance with the defined maximum stocks determined by the number of all circulating kanban (# K).

Denomination Store; Stock Location
Denomination Material
 # Var Number of Kanban Parts
 # K Number of Kanban
 # P Number of Parts per Product
RC Range of Coverage

Fig. 3.34 Data box for supermarket store

The range of coverage of the supermarket store may be calculated similar to that of the buffer stock (Eq. 2.13). In order to assess the mean inventory, we shall assume that – similar to the inventory log of demand-oriented disposition (cf. Fig. 2.37) – the mean inventory equals half the maximum stock. The maximum stock depends on the total number of circulating kanban; taking into account the daily demand, the range of coverage is as follows:

$$RC = \frac{1}{2} \times \frac{\# K \times CQ}{\# P \times DD} \quad \text{respectively:} \quad RC = \frac{1}{2} \times \frac{\sum_{i=1}^{var} \# K_{Var} \times CQ_{Var}}{\# P \times DD}$$

with: # K number of kanban in circulation

RC supermarkets range of coverage [d] (3.7)

CQ container quantity of kanban [pcs.]

DD daily demand [pcs./d]

P number of identical parts per finished product [pcs.]

$\# K_{Var}$ number of kanban of a variant

CQ_{Var} container quantity of a variant [pcs.]

The quantities stored in the supermarket store may be differently defined for each variant – but must always be an integer multiple of the respective container quantity. Thus, the number of kanban corresponds to the number of circulating containers. In the numerical example, four variants are stored in the supermarket store in six containers of ten parts each (Fig. 3.33). In addition, one container is in re-production and five kanban are waiting to be processed at the supplier process.

Over and above that and depending on the respective production procedure design, further containers may already have been made available to the customer process. In the numerical example, all four variants are readily provided, allowing the customer process to retrieve each variant individually. If the customer process was working with lots as well, only one container would have to be provided. In simplified terms, the *production lead time* for one part corresponds to the inventory in the supermarket store, i.e. its range of coverage, calculated as per equation 3.7 – or 60 minutes in our numerical example (cf. Fig. 3.33). On closer reflection, though, the provided containers and the kanban lot in progress must be taken into account as well. This increases the production lead time in the numerical example to a total of 110 minutes.

For simple *dimensioning* of the kanban system, trial and error is recommended, initially starting off with generous assumptions to be on the safe side with a view to possible breaks in the material flow. To begin with, the inventory may still be rather high. After kanban implementation, the relatively high number of work in progress cards and thus the corresponding material is gradually decreased through removal of individual cards until the first disturbances in material availability occur. At this point, possible process improvements may be investigated – or, alternatively, one card may be returned.

Irregular demand. In cases of parts with large demand fluctuations and medium to large piece numbers, adequate supermarket layout is virtually impossible, as this would either result in huge inventories or partially prevent delivery on demand, as the stocks would fall below the possible maximum order quantities. Even prognosis-oriented production planning would find it difficult to reliably forecast such extreme fluctuations. Typical examples would be annual orders placed by foreign branches, large orders by individual customers at special conditions, or seasonal or specially promoted products.

These special cases may be handled by *emergency kanban*. An emergency kanban may only be used once and needs to be specially marked, for instance by a different colour or by way of a horizontal bar printed in the background. Major orders are divided into container quantities. An individual emergency kanban is created for each partial quantity and thus triggered in the production. The splitting up of course increases the delivery time in comparison with small order quantities. In anticipation of expected but not yet received annual orders or temporary promotion-related sales increases, emergency kanban may be used for prognosis-oriented pre-fabrication. Likewise, seasonal kanban may be defined and triggered, avoiding an adjustment of delivery times. Thus, the kanban system may be applied highly flexible.

To consider changes of customer demand by trend (cf. Fig. 2.10, cases B), the kanban system must be consistently *maintained*. For that the number of circulating kanban must be checked regularly (i.e. on a monthly basis). In growing markets, the number of cards needs to be regularly increased, as otherwise the material flow would break. In shrinking markets, however, the range of coverage increases if no cards are taken out of circulation – in order to keep the range of coverage consistent, cards must be removed gradually.

Kanban rules:

1. No production without production kanban and only in accordance with stipulated lot sizes.
2. Production exclusively in the order of retrieval by the customer process unless deviating priorities have been defined.
3. Each filled container must be allocated a kanban.
4. Kanban containers are only ever put down at defined and marked locations.
5. The logistician is responsible for the transport of material and kanban according to schedule on defined routes.
6. The number of circulating kanban must be checked regularly (monthly).

Special kanban characteristics:

1. An accumulation of kanban in front of a production process indicates that the process is not working fast enough.
2. Contrary to FIFO coupling, quality defects of individual parts do not disrupt the entire production process, as the exchangeability of parts makes supplemental deliveries unnecessary.

Material logistics. There are numerous possibilities to implement the control logic described above by way of the two control pulses. The classic solution, so to speak, uses both types of kanban in the form of laminated cardboard cards commuting between the production processes and supermarket store. The kanban contain all relevant control information with a view to material, customer process, supplier process, containers and the actual cards (Fig. 3.35). In the case of material transport, they accompany the containers. These *information carriers* are mainly needed because production and transport tasks are strictly separated to allow independent optimization for each of them.

1 Withdrawal Kanban

Supermarket Store	Part Information	Place of Usage
• Store Location • Supplier Process Card-Number (Total Number)	• Identity Number • Denomination of Part • Sketch of Part • Denom. Container • Quantity of Parts per Container	• Name and Location Customer Process • Workplace (Address) • Staging Place

2 Production Kanban

Supermarket Store	Part Information	Production Process
• Store Location • Customer Process Card-Number (Total Number)	• Identity Number • Denomination of Part • Sketch of Part • Denom. Container • Quantity of Parts per Container	• Name and Location of Supplier Process • Number of Operation Sheet • Raw Material

Fig. 3.35 Information affixed to withdrawal and production kanban

The transfer of information from the production worker to the logistician is done by way of said cards. As soon as an employee from the customer process removes the first part from a provided container, he puts aside the respective kanban for the logistician who regularly and according to a fixed *timetable* passes workplaces on a given route supplying material and collecting withdrawal kanban. In order to prevent too much material accumulating in the customer process, container quantities should be defined for container coverages to roughly match one shift, requiring transport cycles of one to two hours' duration at most.

As if equipped with a shopping list, the logistician withdraws containers with the ordered parts from the supermarket, removes the production kanban affixed to or included in the container and replaces the same by the withdrawal kanban instead. The logistician then transports the thus withdrawn containers to their place of usage, where in turn he picks up new withdrawal kanban, thus always conducting *mixed transports*. The production kanban placed in the supermarket store are in turn taken to the supplier process where they serve as a manufacturing specification. Upon completion, the replenished containers are taken to the supermarket store together with the production kanban, thus completing the cycle.

The material logistics would not necessarily need the kanban in the form of physical cards. The entire procedure could also be realized by way of electronic data transfer in the form of 'faxban' or scanned in as 'e-ban'. But even without EDP, simplifications are possible depending on the specific circumstances. If for instance the supplier process is in direct visual contact with the supermarket store and can physically observe all material withdrawals, it may work on a visual basis and simply refill empty storage places based on the respective labels with item numbers – production kanban would not be required in this case. If the supermarket is located close to the customer process, the withdrawal kanban may not be necessary – if there is only one place of utilization, the supermarket could actually constitute the retrieval area for the customer process. *Container kanban* with returned empties taking over the functions of both withdrawal and production kanban are often the most practical solution. The returning empty containers are equipped with all relevant control information and automatically trigger the required re-production, at the same time taking care of the problem of returning empties.

Specification of Production Kanban

The kanban system is designed through determination of the maximum permissible supermarket inventory and the resulting card requirement. This is achieved with the aid of a simple *calculation formula* based on three summands, the factors of which are not always easily determined. The supermarket inventory consists of the work in progress, a buffer inventory to cushion demand fluctuations and a safety stock to compensate breakdowns and quality problems (Eq. 3.9). These different types of inventory are going to be investigated further in the following.

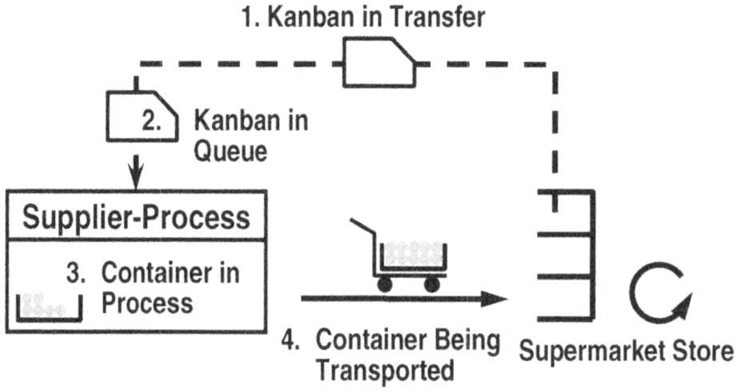

Fig. 3.36 Replenishment cycle

Replenishment. *Work in progress* serves to cover the mean consumption during the time needed for re-production. It results from the multiplication of the replenishment lead time in days and the mean daily consumption (Eq. 3.3). The replenishment lead time is of course difficult to assess.

The replenishment lead time consists of four components: the kanban transfer time from the supermarket to the supplier process, the kanban's waiting time before the supplier process, the operation time per lot and the move time (Fig. 3.36). In an ideal scenario, the logistician's *transport cycle time* determines the replenishment lead time. Let us picture the replenishment advancing one step further with each cycle completed by the logistician. In the first cycle, the logistician delivers the production kanban to the supplier process. During the second cycle, the respective kanban card is waiting in a queue, attached to a kanban board preceding the supplier process. In the third cycle, the required parts are re-produced in the respective kanban quantity and finally taken to the supermarket store in the fourth cycle. Slightly simplified, the required work in progress (WIP) may thus be assessed as follows:

$$WIP = RLT \times DD \approx 4 \times TCT \times DD \tag{3.8}$$

with: WIP work in process [pcs.]
RLT replenishment lead time [d]
DD daily demand [pcs./d]
TCT transport cycle time [time unit]

Since we are exclusively looking at small lot sizes here, we may assume the transport cycle to be significantly longer than one kanban's operation time in our assessment. The logistician will always be transporting several containers at a time, so that several lots will need to be reproduced per cycle. The main value to be verified here is the length of the kanban *queue*, which may become significantly longer than one transport cycle and thus increase the replenishment lead time accordingly. Well-balanced production processes and kanban of uniform time intervals, though, should keep the production flowing evenly.

Demand fluctuations. The *buffer stock* compensates demand fluctuations of the downstream process. In the case of a supermarket store of finished goods, it serves to balance fluctuations in customer demand and accordingly should be amply designed (Sect. 3.4.1). For supermarket stores within the production, the two connected processes should generally be capacitively balanced to prevent volume-related fluctuations of the total demand. However, fluctuations resulting from different variant mixes are to be expected whenever one variant is in greater demand than on average expected. In order to keep these fluctuations at a minimum, the planning concept must try to spread the increased demand as evenly as possible across the production mix (Sect. 3.4.2). As such fluctuations will not be able to be entirely eliminated through planning, though, they should be taken into account in the form of a mark-up (Eq. 3.9). To prevent material flow breaks, also in cases of rare variants, at least two kanban each are required to have parts ready from the second container during replenishment.

Breakdowns. The *safety stock* serves to compensate replenishment fluctuations caused by breakdowns and decreased volumes as a result of rejections. In this context, the percentaged part often cited for the capacitive dimensioning is less significant; rather, the downtime periods resulting in production standstill, or the lengths of time required for re-production of the maximum quantities of rejections are the decisive factors. Both of these increase the following process's waiting times by exactly this time factor. Accordingly, the required additional range of coverage of the safety stock (RC_{SS}) needs to be identified to determine the respective inventory with the aid of the daily demand (Eq. 3.9).

The *maximum supermarket inventory of one variant* is thus calculated by summation of the three factors described above, i.e. replenishment lead time, demand fluctuations and downtimes:

$$SQ_{Var} = WIP + BS + SS$$
$$\approx 4 \times TCT \times DD_{Var} + 4 \times TCT \times \Delta DD_{Var} + RC_{SS} \times DD_{var} \tag{3.9}$$

with: SQ_{Var} maximal stock in supermarket of a variant [pcs.]

- WIP work in process [pcs.]
- BS buffer stock [pcs.]
- SS safety stock [pcs.]
- TCT transport cycle time [time unit]
- DD_{Var} daily demand of a variant [pcs./d]
- ΔDD_{Var} maximal additional daily demand of a variant [pcs./d]
- RC_{SS} range of coverage of safety stock [d]

The *number of kanban per variant* results from the division of the maximum inventory of a variant by the respective (possibly variant-specific) container quantity (to be rounded up). However, in daily factory application, it is recommended to constantly verify this result in accordance with the sixth kanban rule.

$$\# K_{Var} = \text{ROUNDING UP}\left[\frac{SQ_{Var}}{CQ_{Var}}\right] \tag{3.10}$$

with: $\# K_{Var}$ number of kanban of a variant

- SQ_{Var} maximal stock in supermarket of a variant [pcs.]
- CQ_{Var} container quantity of a variant [pcs.]

Signal Kanban

With a regular kanban, the supplier process produces lot sizes identical to those removed from the supermarket store by the customer process. The lot sizes conform to the respective, relatively small variant-specific container quantities and thus require short changeover times. Once processes with long and short changeover times are coupled, uniform lot sizes are not required anymore, because this would either enforce excessive changeover times in the supplier process or unnecessarily increase the customer process's lot sizes – which in turn would increase inventories and decrease flexibility. The solution to this balancing problem is a relatively simple one: Several production kanban are accumulated before they are passed on to the supplier process as one lot. This approach is called *signal kanban*.

Function logic. The signal kanban is symbolized by an upside down triangle, replacing the respective production kanban (Fig. 3.37), inside which the *lot size of kanban* is given – in the numerical example one signal kanban stands for nine production kanban of ten parts each, i.e. the lot size in the supply process amounts to 90 parts. Whenever nine withdrawal kanban of one variant have accumulated, another lot is ordered from the supply process. Otherwise, procedure and symbolism conform to that of the production kanban.

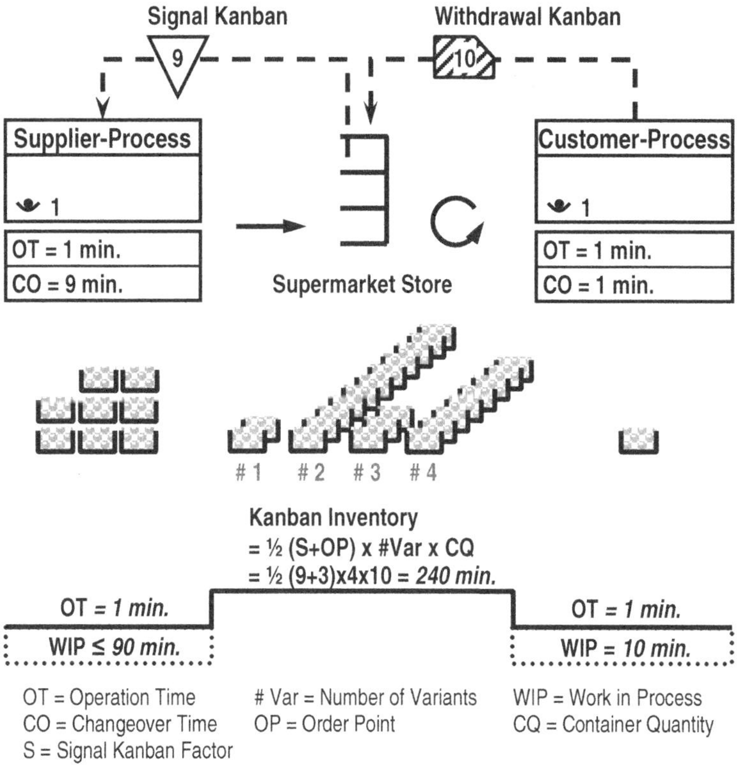

Fig. 3.37 Signal kanban with numerical example

Larger lot sizes, however, increase the inventories accordingly. The supermarket store needs to accommodate bigger lots, which in addition to buffer and safety stocks (cf. Eq. 3.14) determine the maximum permissible stock (cf. Eq. 3.15). In our example, the *production lead time* now amounts to five and a half hours per part – calculated from ninety plus ten minutes for the two lots in progress plus the supermarket inventory of 240 minutes.

Triangle kanban. The signal kanban was originally applied in the form of a paper sheet placed on top of a container at such an angle that one corner was sticking out of the stack of containers like a triangle (Fig. 3.38), which lead to the symbol and the name of 'triangle kanban'. Gradual removal of containers from the stack reduces each variant's inventory step by step until the container with the signal kanban is finally reached. This is the *order point* at which re-production is triggered by taking the signal kanban to the supplier process. The order point must be chosen in such a way as to ensure that the remainder of stock contained in the containers underneath the signal kanban will roughly cover the demand until the newly produced lot arrives. The remaining stock should be slightly higher than the mean demand expected for the duration of the replenishment lead time to cushion demand fluctuations and delivery problems. In addition to the Work in Progress, buffer inventories and safety stocks must be provided.

At the supplier process, the signal kanban are affixed to the signal kanban board in the exact sequence of their arrival. In order to prevent *sequence transpositions*, the exact trigger times may be wipably recorded on the cards. Under no circumstances may the lot sequence be altered on the kanban board, as this would lead to unforeseeable fluctuations in replenishment lead times, which would in turn cause the material flow to break or necessitate higher safety inventories.

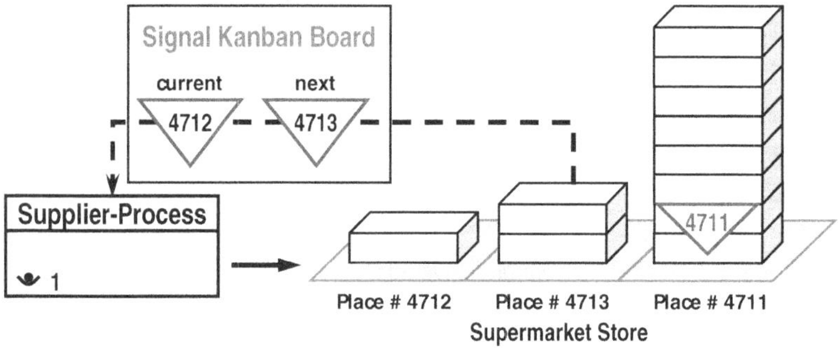

Fig. 3.38 Application of a triangle kanban and signal kanban board

The triangle of the triangle kanban is an *order document* which contains all significant information on material, supplier process and order processing. Order processing requires information on order point and lot size. Material information includes part ID numbers, material denominations and the exact storage location in the supermarket store. Supplier process information may include operation sheet numbers, machine numbers, tool numbers and/or the storage locations of tools. There will be exactly one signal kanban per part ID no., no further grouping together of lots will be permitted. The signal kanban may also be electronically integrated into control software, in which case data evaluations will facilitate the specification of order quantities and points.

Kanban lots. An alternative application of the signal kanban is provided by the *control board*, where kanban are collected and grouped together into lots. Individually arriving kanban are sorted by part ID numbers and attached to the board from left to right (Fig. 3.39). There will be one designated field on the board for each card in circulation. In the case of few kanban, certain fields of the rectangular board will be discarded. The fields are then allocated one of three coloured zones, 'green', 'yellow' or 'red', which symbolize the respective processing urgency.

Ident No.	green				yellow		red
4711	10	10	10	10	10	10	
4712		10	10	10			
4713		10					
4714							
4715		10	10	10	10		
4716	10	10	10				
4717	10						
4718	10	10	10	10			
4719			10	10	10		

Fig. 3.39 Control board kanban lot production

As long as the number of returned kanban is lower than required for a new lot to be re-produced, the cards will be put up in the green area; production may not be commenced yet. When the *minimum order quantity* limit is exceeded, the most recent card is attached in the yellow area and re-production may begin – all kanban accumulated so far will be produced together as one lot. Accordingly, parts 4711, 4715 and 4719 may be re-produced in the illustrated example above. However, depending on the number of yellow fields occupied and the urgency of other parts to be re-produced, the worker may put off the re-production a little while – at most though until the first card reaches the red area, by which time it is really already too late, because the safety stock will have been dipped into by then. This should really only ever happen in cases of breakdowns – and would turn the respective re-production into an urgent order. Depending on the number of yellow fields occupied, this method will lead to differing lot sizes for one and the same part. Besides, the actual sequence also largely depends on the respective worker's preferences. Both can positively increase flexibility and enable local changeover optimization. This however, decreases the level of standardization and accordingly the predictability of the behaviour of the kanban system.

Lot Size Determination for Signal Kanban

Changeover portfolio. One of the major differences between the goals of the kanban system and those of a production plan-controlled production lies in the higher *changeover frequency*. The changeover portfolio shows four possible target values and measures for kanban realization (Fig. 3.40). Production plan-based control uses lot size formulas to determine lot sizes optimized as to cost effectiveness in cases of long changeover times. With short changeover times, cost-effectiveness requires high capacity utilization. By trend, resources are laid out for short changeover times, thus necessitating large lots.

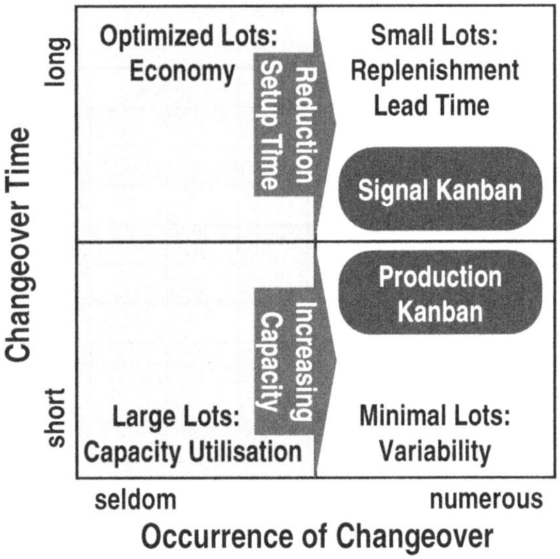

Fig. 3.40 Changeover portfolio

By kanban control, however, the other two target values are pursued. The production kanban wants minimal lot sizes based on container sizes to achieve maximum flexibility in variant mixes. This requires high capacities and short changeover times. Besides, in order to allow small lot sizes to have a decreasing effect on inventories, demand fluctuations must be cushioned by flexibility of capacities which in turn not allows maximum capacity utilization. Accordingly, *increase of capacities* is required compared with prognosis-oriented production control. For the signal kanban, however, measures for the *reduction of changeover times* are of vital importance to significantly lower replenishment lead times at cost-effective changeover effort. However, decreased changeover time quota are not prioritized.

EPEI value. The EPEI value (every part every interval) is highly suited for the determination of lot sizes in a kanban system in cases of fixed *changeover time quota*. The EPEI value indicates the period of time required by a production process to produce exactly one lot of each variant (cf. Sect. 2.3.1). This approach does not use minimized changeover times to reduce changeover time quota, but methodologically converts the same into reduced lot sizes wherever possible with a view to capacity. According to experience, the aspired changeover portion of 10 per cent usually constitutes a reasonable compromise between the minimization of changeover effort and lead time minimization.

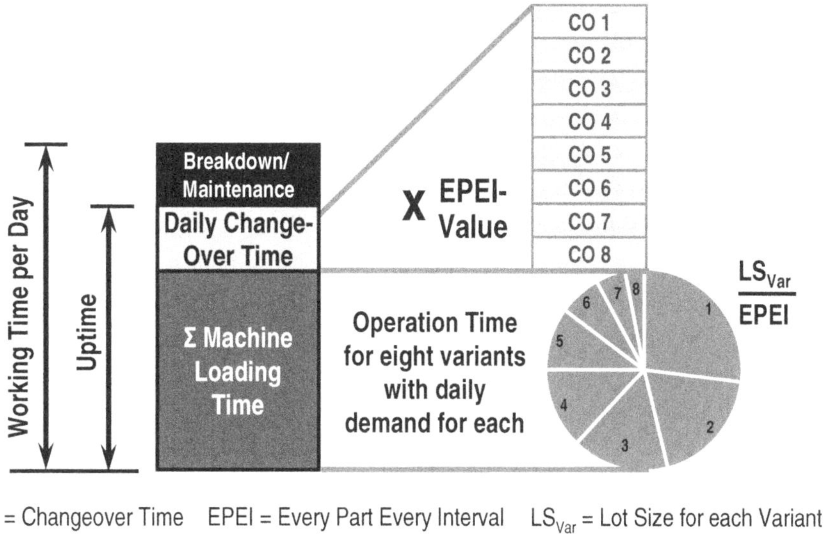

Fig. 3.41 Model for EPEI value calculation

The *determination of lot sizes* assumes resources to be utilized over their entire available life – either for production or changeover – and from there rather works the other way round compared with the regular approach. The intended capacity utilization is set at 100 per cent, then the operation time required to produce the mean daily demand is subtracted and the entire remaining time is used for changeovers. The resulting changeover time thus determines the number of possible changeovers per day, which with the aid of the number of variants leads to the respective EPEI value (Fig. 3.41). Accordingly, the EPEI value is the quotient of the total changeover time required to mount each variant once and the period of time available for changeovers per day (Eq. 3.11). The lot size thus equals the mean piece number per variant used in one EPEI cycle.

Accordingly, two time-related values need to be determined, namely the *total changeover time* and the *daily changeover time*. In the case of identical changeover times, the total changeover time equals the product of one individual changeover time and the number of variants. In cases of differing changeover times, the different changeover times are simply added. The daily available changeover time then results from the difference between the daily available capacity and the actual machine loading time. The available capacity is calculated by multiplication of the working time and the number of available resources. Downtimes and maintenance reduce the available capacity to the percentaged value of uptime. The machine loading time, however, equals the total operation time for all product variants needed in one average day. The EPEI value is thus calculated as follows:

$$EPEI_{\min} = \frac{\sum CO}{\text{daily CO}} = \frac{\#Var \times CO}{(WT \times \#Res) \times UP - MLT}$$

$$= \frac{\#Var \times CO}{(WT \times \#Res) \times UP - \sum_{i=1}^{i=\text{var}} OT_i \times DD_i}$$

with: $EPEI_{min}$ EPEI-value to determine minimal lot size [d] (3.11)

CO changeover time [time unit]

Var number of variants

WT daily working time [time unit/d]

Res number of resources

UP uptime [%]

MLT machine loading time [time unit]

OT operation time [time unit]

DD daily demand [pcs./d]

EPEI calculation example. A simple numerical example will illustrate the above. For eight variants with a changeover time of 1.5 hours each, the total changeover time amounts to 12 hours. The mean daily demand totals 1,170 pieces of a uniform operation time of one minute each. This results in a daily capacity requirement of 19.5 hours. The working time in a three-shift model amounts to 22.5 hours without break times. Accordingly, equation 3.11 results in an EPEI value of four days:

$$EPEI = \frac{8 \times 1{,}5\text{ h}}{22{,}5\text{ h/d} - (1.170/\text{d} \times 1\text{ min.})} = \frac{12\text{ h}}{22{,}5\text{ h/d} - 19{,}5\text{ h/d}} = \frac{12\text{ h}}{3\text{ h/d}} = 4\text{ d} \quad (3.12)$$

However, uptime restrictions have not been taken into account yet. This does not require the EPEI value to be divided by the uptime, though; instead the available capacity is reduced accordingly. This has no significant influence on the EPEI value, because the entire percentaged reduction will be subtracted from the time available for changeovers. In the numerical example, the rather high machine uptime of 90 per cent actually manages to quadruple the EPEI value to 16 days:

$$EPEI_{\min} = \frac{8 \times 1{,}5\text{ h}}{(22{,}5\text{ h/d} \times 90\%) - 1.170\text{ min./d}} = \frac{12}{20{,}25 - 19{,}5}\text{d} = \frac{12}{0{,}75}\text{d} = 16\text{d} \quad (3.13)$$

The EPEI value may be reduced through a reduction of changeover time. A reduction to one third in our numeric example for instance would also reduce the EPEI value to one third, i.e. 5.3 days.

Lot sizes. The EPEI value corresponds to the minimum possible coverage of a lot. The lot size of each variant is at least as large as its respective demand during the period of time corresponding to its EPEI value. Strict conformation with this calculation method determines the *variant-specific lot size* as the result of the respective daily demand multiplied with the EPEI value. As an advantage, this leads to the exclusive mounting of lots of identical coverage. The lots simply need to be adjusted to the respective step-fixed container quantities.

In cases of exotic, rarely needed variants, however, the variant-specific changeover portion may be exceptionally high. For *variants with low piece numbers* it is therefore recommendable to group several lots together rather than produce the same with each cycle. This will significantly increase the respective inventories for the respective variants, but will have little effect on the overall stock, as it only concerns parts with low piece numbers. In the numerical example (cf. Fig. 3.41), for instance, variants 7 and 8 could be mounted alternately in order to allow for production in bigger lot sizes and double coverage.

Changeover reduction. Excessive changeover times are often due to poor organizational and/or technical preconditions. Two basic approaches for the systematic reduction of changeover times, namely *elimination* of waste and *externalization* of ancillary tasks enable a reduction of changeover times somehow historically grown out of proportion to a mere 10 per cent of their original lengths (Fig. 3.42).

The first approach eliminates waste occurring in the production process, which as a result of poor *changeover preparation* and lack of *changeover standards* amounts to around 50 per cent of the changeover time. In order to identify such waste, changeovers are mapped by drawing all movements of the tool setter in the machine layout. This results in a so-called spaghetti diagram and illustrates all walks back, forth and in circles between the various control sections of the machine and the storage places for required fixtures and tools. In addition, a video analysis helps record the time portions used for the searching for tools, the reading of poorly visualized instructions or the fiddly setting of adjustable elements. Adjusting tasks should be rendered superfluous through minute technical measures such as final sets and markings, i.e. mechanical saving of changeover adjustments to make them easier reproducible. Additional waste of time results from poor uptimes or lack of readiness for operation of the required equipment and tools as well as deficient tool setting qualifications.

The second method externalizes as much of the remaining changeover effort as possible. If approx. 80 per cent of the now low-waste changeover procedures can be conducted parallel to the operation time, i.e. before or after the actual *changeover*, the changeover-related machine downtime will be reduced to 10 per cent of the original value. All efforts connected to the provision of materials, fixtures and tools may be externalized, providing this is done early enough and without fault before the previous lot is completed. Trouble-free restarts are best guaranteed if cleaning, servicing and maintenance are always conducted immediately upon demounting and if functionality is verified before remounting.

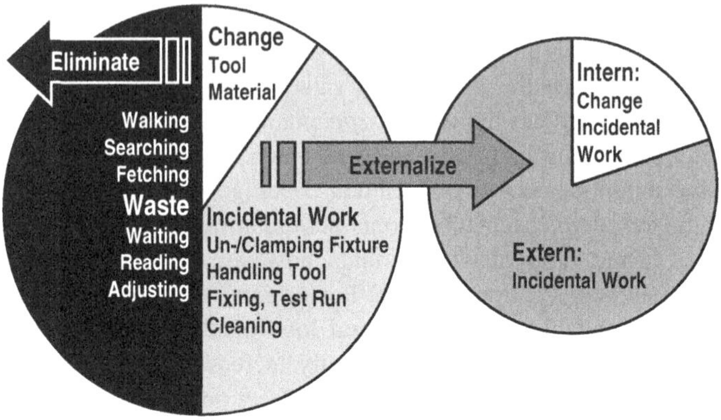

Fig. 3.42 The two approaches for changeover reduction

Reordering with Signal Kanban

The EPEI value is the decisive factor for the *range of coverage* of the supermarket stores behind each production process, which must hold at least one entire lot per variant. Assuming a simplified, uniform consumption of all variants, there will be an average stock of half a lot size per variant. Accordingly, the supermarket coverage then corresponds to one half of the EPEI value. In reality, of course, there will be demand fluctuations as to volumes and product mix, which will need to be balanced with the aid of buffer inventories. In addition, safety stocks must be provided for each variant to cushion possible supply problems. Both of these will increase the range of coverage required to assure an uninterrupted value stream. They are best dimensioned in connection with the determination of the order point for the signal kanban.

As soon as the inventory of a variant falls under a certain predetermined limit, re-production must be triggered. This *order point* must be at least as high as the expected demand during the replenishment lead time of that particular variant in the supplier process. In addition, buffer inventories and safety stocks must be determined to balance demand fluctuations and supply problems. Analogue to the determination of inventories in the production kanban, three different factors are added (cf. Eq. 3.9), though it is not the maximum permissible stock that is determined in this case, but the *variant-related reorder level*. Its division by the respective container quantity then results in the *order point expressed by the number of containers* as per equation 3.10:

$$RL_{Var} = WIP + BS + SS = RLT \times (DD_{Var} + \Delta DD_{Var}) + RC_{SB} \times DD_{var}$$

$$OP_{Var} = \text{ROUNDING UP}\left[\frac{RL_{Var}}{CQ_{Var}}\right]$$

with: RL_{Var} variant-oriented reorder level [pcs.]

OP_{Var} order point of a variant [number of containers]

WIP work in process [pcs.] (3.14)

BS buffer stock [pcs.]

SS safety stock [pcs.]

RLT replenishment lead time [d]

DD_{Var} daily demand of a variant [pcs./d]

ΔDD_{Var} maximal additional daily demand of a variant [pcs./d]

RC_{SB} range of coverage of safety stock [d]

CQ_{Var} container quantity of a variant [pcs.]

It should be noted, though, in the determination of the mean range of coverage, that the number of kanban (# K) depends on the total reorder level at the order point and the signal kanban volume. Inserted in equation 3.7, this accordingly results in:

$$RC = \frac{1}{2} \times \frac{\sum_{i=1}^{var} (OP_{Var} \times CQ_{Var} + S_{Var} \times CQ_{Var})}{\# P \times DD}$$

with: RC supermarkets range of coverage [d] (3.15)

OP_{Var} order point of a variant [number of containers]

CQ_{Var} container quantity of a variant [pcs.]

S signal kanban factor

DD daily demand [pcs./d]

P number of identical parts per finished product [pcs.]

The signal kanban *replenishment lead time* (Eq. 3.14) is composed of the signal kanban's transfer time, the waiting time before the supplier process, one lot's operation time incl. changeover and the respective move time to the supermarket store. Unless move and/or transfer times are negligibly small compared to the operation times, they may – analogue to the production kanban (cf. Eq. 3.8) – be assumed to be twice as long as the logistician's transport cycle time (TCT). In cases of differently sized variant-specific signal kanban, mean values are used to determine the operation time per lot. The general waiting time before a supply process strongly depends on the quality of the kanban layout and its performance within the overall factory operation. In a well-adjusted system which enables fast reactions to the requirements of the customer process and at the same time prevents the supply process from running dry as a result of lack of orders, there will an average of one signal kanban in the queue before each machine. Assuming that half of each lot in progress has just been completed, the total mean replenishment time is:

$$RLT_{\emptyset} = 2{,}5 \times (OT \times S_{\emptyset} \times CQ + CO) + 2 \times TCT \tag{3.16}$$

with: $RLT_{\emptyset}$ average replenishment lead time [d]
OT operation time [time unit]
CQ container quantity of kanban [pcs.]
$S_{\emptyset}$ average signal kanban factor
CO changeover time [time unit]
TCT transport cycle time [time unit]

The calculation formulae given here serve mainly the initial layout and quantity evaluation for a kanban control system and in particular point out the influencing factors to be taken into account. Continuous adjustment of the determined values will then be up to the daily factory operation.

Supplier Kanban

The kanban system is also suitable for external procurement of raw materials or regularly ordered bough-in parts. Depending on the respective types of parts, there are two possibilities, differing mainly from an organizational and commercial point of view, but also with respect to the logistics involved. Low-priced C-parts are handled by way of the *C-part management* easily available from the respective suppliers in the form of complete solutions. For more expensive, product-specific parts, however, as well as component assemblies and raw materials to be procured in large quantities, *supplier kanban* are developed in close cooperation with the respective suppliers. The procurement of customer-specific parts and/or materials, however, is not possible in a kanban system; these may be better handled in the form of individual or just in sequence orders (Sect. 3.3.1).

Purchased parts portfolio. Purchased parts and raw materials may be entered in a purchased parts portfolio with their respective *procurement volumes* – expressed according to ABC classification – and their pertaining *supply security* – expressed through the number of competing suppliers in the market (Fig. 3.43). Each portfolio square is allocated a standard purchase strategy.

Standard parts can be obtained easily and at low prices from numerous competing suppliers. The partial logistics-related costs for transport, handling, inventory management, order processing and invoicing are relatively high for these small parts procured in large quantities (e.g. standardized parts and designed components), compared to their respective unit prices. The decisive target criterion in this context is the *efficiency* of the procurement process. Efficient procurement takes into account not just unit prices but also internal procurement and logistics-related costs. C-part kanban for instance, where the purchase of the parts usually includes the logistical solution for the respective C-part management, are highly suited for this purpose. The supplier guarantees constant material availability and

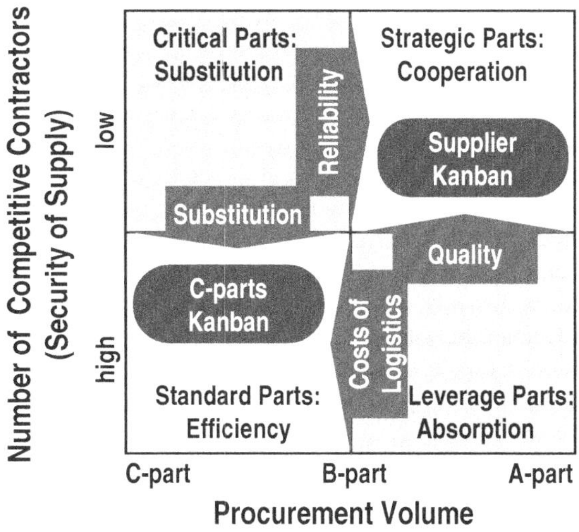

Fig. 3.43 Purchased parts portfolio

handles all aspects of the part supply to the shop floor, possibly even to the respective places of utilization in the production process. No orders are required for such parts.

Expensive materials, parts and component assemblies with high procurement volume, on the other hand, are often highly specific and only obtainable from selective niche suppliers. The standard purchase strategy in cases of such *strategic parts* is to make virtue out of necessity and minimize the pertaining supply risk through long-term agreements with suppliers. Such strategic *cooperation* often enables the development of close logistical supplier connections, namely the supplier kanban explained in detail further down. Again, the unit price is not the only decisive factor, but the minimization of logistical effort and increase of delivery frequency are taken into account as well.

Mass-produced parts with a narrow procurement market – often electronic components – are considered *critical parts*. In spite of their low costs, their lack may well bring production to a standstill. The respective standard strategy is to minimize the supply risk by *substitution* of parts or unreliable suppliers. Ideally, this portfolio square should include no entries all. This may be achieved by substitution of critical parts by easily obtainable standard parts to be procured via C-part kanban. Should this not be possible, a long-term agreement with a suitable supplier should be aimed at in order to minimize the supply risk – which would move the respective parts to the square for supplier kanban cooperation.

High-end A-parts easily obtainable in the market provide major *savings leverage* because they are offered by numerous competing suppliers. High procurement volumes make negotiated price reductions highly substantial. However, this does not take into account the increased procurement and logistics-related costs usually accompanying a large number of alternative suppliers. Accordingly, the cost-effectiveness of each leverage part must be checked as to internal process costs, otherwise it may possibly be better procured as a standard part in a C-part kanban with high logistical efficiency. Unfortunately for some production managers, there is another factor often disregarded by cost-conscious purchasers, namely quality – with a view to tolerances or delivery reliability. As a result, some savings may well turn out to be particularly expensive in the end. Accordingly, it must be carefully checked if maybe all A-parts should be procured from cooperating suppliers in well-organized, standardized supplier kanban, thus waiving the advantages of a larger procurement market, though. The strategy of *absorption*, however, which is aimed exclusively at the target of cost effectiveness, cannot be realized by way of a kanban, but on the other hand ensures the necessary reliability as to procedure and part quality required for a smooth flow of the production.

Function logic. In the kanban system, *supplier connections* require standardized trans-factory container concepts with well-matched container quantities for purchased parts. In cases of raw materials such as granulate, fluids, paper rolls, metal sections, sheet metal and wire coil, the matching of volumes requires slightly higher effort. Production processes in need of raw materials stored in the supermarket order the same via withdrawal kanban from the raw material store (Fig. 3.44). The respective raw material is equipped with a material ID slip, the supplier kanban. This card may now be directly transported to the supplier by truck with the next delivery or, as shown in the illustration, considerable transport time may be saved by triggering an electronic order via a procurement process. In the latter case, the card is transferred to the order receipt where it then waits for the receipt of the respective raw material in a *kanban post box*, which is symbolically depicted by an open top rectangle on a stand, into which the kanban is slotted vertically. By allocating an individual post box to each day, delivery defaults are immediately visible.

With the supplier kanban, the processing of external transports needs to be adjusted with the kanban logic to lower inventories without increasing transport costs. The raw material inventory's range of coverage depends directly on the delivery frequency and may never be lower than the demand between two deliveries. A change from weekly to daily deliveries for instance may potentially decrease the respective inventories by 80 per cent. Depending on various inherent restrictions, though, this potential may only partially be realized. In cases of a daily demand of less than one container, the lowest possible coverage will be longer than one day – unless the container size may be decreased accordingly. Another important factor for inventory reduction are the transport costs which may be lowered by way of higher delivery frequencies with only partially filled trucks conducting so-called *milk runs*. In these cases the carrier calls at all the different suppliers' locations one after the other, picking up relatively small delivery quantities. The total distance driven is lower than in a shuttle service, so that each supplier may be visited more often at identical effort. The milk run is depicted as a truck symbol with an arrow like the one depicting external transport – curved with a white inside (Fig. 3.44).

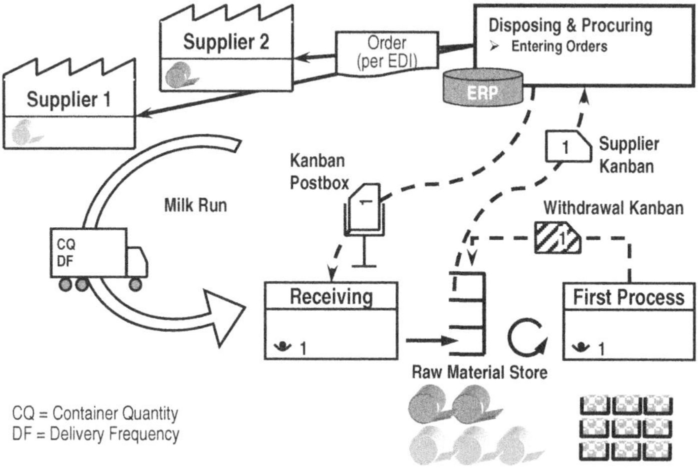

Fig. 3.44 Supplier kanban

Balancing of volumes. Contrary to individual, countable purchased parts, raw materials are usually measured and billed in units of weight or length. Due to differing scrap rates and varying start-up losses after changeover, package sizes expressed by way of such units may only roughly be converted into piece numbers and parts produced therefrom. As a result, raw material inventories cannot be converted into exact coverages but are defined with the aid of mean values of *specific material consumption*, i.e. yield of parts per package. In addition, raw materials are ordered in numbers of packages, but the respective package sizes (weight or length) are subject to significant ranges of tolerance, so that in reality even identical order volumes vary all the time. Accordingly, the order quantity for one supplier kanban of wire coils for instance may be anywhere between 450 and 550 kilograms. All above mentioned values are entered in the data box for the raw material supermarket store (Fig. 3.45).

Denomination, Location	
# Var	Number of Types of Material
PS	Package Size [kg, m, ..]
# B	Number of Raw Material Packs
PY	Yield of Parts per Package
RC	Range of Coverage

Fig. 3.45 Data box for the raw material supermarket store

According to signal kanban logic, production processes utilizing the raw material work with fixed lot sizes, but never need exactly one package (or an integer multiple of the same) of the raw material as required by kanban logic. A little will usually be left over in the production process – or, alternatively, material will run out shortly before a lot is completed. The second case, i.e. slight underdelivery, is usually rather unproblematic, as it merely leads to the next order being triggered slightly earlier. In the first case, the remainder could of course be returned to the store, but this would require additional logistical effort as well as larger storage spaces to hold the respective additional partial quantities for each variant. Besides, this would clearly violate the kanban rules, therefore suitable solutions must be found for each individual case. Minimal overdeliveries could for instance be permitted, the FIFO principle could be violated in the raw material store by allowing individual selection of package sizes best suited for the respective lots to be produced, or such *partial quantities* could be stored directly in the production process. Ideally the plant in question could have a store for all regularly used types of material.

As a rule, though, several variants are processed from one type of raw material, so that one package may be used for several lots, and with suitable variant mix sequences changeover efforts due to *changes of material* may also partially be avoided. However, such alterations of changeover sequence must be determined by production planning by rearranging the signal kanban or modifying the processing logic accordingly (Sect. 3.4.2), paying for the resulting changeover reductions with increased planning effort and longer replenishment lead times, though.

Kanban control

Kanban regulates the consumption-related re-production of parts. The various possible applications may be divided into three groups.

1. The simple *production kanban* enables the decoupling of production processes with small lot sizes – one lot corresponds to one container size. Short changeover times and small lot sizes keep replenishment lead times relatively short. The customer process withdraws parts in lots, but may alternatively process the provided parts individually. The supermarket store has all variants on stock.
2. The *signal kanban* allows the integration of production processes with long changeover times. In a triangle kanban, each variant is re-produced in a fixed lot size as soon as the respective order point is undercut. The application of a control board allows the accumulation of individual production kanban which are then re-produced in lots. According to the changeover portfolio, the kanban system aims to reduce changeover times without changes in changeover portions. The EPEI value determines the smallest possible lot sizes and maximum variant flexibility at optimal resource utilization.
3. The *supplier kanban* enables low-effort procurement of raw materials and purchased parts. The C-part kanban procures small standardized parts and designed components according to criteria of logistical efficiency. The supplier kanban regulates the reliable and balanced supply with strategic parts by way of cooperation. The milk run allows an increase in delivery frequency, which in turn lowers raw material inventories and balances the value stream. The first production process based on container quantities requires volumes to be aligned with the fluctuating package sizes of raw material.

Case study At *Liquipur*, the assembling process is supposed to be connected with the milling process either by production kanban or by signal kanban, depending on the respective variant – across the washing process connected via FIFO coupling. In addition, a supplier kanban is to be implemented for unfinished castings (Fig. 3.46).

Production kanban. The customer requirements with a view to shipping containers remain unchanged, so that in future the finished products may continue to be shipped in containers of 60 pieces each and accordingly use the same lot size in the last production process of the value stream, i.e. the assembly process. With each variant, 60 pieces are therefore withdrawn from the upstream supermarket store. As the parts are not supposed to be re-packed after washing, three wash baskets of twenty pieces each are stacked at each storage location.

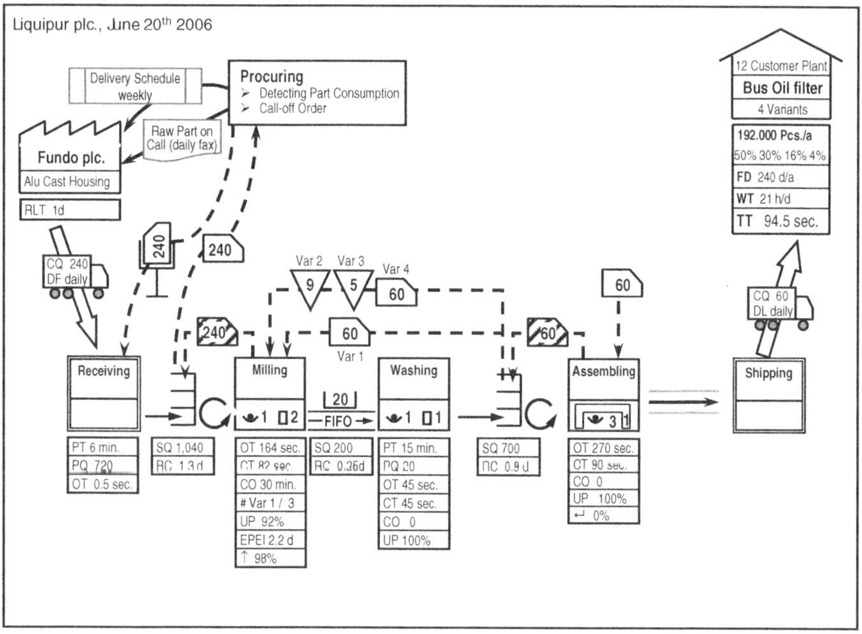

Fig. 3.46 Value stream design at *Liquipur* (3): Kanban control

It turns out that variant no. 1 with a piece number portion of almost exactly 50 per cent, may be continuously processed on one milling machine, which thus does not require any changeovers anymore. For each withdrawn lot, one production kanban of 60 pieces is triggered. The empty wash baskets are returned to the washer independently from the kanban control cycle.

Consistent preventive maintenance is to raise the uptime of the milling machine to 92 per cent. A continued operation time of 164 seconds results in a mean capacity requirement of 178 seconds per piece. The new layout does not assume the machines to work during break times; accordingly the daily available working time amounts to 21 hours The daily demand of 400 pieces of variant no. 1 needs to be marked up by a 2 per cent scrap quota, so that on average 408 pieces need to be processed. According to equation 2.1, this results in a customer takt time of 185 seconds, i.e. 4 per cent above the required cycle time.

Signal kanban. The remaining three variants are processed on the second milling machine. The changeover time is significantly reduced to 30 minutes, which does not suffice for a fixed lot size of 60 pieces, though. Accordingly, the EPEI value needs to be calculated as per equation 3.6 for the determination of the lot size:

$$MLT = OT \times DD = \left(\frac{164\,\text{sec}}{3600} \times 408\frac{\text{Pcs}}{\text{d}}\right) = 18{,}59\frac{\text{h}}{\text{d}}$$

$$EPEI_{\min} = \frac{\#Var \times CO}{(WT \times \#Res) \times UP - MLT} = \frac{3 \times 0{,}5\,\text{h}}{\left(21\frac{\text{h}}{\text{d}} \times 1\right) \times 92\% - MLT} = \frac{1{,}5\,\text{h}}{19{,}32\frac{\text{h}}{\text{d}} - 18{,}59\frac{\text{h}}{\text{d}}} \approx 2{,}05\,\text{d}$$

with: CO changeover time [time unit] (3.17)
\# Var number of variants
WT working time [time unit]
\# Res number of resources
UP uptime [%]
MLT machine loading time [time unit]
OT operation time [time unit]
DD daily demand [pcs./d]

Multiplication with the mean daily demand thus results in the minimal lot size per variant (LS_{Var}) as follows:

$$LS_{Var2} = \frac{58.000\,Pcs}{240\,\text{d}} \times 2{,}05\,\text{d} = 495\,Pcs \cong 540\,Pcs = 9 \times 60\,Pcs$$

$$LS_{Var3} = \frac{30.000\,Pcs}{240\,\text{d}} \times 2{,}05\,\text{d} = 256\,Pcs \cong 300\,Pcs = 5 \times 60\,Pcs \qquad (3.18)$$

$$LS_{Var4} = \frac{8.000\,Pcs}{240\,\text{d}} \times 2{,}05\,\text{d} = 68\,Pcs \cong 60\,Pcs$$

Accordingly, the signal factor for the second variant is chosen as nine, as five for the third variant and as one for variant no. four. The exotic variant is processed analogue to the production kanban in spite of its high changeover portion of more than 15 per cent (30 minutes changeover at an operation time of 164 seconds). In accordance with equation 2.6, rounded up lot sizes thus lead to an EPEI value of:

$$EPEI_{fix} = \frac{\sum OT + \sum CO}{\#Res \times UP \times WT} = \frac{((9+5+1) \times 60 \times 164\,\text{sec}) + 3 \times 30\,\text{min}}{1 \times 92\% \times 21\,\text{h}} = \frac{42{,}5\,\text{h}}{19{,}32\,\text{h}} = 2{,}2\,\text{d} \qquad (3.19)$$

Supermarket inventory for work in progress. In a well-established signal kanban, the WIP supermarket store before the assembly process contains half the coverage of the EPEI value plus buffer inventory and safety stock. At a daily demand of 400 pieces for variants 2 to 4 and an EPEI value of 2.2 days, this amounts to 440 pieces. A fluctuation tolerance of 30 per cent, or approx. 130 pieces, is assumed to balance the production mix. A maximum downtime of one hour per machine must be able to be compensated, which at a daily demand of 800 pieces, equals approx. 40 pieces. All in all, this amounts to a mean inventory of 700 pieces, conforming to a range of coverage of 0.9 days.

Order point. The order point for variant no. 3 is calculated as an example. At a lot size of 540 pieces, a cycle time of 164 seconds and a 30 minute changeover time, the operation time for variant no. 2 amounts to 25.1 hours As there are only three variants in total, the determination of the order point should take into account the fact that the previous lot has already been half completed, the waiting time will thus be 12.5 hours plus the operation time for variant no. 3 (cf. Eq. 3.16). As the material will be taken to the supermarket in the wash baskets via FIFO line and the wash process in stacks of three baskets each immediately after completion, the replenishment lead time needs to include 15 minutes for washing, a FIFO buffer of 200 pieces equalling 5.25 hours plus the operation time for 60 pieces, i.e. 2.7 hours This amounts to a total of 20.7 hours, i.e. approx. one day. At a daily demand of 125 pieces for variant no. 3, this equals 123 pieces. Another 30 per cent need to be added for the buffer, totalling 160 pieces. A safety stock of 6 pieces per hour is required, which determines the order point at 2.7 lots of 60 pieces each; the signal kanban is thus placed on top of the third stack of containers.

Supplier kanban. The supplier connection is already very good in the current state. With the implementation of the supplier kanban there is no need to increase the current delivery frequency of one day. It is assumed that no more inspection of goods received will be required. The recorded working time of 6 minutes per pallet of 240 pieces each results from unloading. The inventory is determined as a demand buffer of 30 per cent, resulting in a raw material inventory of 1,040 pieces, i.e. 1 day and 1 container.

3.3.3 Production Control at the Pacemaker Process

Three design guidelines are needed to achieve a lean material flow: continuous flow production, FIFO coupling and kanban control. With their help, the *entire* material flow of *any* given piece production may be designed; no other guiding principle will be required. Individual processes or flow productions are the components of the value stream which will be connected by FIFO lines or different kanban controls. However, the following rules must be observed in the combination of these connections.

Pacemaker. Central importance is attached to the decision as to how to control the entire value stream. In contrast to prognosis-oriented control, there must be exactly *one trigger point* for the entire processing of a manufactured product. From this trigger point onwards, the pertaining production processes are regulated by way of above mentioned coupling principles.

As a crucial advantage, explicit determination of the trigger point prevents contradictory control instructions which would inevitably result in higher inventories and shortfall quantities. Accordingly, production control should only ever intervene at precisely one point of the value stream. Normally, i.e. when all products of a product family are subject to identical control, exactly one production process will be controlled in each value stream, all other production processes of the same value stream will be regulated accordingly. The thus directly controlled process is called *pacemaker process*. This process sets the pace and thus the production rhythm for all other processes of the value stream. Accordingly, value stream control is subject to the following design guideline:

> **Design Guideline 5:** *Pacemaker process*
> Each value stream should be triggered in customer takt time at one definite point, the pacemaker process.

The pacemaker process adopts the customer takt time for the entire production. All orders should be triggered in conformance with the customer takt time, therefore the pacemaker should take particular care to stay in tune with the customer takt time and make the same visible. Only in this way the customer demand can trigger as *rhythm* the entire production. Deviations from the customer takt time originating from the pacemaker will spread over the entire value stream and thus impede the target achievement for the entire production. Therefore, the pacemaker should be equipped with measures to *visualize* possible timeouts, for instance in the form of illuminated displays, acoustic signals or maybe display panels showing targeted and actual output.

Production types. Production planning converts customer orders into production orders (Sect. 3.4), which are then triggered at the pacemaker process, which is also the location of the *customer decoupling point*. This point separates the customer-anonymous pre-production from the customer order-related downstream production. Accordingly, products with customer-specific characteristics can only be produced at the pacemaker process or further downstream. Depending on the location of the pacemaker in the value stream, different production types may be realized.

The logistical linkage between production processes and the pacemaker process is subject to the observance of certain rules. In general, the customer-anonymous pre-processes are connected by kanban control cycles, while the customer-related subsequent processes are linked by FIFO coupling. Accordingly, there may be a sequence of several separate downstream flow productions, or individual production processes, respectively, each connected to the pacemaker by a FIFO line. In these cases, the pacemaker is located relatively far up in the value stream, which indicates that most production processes will be working in a customer order-related manner. As these solutions also enable customer specific part production, this is also referred to as *make-to-order*-production (Fig. 3.47, case 1). There will be no kanban control cycle downstream of the pacemaker process, as this would result in decoupling and shifting of the pacemaker.

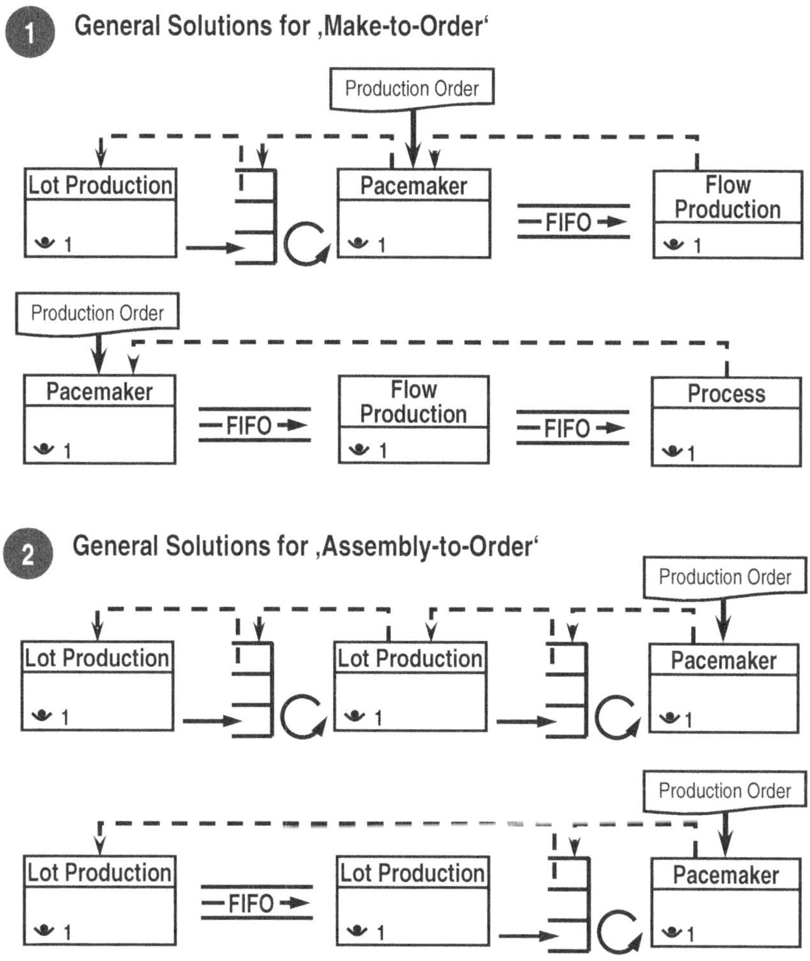

Fig. 3.47 Location of the pacemaker in the value stream

Upstream, however, there may well be a sequence of several production processes – possibly working with different lot sizes – connected via kanban. Customer-anonymous part production where only the very last process, normally an assembly process, is triggered in a customer-related manner is called *assembly-to-order* (Fig. 3.47, case 2). As a rule, kanban control is implemented through a number of kanban control cycles corresponding to the number of integrated production processes. Alternatively, two lot-producing processes may be directly coupled via FIFO line and jointly controlled by a mutual kanban cycle.

Shipping principle. The shipping process has deliberately been disregarded so far, as for numerous reasons it is not suitable as a pacemaker at all. Among other things, this is due to the fact that the only way to send off products in accordance with the customer takt time would be to ship each product in a truck of its own. At best, loading could be aligned with the customer takt, but then the trucks would have to make regular deliveries corresponding to the respective shift model – only feasible, even necessary, with just in sequence deliveries. The dispatch must be adjusted to shipping routes and carriers' time schedules, wait for customers' trucks, and take into account parcel service providers' pickup times. As a result, shipping usually uses shift models different from those of the production. Besides, the shipping end is usually in charge of several value streams and would need to act as a pacemaker for different takt times. However, shipments containing different products to be packed and dispatched together in packages or pallets are based on lot productions with greatly differing lot sizes. Besides, time requirements don't necessarily linearly dependent on piece numbers, which also rules out the pacemaker function, which would align the production to coincidences and external restrictions rather than actual customer takt times.

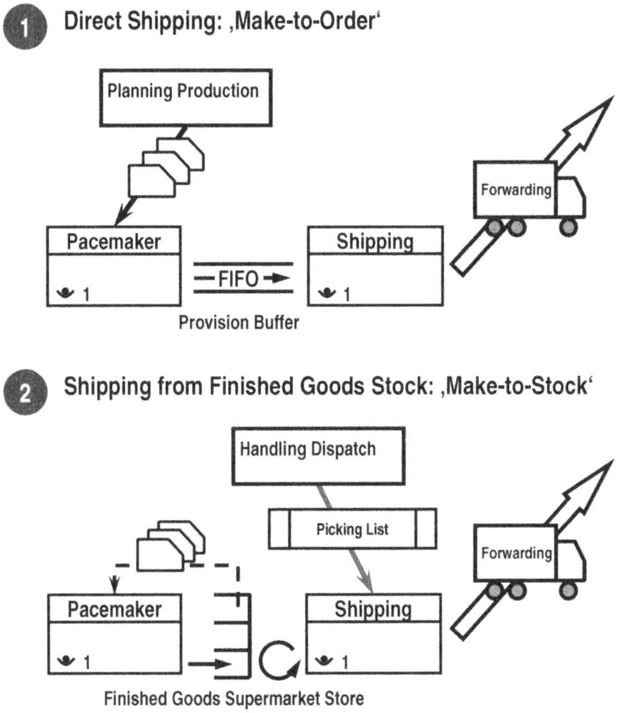

Fig. 3.48 Linkage between pacemaker and shipping end

Accordingly, the pacemaker process will always be located upstream from the shipping end, which may be logistically linked to the pacemaker in two possible ways: If the design guideline is implemented to directly trigger the pacemaker by way of the customers' orders, this will result in *direct shipping*. From the pacemaker onwards, all products are manufactured, prepared for shipping and dispatched as fast as possible in a customer-specific manner. Pacemaker and shipping end are connected by a FIFO line (Fig. 3.48, case 1). This solution suits all *make-to-order* variations and is the only possible way to handle customer-specific production of small piece numbers. However, direct shipping is also highly suitable for serial productions of low variance, if the last production process produces the correct amounts straight into the respective shipping containers.

The second method indirectly triggers the pacemaker via a finished goods storage. The shipping end is given a picking list, withdrawal list or similar document and withdraws the respective goods from the inventory accordingly. Reproduction of the removed products is then conducted via kanban rule in accordance with the fourth design guideline. In this case, the goods are *shipped from the supermarket store*. This *make-to-stock* solution is best suited for the primarily customer-anonymous production of serial products with very little product variance (Fig. 3.48, case 2). Thanks to the connection with the store, products manufactured in lots may also be shipped as partial commissions, whereas direct dispatch requires customer order lots to match production lots. If an entire production is made to stock, the finished goods store protects the production from disturbances originating from the shipping end.

Order lead time. Strict conformance with the design guidelines concerning flow production, FIFO coupling and kanban rules irrevocably determines the *location of the pacemaker process* as the first production process looking downstream not requiring large lots. It is thus located as far upstream as technically feasible, enabling a large part of the production to work in a customer order-related manner, but also customer-specifically, if required. The first prevents overproduction in excess of the actual customer demand while the latter allows a high variance in product characteristics to meet varying customer wishes at minimal effort.

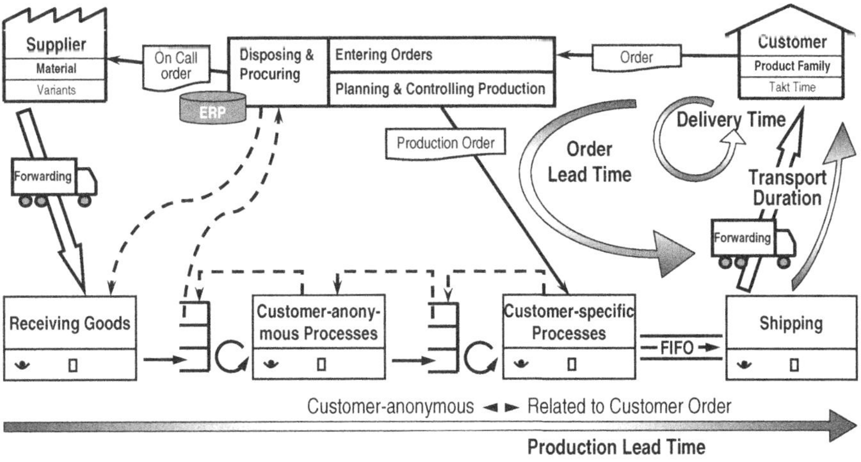

Fig. 3.49 Schematic representation of the ideal future state of a production

Sufficiently long *delivery times* offered in the market are a vital prerequisite for the conformance the design guidelines no. one to five. The delivery time is composed of the order lead time and the transport time. The respective periods of time in relation to the value stream are shown in the schematic representation of the ideal future state of a piece production with direct shipping (Fig. 3.49). For the achievement of the aspired ideal production procedure, the total time must be lower than the desired delivery time – otherwise different possible ways of acceleration must be looked into. In general, it is not possible to shorten *transport periods* without (costly) changes to the means of transport. In addition, delivery schedules and their rigid commitment to defined shipping dates may lead to additional waiting times. The only feasible approach left for production optimization therefore lies in the reduction of order lead times.

The *order lead time* is made up of two compounds: administrative and planning-related order processing on the one hand and the customer order-related portion of the *production lead time* on the other. The latter could theoretically be reduced through lower maximum FIFO inventories or shorter operation times, both of which will hardly be possible in view of the already implemented improvements, though. As an alternative, the pacemaker process could be moved further downstream, provided customer-anonymous pre-production is at all possible. This would shorten delivery times at the cost of increased stocks in prefabricated parts. Unless we are looking at customer-specific products, the implementation of a finished goods store to trigger the pacemaker would also be possible. Accordingly, the shipping kanban (Fig. 3.48) would enable the realization of the shortest possible delivery time, but increase finished goods inventories to make up for the production being too slow for the shorter delivery times.

Another possible toehold to achieve a reduction in order lead time is *order processing*. Unless the customer orders containing customer-specific information are received by letter post, order processing is exclusively determined by internal business processes, i.e. order entry, order dispatching and shipping management. The possibility of speeding up administrative processes such as order entry and shipping management through improved electronic tools, parallelization, partial automation, elimination of duplicate work und coordination procedures, shall be disregarded herein, though. More importantly, the application of the FIFO principle constitutes a vital prerequisite for the implementation of lean production, as this is the only possible way to process all orders in an identical manner. With optimal implementation and utilization of suitable software, the time effort for order processing and product configuration may be lowered to a negligible minimum, at least as far as low-variance standard products are concerned.

Order disposition converts customer orders into production orders and fairly precisely determines at which point in time to start the production, which falls into the responsibility of production planning (Sect. 3.4). The *waiting times* ensuing for the customer orders make up a substantial time portion of order processing. Only in cases of primarily customer-specific products with large construction-related time portions, such as in plant construction, the waiting times are less significant. Accordingly, waiting times constitute a decisive factor for the achievable delivery times of all products of not entirely customer-specific manufacture. They result from the planning logic introduced in the following section and developed with the aid of further design guidelines to determine the overall production procedure as such, which may then be adjusted to match the desired delivery time.

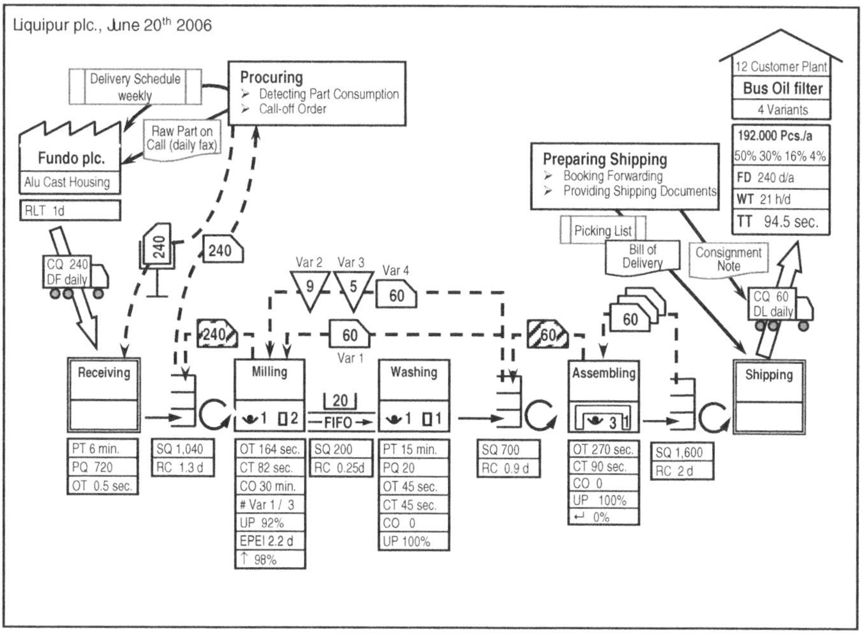

Fig. 3.50 Value stream design at *Liquipur* (4): Shipping kanban

Pacemaker Process

Goal Each value stream should be triggered at exactly one point. The triggered production process acts as the pacemaker for the entire production. All other processes are regulated accordingly.

Implementation in the value stream The customer decoupling point is located at the pacemaker. Customer-specific production can only be carried out at the pacemaker process and downstream from there. All upstream processes are customer-anonymous ones. The location of the pacemaker determines whether the production, or assembly, respectively, is conducted in a customer-specific manner or as a make-to-stock production.

The shipping process can never be the pacemaker, as it needs to consistently shield the smooth production flow from customer demand fluctuations and shipping logistics-related restrictions. In a make-to-stock production, the pacemaker is completely isolated from the dispatch by the shipping kanban. In direct shipping, the rhythm of the production is determined by the pacemaker, not by the shipping process which in turn is supplied from a provision buffer.

Impacts The production lead time results from the application of the first five design guidelines. The location of the pacemaker is decisive for the customer order-related portion of the production lead time, which forms part of the achievable delivery time. The remaining portions result from transport times and production planning.

Case study. At *Liquipur*, the shipping principle of choice is the shipping kanban (Fig. 3.50). Due to the very short delivery time of only one day, the assembly process cannot wait for binding orders, nor is it possible to reliably produce one day in advance, as demand forecasts are rarely totally correct. A finished goods store with a range of coverage of two days is therefore created to buffer demand fluctuations (cf. Fig. 3.52) and instantly meet customers' wishes. Demand forecasts helps react to possible changes in the variant mix in good time.

3.4 Production Planning

The guidelines introduced above define control rules for optimal management and regulation of the material flow in a production. In the following section on *production planning,* relevant design guidelines for the conception of the planning logic up to point of order release will be developed. For the smooth functioning of the production control devised above, all production orders were assumed to have been appropriately released by production planning. Accordingly, the *order release* must be triggered in such a way as to enable a smoothly flowing value stream with the aid of said control rules, for the purpose of which production planning needs to complete the following three tasks:

First of all, the customer orders need to be converted into production orders for the pacemaker process. The best solution for easier control is to create equally sized release units. To enable an even production the load has to be carried out evenly. When customer demands fluctuate, by this volume-related *production smoothing* orders are accumulated in *queues* (Sect. 3.4.1). The thus created production orders must secondly be put in the correct *sequence* through *production levelling* across all variants considering their respective different production requirements (Sect. 3.4.2). Thirdly and finally, the time of release and the order sequence will sometimes also be dependent on capacitive and technological *restrictions* caused by individual production processes which will also need to be taken into account in the planning process (Sect. 3.4.3). This simple system of rules enables the planning of value streams, including those which manufacture very complex products in highly sophisticated production processes. Contrary to traditional PPS approaches, however, this is not done by way of mean value-based prognosis-oriented data, but each order is planned individually.

Production planning targets

To facilitate production control, production planning needs to pre-process the customer demand to keep the production load even and thus enable a smooth value stream. This requires uniform production order units to be released in a homogenous rhythm. Production order sequences must enable the variants to create consistent demand for parts and capacities and at the same time take into account other restrictions affecting the production processes.

3.4.1 Smoothing the Production Volume

The main objective of production planning is the creation of a smoothly flowing production. Accordingly, the pacemaker process should only ever release clearly defined volumes of work in a steady rhythm. The *smoothing* of production volume is achieved by consistent triggering of identical lot sizes at identical intervals rather than different lot sizes varying with each production order. In simple cases with identical operation times, the respective volumes may be determined by piece numbers, otherwise the capacity demand at the pacemaker process will be used as a basis.

With the definition of standardized *release units* as a means of value stream control, the decisive time frame for production planning and control is determined. In lean literature, this is also referred to as 'pitch', named after the act of throwing a baseball toward home plate to start a play in accordance with the ninth Knickerbockers baseball rule. The release units regularly released at the pacemaker process to define the start of production should be chosen as small as possible to encompass a proper 'feeling' of customer takt time. The design guideline pertaining to smoothing of the production is as follows:

> **Design Guideline 6:** *Definition of release units*
> The release of production orders has to be done in small, standardized dimensioned amounts to reach an even production volume.

Determination of release units. The correct determination of release units is of vital importance for the achievable value stream flexibility as well as for planning effort and efficiency and for transparent procedures. Release units should not be defined too small to keep the related planning and control effort manageable. Overly large *release quantities*, on the other hand, have numerous disadvantages: The release of production orders containing several days' work results in excessive material volumes at the pacemaker process, cluttering the production and generally disorganizing the shop floor. Besides, product lead times will accordingly rise, so that short-term flexible reactions to changes in customer demand require complicated alterations and/or urgent orders. In addition, overlarge production lots obscure the current market demand, because the demands are combined in lots.

In order to be able to not just use the customer takt time determined in the value stream analysis for the purpose of capacity dimensioning of the flow production, but to also make it the *pacer* for the value stream, the respective *release interval* must be small enough to keep the takt time discernible – the only way to introduce any rhythm at all into the production! Thus, the end of each release interval indicates whether or not the production is still running smoothly. In cases of release intervals of one day each, there will only be one chance per day to verify whether or not the production goal has been achieved. In cases of release units of 30 minutes each, however, the conformation with the intended rhythm, or the occurrence of deviations resulting from disturbances, respectively, is regularly verified in short intervals throughout the entire production procedure.

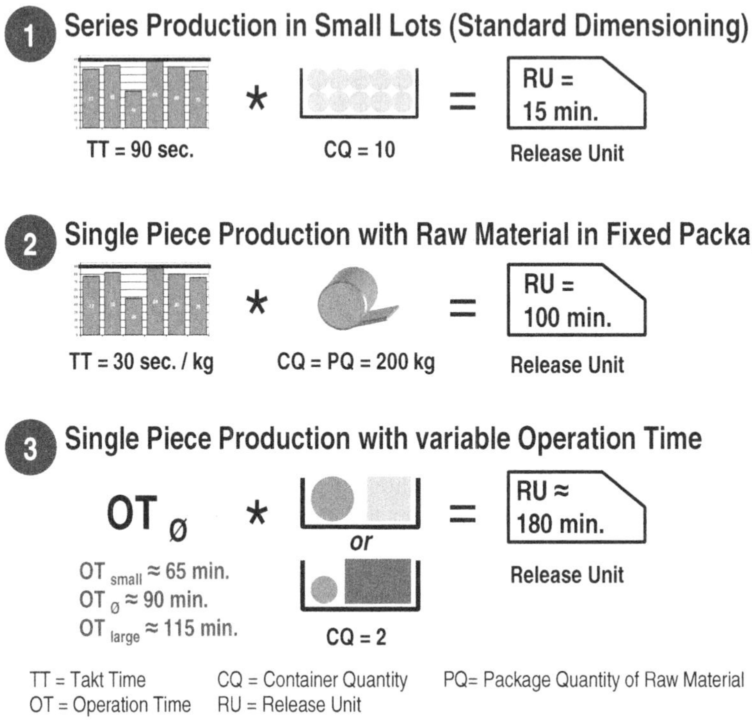

Fig. 3.51 Calculation of release units

For appropriate linkage with the customer demand, release units are determined as integer multiples of the customer takt. The suitable size of a release unit (RU) is best determined by multiplication of customer takt time and *container quantity* (Fig. 3.51, case 1), as is generally done in serial productions.

In the numerical example, a customer takt time of 90 seconds and a release unit of 10 pieces, i.e. exactly one container, results in a release interval of 15 minutes. In the most straightforward and simple case, a container size defining one release unit corresponds to the number of pieces in one such container for finished goods used at the pacemaker process. Customer-specific, product-related containers for finished goods are mainly used in cases of serial products supplied to the processing industry. Retail shipments are usually conducted in commissions, often via finished good storages with homogeneous pallets. Their container size may be used for the determination of suitable release units, or, alternatively, factory-internal transport containers for products or parts may form the basis for the release units. Should a release unit formed on the basis of such a container seem too small, several containers (quantity b in the equation below) may be used as a basis. In cases of pacemakers or downstream processes comprising lots of several container quantities each, the respective lot size will be decisive for the size of the release unit. Accordingly, release units in serial productions are determined as follows:

$$RU_{Series} = TT \times b \times CQ$$

$$RU_{Variety} = TT \times \frac{\sum_{i=1}^{var} b_i \times CQ_i}{\#Var} \qquad \langle b_i \times CQ_i = const.\rangle \tag{3.20}$$

with: RU release unit [time unit]
TT takt time [time unit/pcs.] *or* [time unit/quantity unit]
b integer factor for containers to be released simultaneously
CQ container quantity [pcs.]
#Var number of variants

Above formula, of course, should be considered a rule of thumb only and will need to be individually adapted to the respective production circumstances. In high-variance product families for instance, container quantities often vary considerably – the obvious solution would be to use the lowest common multiple, though this would lead to unrealistically large quantities. Instead, container quantities will have to be adjusted to achieve uniformity with small quantities (Eq. 3.20, bottom). Besides, for practical reasons, round numbers should be chosen, which will only approximately correspond to the customer demand, though. In the numerical example, for instance, the customer takt time of 95 seconds would lead to a release unit of 15.83 minutes. The pacemaker may then be laid out somewhat more powerful than required by the customer takt time, enabling the release unit to be determined at 15 minutes based on a customer takt time of 90 seconds.

In customer-specific productions of significantly varying or very long operation times per piece, it will be difficult to find a suitable piece number as a basis for the determination of the release unit. In the first case, the raw material demand may be identified as a common reference parameter – with piece productions of little complexity, consisting mainly of raw materials delivered in fix *package quantity* (PQ), the respective volumes may be used as a basis (Fig. 3.51, case 2). This simple mathematical relationship may be formulated as follows (top equation):

$$RU_{Single-Item} = TT \times b \times PQ$$

$$RU_{OT} = CQ \times OT_{\emptyset} \tag{3.21}$$

with: RU release unit [time unit]
TT takt time [time unit/pcs.] *or* [time unit/quantity unit]
b integer factor for containers to be released simultaneously
PQ package quantity raw material [quantity unit]
CQ container quantity: no. of products combined by size group
$OT_{\emptyset}$ operation time [time unit]

In cases of more complex products, though, a reference parameter is not found so readily, and besides, the operation time would increase to such an extent, that any release unit of reasonable size could only include very few products. Instead, varying numbers of products are combined in such a way that their total work schedule-based operation times result in the (relatively) constant value of the determined release unit. To simplify matters, operation time-based *size groups* may be created and different combinations of products pertaining to said size groups may then be allocated to a mutual release unit (Fig. 3.51, case 3). The respective release unit will be an integer multiple (CQ) of the mean operation time ($OT_{\varnothing}$), with the resulting factor corresponding to the number of simultaneously released products (Eq. 3.21 bottom). With simultaneous release of one pair at a time, a smoothed out production volume may easily be achieved by pairing large with small products or medium-sized with medium-sized ones. Any deviations still remaining after this rough smoothing may then be balanced with the aid of a little capacitive flexibility of the respective production processes, e.g. production-related overtime or stand-by personnel.

In cases of low piece numbers, typical in plant construction, it makes no sense at all to pool products for the purpose of mutual release. On the contrary, the production can only be rhythmically structured to match the pace of the production if broken up into individual *process steps*. The release units will then be defined on the basis of individual work processes of different extent which will be combined to release units of identical size. In these cases, the release units will remain lower than the customer takt time.

Smoothing the production. Fluctuations in customer demand may either be smoothed with the aid of the time buffers of the production orders queuing up before the pacemaker or by way of the finished goods inventories in the supermarket store. Direct shipping (cf. Fig. 3.48) requires production smoothing in the queue. Regular release of standardized production volumes at fluctuating customer causes different lengths of order *queues*. As long as the customer demand remains lower than the mean demand, the queue will get shorter as a result of the faster pacemaker. In the case of strong customer demand, though, orders will queue up and the line will accordingly grow longer. In this manner, the production load may be kept at constant mean demand level (Fig. 3.52, case 1). In spite of consistent production lead times, though, order lead times will still fluctuate: If the queue never grows too long, only part of the products will be completed on time, others will be completed early. The *bringing forward* of orders to a point in time prior to the actual demand is the price to be paid here for the homogenization achieved.

Alternatively, and taking into account the length of the queue, the production capacity may be adjusted to the fluctuating demand starting from the pacemaker in a downstream direction. Depending on the relative capacitive flexibility, either the *release rhythm* of the respective production orders may be changed – possibly through allocation of additional staff to the pacemaker process – or, without changing the rhythm, the daily *available period of time* may be altered – for instance through overtime or shorter shifts. Mostly, though, the capacitive flexibility will be lower than the customers' fluctuation range, enabling the mutual application of both tools, i.e. the bringing forward of orders and the simultaneous adjusting of capacities. *Delivery defaults* must be expected whenever the customer demand exceeds the maximum permissible fluctuation range, thus causing queues to overrun their lead time-related maximum values.

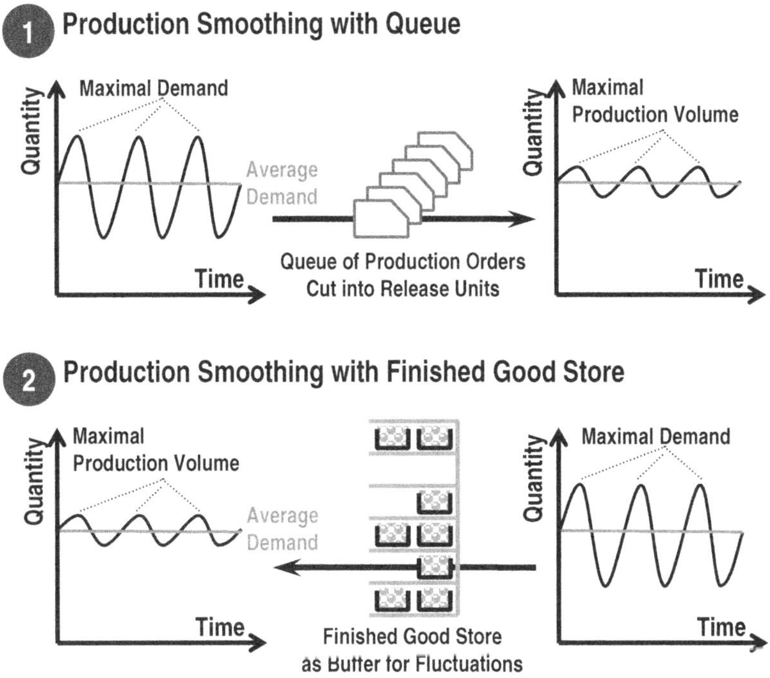

Fig. 3.52 Smoothing of the customer demand by way of queues or finished goods storages

If goods are shipped from the finished goods storage, the store also handles the smoothing of the production by buffering demand fluctuations (Fig. 3.52, case 2). As long as the *finished goods inventory* is not exceeded, fluctuations in customer demand are immediately compensated for. According to kanban logic, shipping triggers the respective production orders. These, however, are not taken to the pacemaker immediately upon withdrawal from the store, but are collected in a queue and then released in fixed intervals, thus ensuring an even strain on the production. Similar to the first case (Fig. 3.52, case 1), the capacity may be slightly adjusted to the fluctuations to reduce inventories.

Three types of inventories are generally differentiated in the *dimensioning* of a finished goods supermarket store, namely work in progress, buffer inventory and safety stock (cf. Eq. 3.9). The work in progress corresponds to the mean volume required to cover the replenishment lead time. The buffer inventory serves to balance fluctuations in customer demand in excess of the pacemakers' capacitive flexibility. Thus the finished goods store shields the production from medium-sized fluctuations in customer demand. The safety stock is required in cases of internal problems, such as machine breakdowns or increased scrap rates interrupting the production process. The safety stock should not be easily accessible to the staff to ensure that production problems are really taken notice of, which is the only way to create sufficient pressure for constant improvement. The three different types of inventory are required for different reasons which may over time be subject to change and should therefore be checked for suitable size from time to time.

3.4.2 Levelling the Production Mix

Production smoothing homogenizes the production flow volumes (Sect. 3.4.1) with the aid of release units. As soon as a production includes more than one variant, production planning has a second task, namely the responsibility for the determination of order sequences. In other words, unless all product orders are identical, their sequence must be planned. This *sequencing* of release units generally levels the production through determination of the sequence in which different product variants are produced. Target at this is to achieve a levelled production mix by sequencing. Said *levelling of the production mix* means to change the variant after each release interval. Accordingly, the design guideline for sequencing is as follows:

Design Guideline 7: *Levelling the production mix*
The sequence of production orders has to be thoroughly intermixed concerning the variants to reach a balanced production variety.

Function logic. The immediately effective advantage of production levelling is the prompt satisfaction of demands for variants other than the currently produced ones. Basically, a different variant could be produced after completion of each release unit, which should be kept fairly short in any case (Sect. 4.3.1). The ideal highly balanced *intermixture of variants* does not permit production orders for identical variants to be accumulated for lot formation (Fig. 3.53). Each release unit conforms to one variant's fix lot size as determined by the pacemaker process. After each release, variants must be changed. As shown in the schematic diagram, the identical variants symbolized by identical geometrical shapes may not be arranged next to each other but must alternate in a balanced intermix which at the same time defines the requirements with respect to changeover times, which must be adjusted to the release interval.

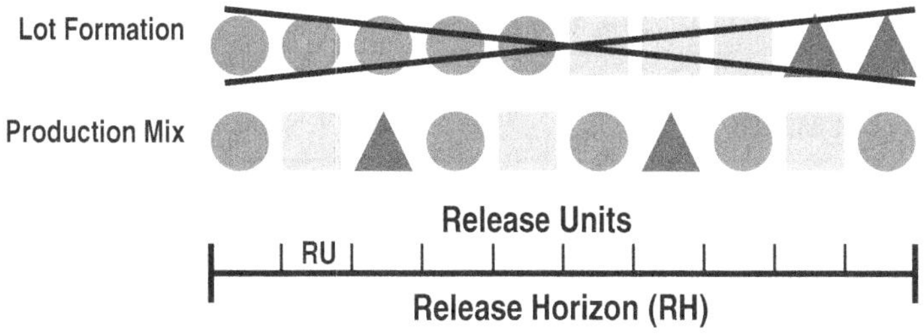

Fig. 3.53 Levelling of the production mix

Levelling of the production mix serves not only to align the production with the customer demand, though, but also enables a *homogenized load* on the production processes. First of all, the intermixture of variants automatically compensates slightly fluctuating operation times at the pacemaker process. The same applies to all following processes connected by FIFO. Secondly, variant-related demands for parts are homogenized in the respective supply processes, thus counteracting demand swings otherwise resulting in the bullwhip effect (cf. Sect. 3.3). Different materials are alternately required and different supply chains are drawn on to satisfy varying demand. Thirdly, should the value stream branch out behind the pacemaker, the relative strain on the production could be regulated by way of a suitable variant mix. In cases of assembly lines followed by a choice of different test devices with test periods in excess of the assembly takt time, a fix variant mix may achieve an even load.

The production order sequence may be homogenized with the aid of the *load levelling box*, 'heijunka box' in Japanese. A real-life levelling box for instance may consist of compartments, with vertical columns determining the respective release units and horizontal rows defining the product variants. These compartments may hold evenly distributed production order cards, possibly kanban (Fig. 3.54). The heijunka box is a simple tool for the smoothing of a production mix. Due to its dependency on the respective release interval, the production volume is subject to consistent smoothing (Sect. 3.4.1). The order in which the production orders are sorted into the levelling box determines their sequence and release times. They are generally implemented in the production by the logistician, who at regular intervals – the release intervals – withdraws the next following orders and transfers the same to the pacemaker process.

The application of a levelling box or respective planning logic requires the specific reference to a certain time period, the so called *release horizon* (Fig. 3.53). The production is levelled over this particular time period, for instance one shift. For each release unit within this defined planning horizon, one production order is scheduled, ensuring a distribution of variants as evenly as possible across said shift. The release horizon forms a so-called 'frozen zone' within an order sequence, which should not be altered anymore. The symbol for the levelling of a

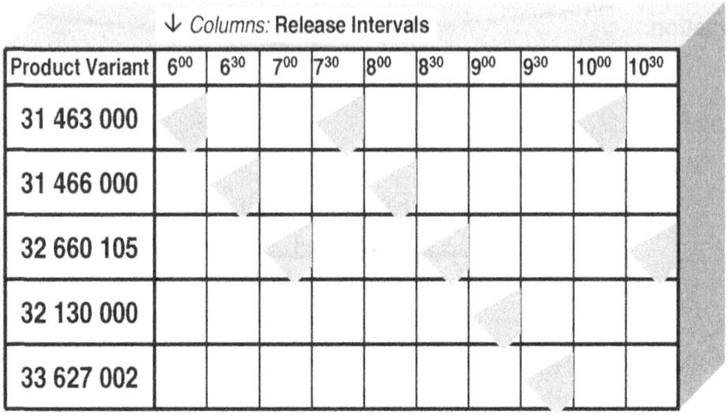

Fig. 3.54 Load levelling box

production mix consists of the letters OXOX framed by a rectangle and combined with an information flow arrow (Fig. 3.55). Immediately adjacent, the size of the release horizon (RH) is stated, expressed by a period of time, such as one shift or one day, or alternatively by the number of scheduled release units.

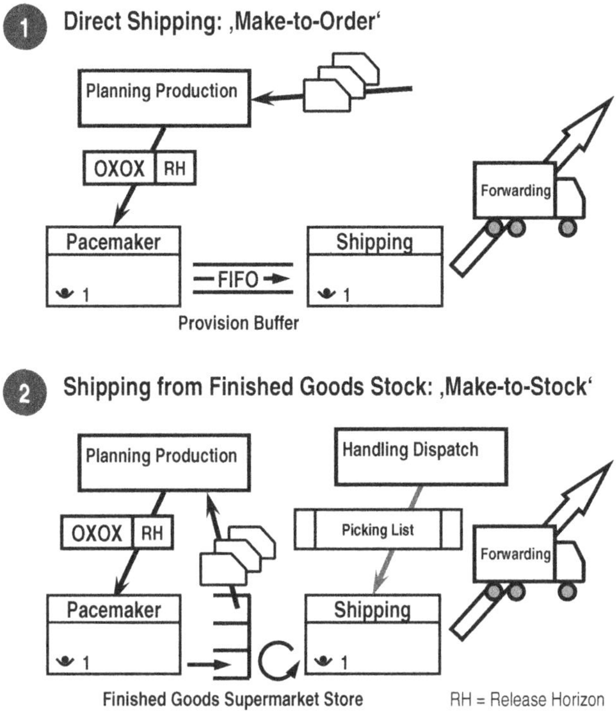

Fig. 3.55 Production levelling at the shipping end

Generally, sudden urgent demand for a certain product variant may easily be scheduled into the next following release horizon. As the queue will normally be somewhat longer than the release horizon, the planning principle at hand simply handles *urgent orders* by allowing them to jump the queue. Within the production itself, there are no urgent orders nor any progress chasers. Within the release horizon, only one order is released at a time, so particularly urgent demand may be immediately slotted into the very next release, which will cause additional planning effort, though, as another ready order would need to be returned to the queue. But even in such extreme cases normal operation continues on the shop floor, because any possible turbulences will have been straightened out at planning level already.

Depending on the respective *shipping principle* (cf. Fig. 3.48), the levelling of the product mix affects different points of the production. Customer-related product orders need to be allocated to the correct release horizons depending on scheduled and/or customer-specific delivery dates. The sequence developed by production planning in accordance with the related urgency is triggered at the pacemaker process. The products are directly shipped immediately after completion (Fig. 3.55, case 1). Production orders for products shipped from the finished goods storage, however, constitute shipping kanban (Fig. 3.55, case 2). The cards transmitted by the dispatching store are collected in production planning. For production levelling, a number of kanban corresponding to the size of the release horizon are removed from the queue; the different variants are chosen in accordance with their relative frequency in the queue. The thus determined kanban sequence is then triggered at the pacemaker process which in turn provides the products for the finished goods supermarket store.

Bulk orders. A typical manufacturer of variants will produce a mixture of numerous small customer orders suitable for processing in defined release intervals but also bulk orders. These will comprise customer orders for variants corresponding to the size of the chosen release horizon. Variants will not be mixed, as this would block the production for a considerable length of time as far as other orders are concerned and as an added disadvantage would make the entire production load lopsided and uneven. A more suitable horizon for bulk orders must therefore be determined. In cases of *extended delivery periods* for bulk orders, partial customer orders may be held back to create sufficient slots for small orders of other variants. Accordingly, either the queue – and with it the delivery times – must be extended in order to be able to once again mix customer orders or the production will not be levelled.

Bulk orders supplied from the finished goods storage, however, require a different solution. Direct withdrawal from the store would not level the production at all. Order volumes higher than the inventory kept for the usual small and medium-sized orders would be too small to complete the order. As a rule, customers do not appreciate partial deliveries of orders split into smaller portions, which, in addition, would greatly increase transport costs. Instead, however, orders could be provided from the finished goods store in small portions and then loaded for shipping on the intended delivery date as planned. This of course requires sufficient retrieval areas and excellent organization. A more elegant solution, the so-called emergency kanban (Sect. 3.3.2.1), slot additional kanban into the queue, each of which may only be used once. This gradually creates *special stock* in the finished goods storage which will be automatically exhausted upon delivery.

Campaign formation. If long changeover times in the pacemaker process cannot be avoided, simple production mix levelling as per design guideline no. 6 may prevent large changeover portions – at the cost of very large release units, though. In connection with high numbers of variants, this makes the value stream way too sluggish. Regular control with uniform release quantities will not be as effective as conventional, demand-related production planning. As a rule, changeover times vary greatly depending on the respective *variant changes*. Accordingly, a restriction of the *level of freedom* in sequencing may decrease the accumulated changeover effort, similar to the application of a changeover matrix (cf. Sect. 2.3.1).

Generally, with high numbers of variants at the pacemaker, two types of changeover procedures are differentiated depending on the type of variant change. The first type requires extensive plant modifications resulting in longer changeovers and is thus only justifiable with large lots. The second type needs only minimal changes, such as vernier adjustment, idle cycles, container changes etc., keeping changeovers roughly as short as one part's operation time, thus enabling small lots. In this case, simple production levelling may be achieved by way of a two-step approach: First, the size of a release unit is determined based on a short changeover time, enabling production of any given variant in small lot sizes. Subsequently, a certain number (c) of release units are combined to form a *campaign*, containing only variants would require only minimal changeover times in relation to each other. A change of campaign, however, may well necessitate a large changeover time to allow fabrication of a different changeover type.

The formation of campaigns is a simple control method to enable smooth production volumes and levelled production mixes in the production of small lot sizes per variant in spite of partially substantial changeover times (Fig. 3.56). Order sequencing is done in two steps: The first step determines a consistent time interval, the campaign duration (CD), into which all variants featuring a certain changeover-related *group characteristic*, such as size or diameter are slotted. Figure 3.56 schematically depicts said group characteristic as the edge length of the respective squares. The second step arranges release units of variants with different *secondary characteristics* within a campaign, such as precise dimension settings within the overall diameter dimensions of the campaign, or maybe differently coloured identical material, which would not require changed settings, but would still require an interruption to ensure clean material separation. In our illustration, this is symbolized by variants depicted in different shades of grey.

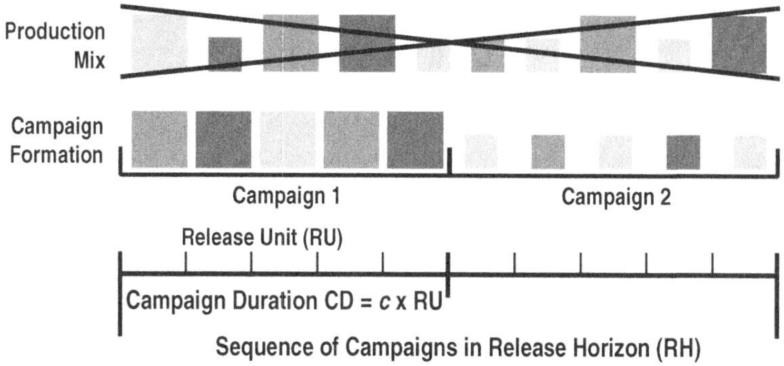

Fig. 3.56 Campaign formation for production mix levelling in two-steps

Within a campaign, the various release units may be intermixed according to the logic of a levelling box. The determination of campaign sequences, i.e. of the changes in variant groups, depends on the size of the respective release horizon (RH). For methodological reasons, campaign durations are comparatively long, therefore release horizons of reasonable duration, i.e. roughly matching the respective delivery times, will rarely comprise significantly more campaigns than there are groups of variants, which would be the only case where the application of a levelling box on campaign level really made sense. As a rule, an individual campaign of predetermined *campaign sequence* is planned for each variant group. Thanks to the resulting long release horizon, the individual campaigns do not initially need to be completely filled with orders. Unfortunately, though, this allows customers' order patterns to influence campaign durations, triggering campaign changes at irregular intervals and thus unsettling the production. Decisions as to whether or not to include a new order into an existing campaign that's already full or whether to shift that particular order to the follow-up campaign will repeatedly be required. The first option would extend the campaign and push all others back accordingly. Such optimization measures may easily make a production skittish. Besides, the resulting increased effort is not really compatible with the principle of simple production control.

Alternatively, the release horizon may be sized to match the campaign, in other words be amalgamated into it. This highly flexible solution leaves all options open as to which variant to produce next, allowing spontaneous decisions concerning group priority up to the moment of release. It also ensures that only campaigns of identical size are released. Two-step production levelling in campaigns is symbolized by a double sequence of the letters O and X, separated by two vertical lines (Fig. 3.57). The *campaign duration* (CD) is added to adequately describe the lean campaign control.

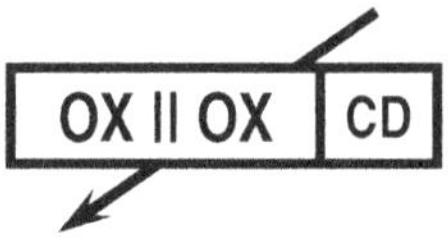

Fig. 3.57 Campaign control symbol

However, contrary to fix campaign sequences, this simplified procedure does not provide any exact *delivery dates*. At best, a latest possible delivery date might be confirmed, resulting from the number of variant groups and the respective fix campaign duration. Not all campaigns would be necessarily filled to capacity in this case.

Combined planning and control. In productions handling well-established as well as exotic variants, *combined control with volume storage* often seems the best possible solution, combining the two previously explained standard cases of direct shipping and shipping from the finished goods store (cf. Fig. 3.55). As a result of make-to-stock and make-to-order production from identical resources, the value stream will have *two trigger points*, which each product variant as per ABC classification (Sect. 3.1.1) being allocated exactly one control procedure. All 'standard' variants required on a regular basis and/or in large quantities are thus shipped from the finished goods store (FG) and re-produced from the semi-finished goods store (SFG) via kanban (Fig. 3.58). 'Exotic' variants hardly ever needed and/or only required in small piece numbers as well as customer-specific items are split up by release units, triggered upstream and shipped directly.

In this combined solution, the supermarket store takes over the function of *volume smoothing*. In cases of high demand, standard products are withdrawn from the finished goods supermarket store, leaving sufficient capacities for the production of all customer-related variants ordered at the same time. On days with low demand, free capacities are used to replenish the supermarket store. This solution is only worthwhile if delivery times are too short to balance volume fluctuations by way of queues, otherwise entirely customer-specific production is generally recommendable.

As the relative frequency of the different order types varies, fix resource allocation by way of *time slices* is not suited for either of the two cases mentioned above. Instead, customer-specific orders always need to be given higher *priority* in order to ensure consistent lead times. The kanban system must be dimensioned to balance possibly resulting time losses incurred by standard products waiting for free capacities.

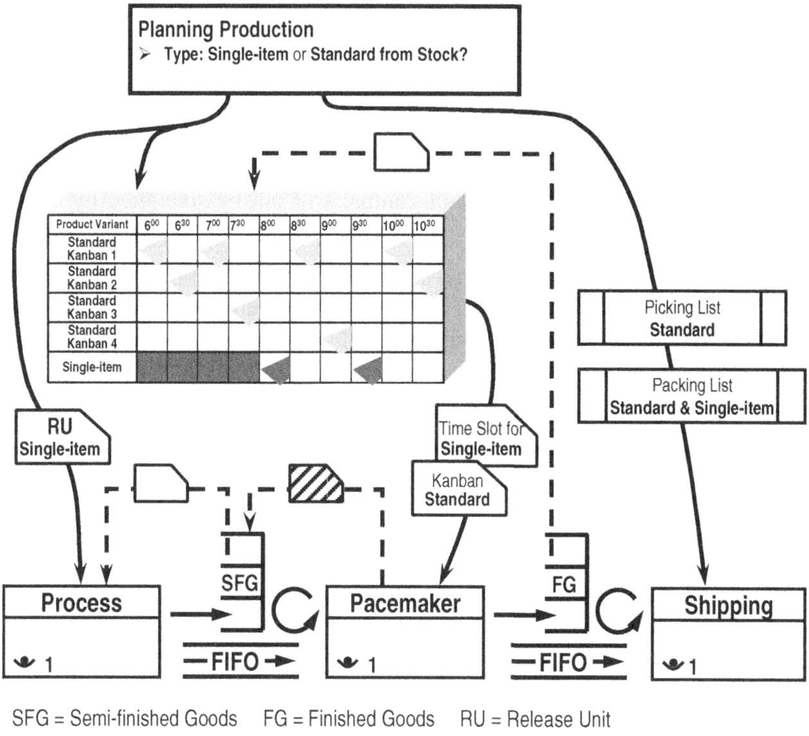

Fig. 3.58 Production subject to combined control

Strictly speaking, the combined solution violates the fifth design guideline which permits only one trigger point. However, since each product variant is allocated an individual trigger point, we are really looking at two value streams with shared resources, hence the design guideline is met after all. The combined system's specific problem of having to *coordinate* two trigger points, remains unsolved, though. Each production process receives control information from two different sources. The pacemaker process receives kanban for customer-anonymous goods from production planning as well as customer-specific orders from the upstream process via FIFO line. Kanban triggering for storage replenishment is done via levelling box containing extra placeholder cards reserving time slots for the processing of exotic orders coming in via FIFO line. The upstream process contains exotic orders which always need to be processed immediately as well as kanban from the semi-finished goods store (SFG), which are processed whenever there are no exotic orders. The number of release units for exotic products at the upstream process of course needs to match the time slot at the pacemaker. When sorting the time slots into the levelling box, a certain time delay for processing by the upstream process as well as transport needs to be taken into account.

In combined control, no residual amounts of exotic variants may be stored, i.e. the ratio between customer order quantities and release units must always be an integer number. Should this not be possible, only variants of low customer demand will be stored while high-volume products will be subject to customer-related production as an alternative solution. In this *combined control with variant storage,* the finished goods store will contain low inventories of numerous different items. As customer order quantities for exotic variants are lower than the respective release units, the finished goods storage contains nothing but residual volumes. Accordingly, the production volume will not be smoothed by the store but by the queue in this case.

Four planning and control systems. Four different planning and control methods for lean production may be derived from different combinations of the two basic shipping principles of direct shipping and shipping kanban, the respective possible applications of which shall be compared/set in contrast in the following:

1. The first standard case processes and ships all products in a customer-order-related manner (cf. Fig. 3.48, case 1). In this case of *control with direct shipping,* the production is smoothed by the respective queue of production orders (cf. Fig. 3.52, case 1), production levelling is achieved through even distribution of the variant mix in the release horizon, within which the production orders are sorted by urgency with respect to deadlines (cf. Fig. 3.55, case 1). This solution leads to *minimal inventories*, consisting of approximate production lot sizes plus any FIFO line lengths possibly required for line balancing. This is the only possible solution for *customer-specific* production. Besides, it is the *ideal solution* for any type of production not subject to restrictions due to changeover times, delivery times or insufficient capacity flexibility.
2. The second standard case pre-fabricates all products in a customer-anonymous manner and then ships the same from the finished goods storage (cf. Fig. 3.48, case 2). In case of this *control with shipping kanban*, the production is smoothed by the finished goods storage (cf. Fig. 3.52, case 2) and levelled through even distribution of the variant mix of the kanban triggered for re-production in the release horizon (cf. Fig. 3.55, case 2). As a result, inventories are relatively high, thus this solution is better suited for decreasing numbers of product variants. It is usually the only way to reliably implement short delivery times and is also required in cases where the pacemaker's production lot sizes cannot be balanced with the customer order quantities. It is usually the best possible solution for *low-variance* productions.
3. The *combined control with volume storage* customer-anonymously pre-produces products with high piece numbers while products with low piece numbers are manufactured in a customer-specific manner (cf. Fig. 3.58). Similar to the second case, production smoothing is effected by the finished goods storage, while the combination of both order types, i.e. customer order-related production orders and overlapping kanban, levels the production,. This solution keeps finished goods inventories in productions of *high variance* at a reasonable level and at the same time enables short delivery periods, though in view of the problematic control situation the possibility of piece-related segmentation should be considered to enable individual control

in accordance with the two standard cases above. In addition, products with low piece numbers may face problems with respect to the adjustment of customer order volumes and release units, or lot sizes, respectively.

4. The *combined control with variant storage* customer-anonymously pre-produces products with low piece numbers while products with large piece numbers are manufactured in a customer-specific manner. Similar to the first case, the smoothing of the production is effected by the queue, while production levelling results from overlapping combination of both order types. This solution leads to low inventories in the finished goods storage, which can also hold lot size-related residual volumes. However, for the purpose of production smoothing, suitably *long delivery times* are required for standard products. This solution is only practicable in cases where long changeover times prevent the application of the first solution for variants with low piece numbers.

Case study. Due to the small number of variants and short delivery periods, control with kanban shipping is best suited for *Liquipur* (Fig. 3.59). Once a day, the products to be shipped according to customers' call-off orders are withdrawn from the finished goods storage in containers of 60 pieces each. This corresponds to the defined kanban volume and also determines the release quantity. Multiplication with the pacemaker's cycle time of 90 seconds results in a release unit of 90 minutes, i.e. one hour and a half. Accordingly, up to 14 kanban may be completed in a three-shift model, i.e. a maximum of 840 oil filters may

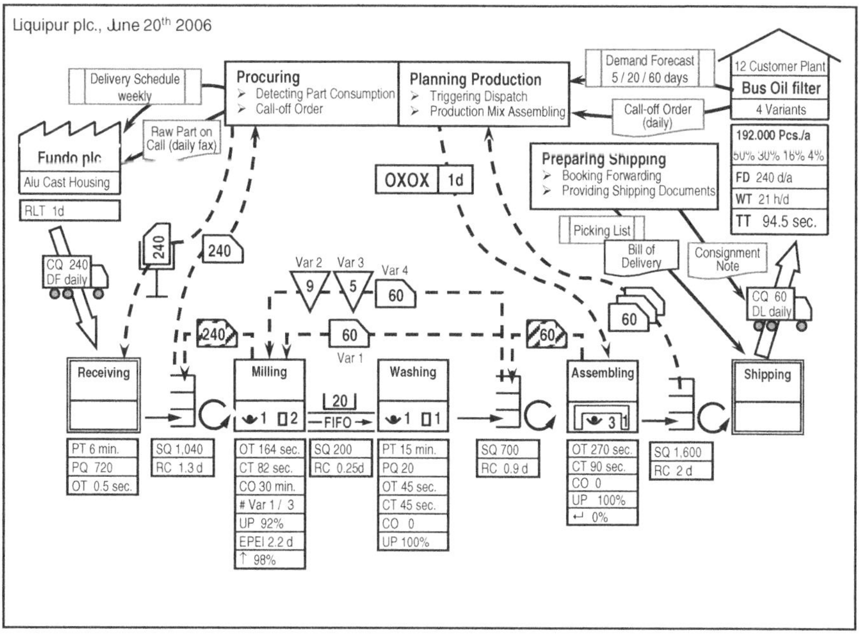

Fig. 3.59 Value stream design at *Liquipur* (5): Production control

be assembled. As this exceeds the mean demand of 800 pieces, days of low call-off order volume may result in short or cancelled shifts. According to the levelling box, the release horizon for the four variants corresponds to one day of 14 release units.

3.4.3 Capacitive and Restrictive Bottleneck Control

With respect to productions with identical available capacities for each process, the planning logic is concluded as soon as the production volume has been duly smoothed and the production mix suitably levelled. In many productions, though, available capacities will vary with each process, thus creating another task to be handled by production planning. The order frequency must be aligned with the production process with the lowest available capacity in order to avoid hold-ups in the production flow. If the respective low-capacity process is the pace-maker, no additional planning tasks arise. In all other cases, though, the resulting *bottleneck* in the value stream will need to be taken into account in the respective control.

Such bottleneck may either purely *capacitively* limit the maximum possible volume in piece numbers in a value stream, or it may *restrictively* define process technique-related sequencing rules, the violation of which would lead to considerable losses in capacity and/or first pass yield. In any case, the bottleneck process may not release an order until the respective release signal has been received. The design guideline concerning bottleneck control is thus as follows:

> **Design Guideline 8:** *Bottleneck control*
> The release of production orders is subject to quantity or sequence-related control depending on possible capacitive and/or restrictive downstream bottlenecks.

Capacity profile For a better understanding of the functionality of a bottleneck, let us take a look at a value stream's capacity profile, which symbolically depicts the available capacity of a production process by the width of the channel available to the respective flow at the bottleneck (Fig. 3.60). Said channel is formed by stringing together the available capacities along the production flow. In an ideal value stream, all processes have identical capacities, thus keeping the channel's border lines parallel. In real factory life, though, the various processes' capacities will very rarely be identical, not least as a result of line balancing difficulties. Under real conditions, the *best capacity profile* will therefore provide the pacemaker process with the lowest capacity of the entire value stream (Fig. 3.60, case 1). This prevents the material flow from breaking, as it enables all upstream processes to provide ample material for all triggered orders, thus allowing the downstream processes to handle all orders without problems.

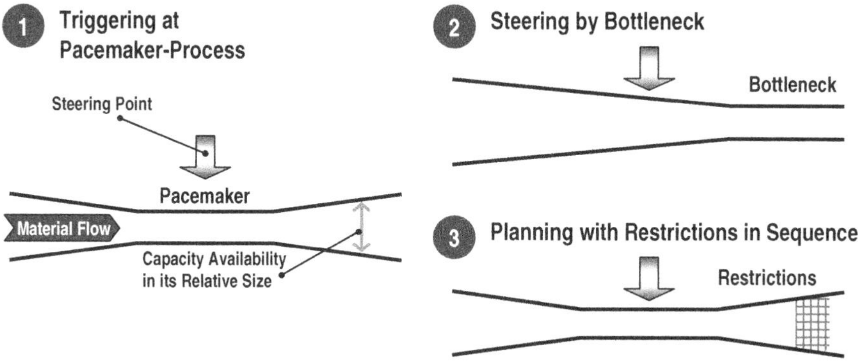

Fig. 3.60 Value stream capacity profiles

However, a trigger point at any other process but the one with the lowest capacity, will lead to a bottleneck at another point of the value stream. If this is located downstream of the pacemaker, *bottleneck control* is applied (Fig. 3.60, case 2). Ignoring the bottleneck would lead to overproduction, because the pacemaker would release more orders than could possibly be handled subsequently. The opposite case, i.e. a bottleneck located upstream from the pacemaker, should be avoided at all cost, because it would inevitably lead to repeated material flow breakages. However, should it not be possible to increase the capacity of an upstream bottleneck, the pacemaker's capacity must be gradually reduced through decrease of process-specific working time, which turns the pacemaker itself into a bottleneck and makes the value stream applicable to the first case as above.

In bottleneck control, the orders released by the pacemaker process are subject to capacitive restrictions defined by one of the subsequent processes, additionally restricting the release quantities levelled in accordance with the sixth design guideline. Subsequent processes, however, may also exert restrictions with a view to the sequencing determined as per design guideline no. 7. Depending on the restricting resources, a certain variant mix may be required to achieve maximum possible capacity. Another typical case would be downstream batch formation through accumulation of matching product variants. The batch formation depends on the respective order backlog as well as the individual orders' urgency and also needs to be taken into account in triggering. In cases of *sequencing by restrictions*, certain excess capacities are required at the restricting bottleneck, as otherwise unforeseeably long waiting times would result for some orders through overlapping of capacitive and production process-related restrictions (Fig. 3.60, case 3). A typical technical restriction upstream from the pacemaker would be order-anonymous pre-production with optimal changeover sequences. Such sequencing restrictions, the violation of which may lead to decided output loss, are only partially compatible with kanban rules, as kanban work best in accordance with the purely chronological arrival sequence of the respective cards (Sect. 3.3.2). Accordingly, changeover losses will have to be weighed up against incalculable replenishment lead times – or, ideally, sequencing restrictions with respect to changeovers will be eliminated.

Accordingly, three types of *controllable* cases may be differentiated with a view to a value stream's capacity profile:

1. Standard case: Triggering at the pacemaker process which is also the bottleneck.
2. Bottleneck control: Upstream bottleneck, to be taken into account in order release.
3. Sequencing by restrictions: Triggering at the pacemaker process, which is also the bottleneck, duly taking into consideration possible sequencing restrictions of downstream processes.

Bottleneck control. Bottleneck control is responsible for the alignment of release quantities with the capacities of capacitive bottlenecks located downstream of a pacemaker process. It is also known as *drum-buffer-rope-control* (Goldratt 1995) and is highly suited for depiction by value stream symbolism (Fig. 3.61). The bottleneck – here highlighted by hachure – takes over the function of the pacer – literally like a drum. Upon completion of each production order, the bottleneck process will send a signal to production planning or directly to the pacemaker, symbolized by a ConWIP circle, which then permits the release or commencement of the following order and thus prevents excess production at the more powerful processes. The order release sequence is synchronized with the cycle time of the bottleneck process. The orders waiting to be released in the queue and the release horizon are being dragged along on a 'rope', so to speak, by bottleneck control. The respective order sequence is depicted by way of the OXOX load levelling symbol stating the pertaining release horizon. The released orders reach the bottleneck process via FIFO lines and queue up in the buffer before the bottleneck, thus ensuring that there will always be sufficient material for the bottleneck.

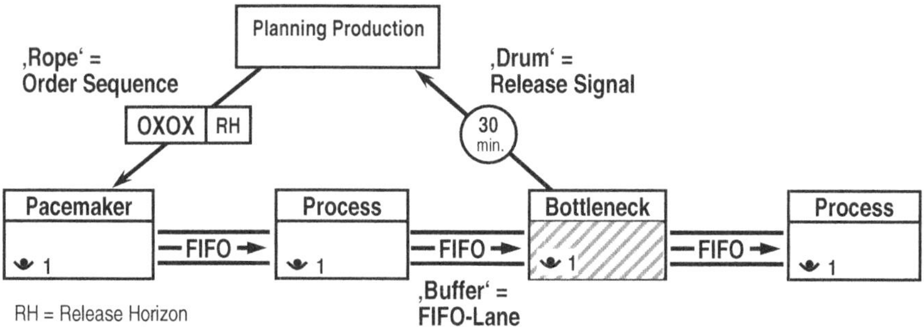

Fig. 3.61 Bottleneck control

Bottleneck control aims at as much *capacity utilization* as possible at the bottleneck. Every single minute lost there will remain lost throughout the entire production. Strictly speaking, any idle time at the bottleneck temporarily cripples the entire factory. Conversely, the value stream does not benefit from any temporary capacity increases at other, non-bottleneck processes at all. Even minute performance improvements at these processes mainly constitute waste, as they do not increase the overall production volume. At best, they can be set off against savings in labour force, but investment at the bottleneck would be disproportionately more valuable.

In cases where the entire *factory output* depends exclusively on the bottleneck, the bottleneck's output in particular must be maximized. This may be supported by efficient failure management, which in maintenance and services always gives top priority to bottleneck resources, even justifying the interruption of ongoing works to this end. It is therefore vital to know all existing bottlenecks. Besides, changeover portions must be decreased – through changeover time reductions, or lot size increase, or both. Increases in lot size, however, may easily disagree with measures applied in accordance with other design guidelines. All in all, the capacity of a value stream is decided at the bottleneck: Its cycle times, changeover times and lot sizes determine the achievable annual piece numbers.

Sequencing by restrictions. Sequencing is done in accordance with the technical restrictions of a downstream process to align the release sequence of the pacemaker process with said restrictions. Similar to capacitive bottleneck control, the *restrictive bottleneck* – here highlighted by a lattice pattern – sends a release signal to production planning (Fig. 3.62), accompanied by additional technological requirements to be considered with the next following order release. The sequencing restrictions are symbolically depicted by the combination of a ConWIP circle with a OXOX-load levelling symbol.

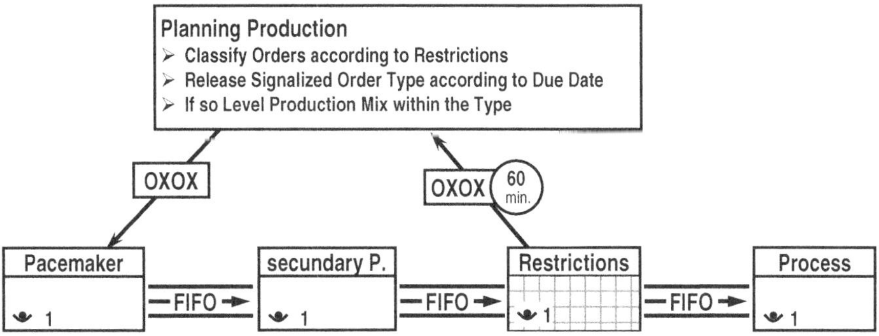

Fig. 3.62 Sequencing by production process-related restrictions

Production planning classifies all incoming orders according to their characteristics with respect to technological restrictions. Therefore, all customer orders need to be converted into production orders in such a way as to enable homogeneous classification. The most urgent orders matching the signalled order type are then released. There may be different variants within one order type, as the initial selection only observes restrictive characteristics. In cases of several variants, there will be another subordinate sequencing process such as simple intermixing. The resulting sequence is then released at the pacemaker process (OXOX symbol). No release horizon needs to be added, as this is already defined by the release signal of the restrictive process.

Sequencing restrictions may either refer to the *permissible variant mix* in the release horizon or be required for downstream *lot formation.* Each of these two may again be split into two types, resulting in four types of sequence-restricted planning as follows: secondary bottlenecks and secondary line balancing losses as well as cyclical production and batch production. These four shall be discussed in the following.

The first case corresponds to bottleneck control applied to average variant mixes within a certain scope. In contrast to variants with high piece numbers, there may be variants whose cycle times are longer at the secondary process than at the actual bottleneck, a disproportionate increase of which would turn the secondary process into a *secondary bottleneck.* This may be prevented by restricting that particular variant's piece numbers in the planned variant mix. This restriction of the permissible mixing ratio of the different variants leads to additional sequencing restrictions in planning. In cases where even minimal fluctuations of the production mix cause secondary bottlenecks, the substantial resulting extra planning effort will inevitably require additional buffer control before the secondary bottleneck. These complications are already well known in job-shop operation, and should ideally be avoided in the capacitive dimensioning of a value stream.

In the second, opposite case, the deviating variants require a decidedly higher capacity in the restrictive process than the majority of the manufactured products. If their portion increases, the pacemaker process will run dry, as the considerably lower piece numbers per release signal (which conforms to a fix release quantity) will decrease the work content provided to the pacemaker accordingly. In an average variant mix, however, the restrictive process (for example a test process), will run faster than the pacemaker process (for instance a flow production). Up to a certain degree^, the FIFO line will be able to balance the resulting takt differences between the two processes. Once the maximum inventory is reached, though, the production will come to a standstill in order to prevent overproduction – resulting in waiting time losses for the operators. This should of course be prevented to happen as a mere result of poor sequencing. The pacemaker incurs *secondary line balancing losses* caused by the restrictive process and successful line balancing may require further restrictions with a view to the permissible variant mix.

In the third case, the restrictive process demands certain optimal production sequences for the different variants, such as for instance changeover sequences with minimized changeover times like the ones normally found in pre-production rather than in downstream processes. With this type of *cyclical production,* conformance with optimal variant sequences guarantees minimal total changeover times. Colour sequences in painting procedures or injection moulding for instance help economize on material as well. In thermal processes, certain sequences may eliminate transition times between variants. In all of these cases, the variant mix sequences are subject to a predefined pattern. The restrictive process handles the variants in a fix cycle and withdraws products from the queue accordingly. As the variant mix in the queue is subject to fluctuations, different amounts of release units will be released per variant. Lot sizes will also vary within the fix sequence pattern depending on the delivery date-related urgency of the respective orders.

In the fourth case, the products of a downstream process are grouped together into lots according to certain criteria and produced in *batches*. The process time of a batch equals one release unit and may contain different numbers of products depending on the variant. These processes are often more flexible with a view to lot sequences than a cyclical production would be. Batch productions are found in surface treatment, for example, where coating processes may include arbitrary sequences of different types of coating, but they will always comprise a certain number of product variants with identical geometrical characteristics.

Planning principles

Goal Value stream production planning aims at the creation of a balanced production flow by way of simple production control. This requires evenly balanced order release in tune with the value stream's capacity profile.

Approaches

1. A release unit is defined as a relatively small planning and release interval which acts as a pacer for the pacemaker process.
2. The production volume is smoothed with the aid of the queue of customer orders, or stock replenishment orders, respectively.
3. The Production mix is homogenized by variant-related intermixing of orders within the respective release horizon with the aid of a levelling box.
4. Changeover-related sequencing restrictions at the pacemaker process may be taken into account through campaign formation.
5. Capacitive or restrictive bottleneck control is applied in cases where production processes located downstream from the trigger point cause restrictions as to volumes and/or sequencing.

Result are four fundamental planning and control solutions: two standard cases applying control via direct shipping or shipping kanban, respectively, as well as two combined control systems with volume, or variant storage, respectively.

3.4.4 Characteristics of Lean Planning und Controlling

The superordinate goal of lean planning and control in accordance with the eight value stream design guidelines is the minimization of overproduction and inventories. As a tangible result, factories are stripped of all superfluous material to create lean material flows. This aids *transparency on the shop floor,* i.e. deviations from the standardized production floor are easily recognizable. Lower inventories mainly result in shorter production lead times. Accordingly, a lean factory is always a *fast factory*; the time line, which has great potential for improvement (Sect. 2.4.1) is particularly important for the redesign of a production in accordance with the value stream design guidelines.

All design guidelines strongly observe the virtue of utmost simplicity in the development of lean planning and control systems. Planning and control tasks are formulated to avoid extensive, possibly untraceable calculative effort, in principle enabling manual solutions without IT-support. This does of course not rule out IT application on the whole, certain tasks like the printing of labels, feedback concerning completed orders, the sorting of large numbers of production orders, the allocation of numbers etc. should not really be carried out manually any more. Above mentioned simplicity ensures easy traceability of specifications and results within the system and keeps the calculative effort low in order to achieve *transparency in planning and control.*

The most important principle of the respective planning and control rules absolutely forbids any alteration of the defined sequence of released production orders in order to secure consistent, highly predictable production lead times. In view of very short production lead times, strict adherence to said sequence is vital, because any alterations of the given sequence would have a major percental effect on the few released orders. All process-related sequencing adjustments are therefore completed during the planning process. This requires a suitably long queue, the dimensioning of which should be taken in to due account in the initial conception. All in all, orders must be triggered into the value stream in such a way as to ensure *even capacity load.*

The accumulated maximum inventories in the value stream are fixedly determined by the dimensioning of the kanban control cycles and the FIFO lines. With the exception of emergency kanban, which enable situational stock increases, inventory sizes may *not* be changed by short-term production planning. On a medium-term basis, inventories may be increased through adjustment of the circulating kanban and/or shifting of the signal point in a FIFO coupling. Minor unplanned inventory fluctuations within a production occur as a result of uneven capacity utilization of FIFO lines and/or fluctuating kanban quantities waiting to be processed. With respect to the different possible value stream design approaches, the following rule of thumb may be applied to the inventories' various *ranges of coverage*: Minutes in flow production correspond to hours in line production, to shifts in kanban control cycles and days in conventional, prognosis-based storage.

In general, the design guidelines are based on the assumption of *failure-free* production. Planning does not explicitly allow for breakdowns – this is done on purpose, to make sure that breakdowns really are taken due notice of, to create transparency. Short-term upstream breakdowns in pre-production are cushioned by safety buffers provided by the kanban inventories. Longer breakdowns, though, invariably lead to *missing parts* in the pacemaker process. Missing parts processes are deliberately not defined, as this would establish the same as standardized waste.

Breakdowns occurring downstream from the pacemaker process are automatically taken into account, as they will delay order release both in FIFO coupling and in bottleneck control. The resulting *delays in delivery*, though, are *not* taken into consideration, because alternative resources from other segments will be scheduled and the resulting backlogs will be balanced by capacitive flexibility. The required tools, such as longer and/or additional shifts, are not specifically codified in the design guidelines, but may be applied in accordance with the conventional solutions.

Essential criteria of lean planning and control

1. Lean factory has a *fixed structure*.
   The extensive freedom of complex planning systems is curbed by way of numerous irrevocably determined predefinitions on the shop floor. Accordingly, the fix allocation of resources is an integral characteristic of value stream oriented segmentation, which eliminates all planning tasks concerning the *situational allocation* of orders to resources with suitable capacities.
2. Planning and control values are *based on threshold values*.
   Accordingly, only low-effort verifications of threshold values (FIFO buffer, kanban quantities) with few direct intervention points (pacemaker, bottleneck) are conducted. No other *singular events* within the production procedure (such as utilization of other production processes) are taken into account or recorded by the production planning and control system.
3. Lean production requires a certain *readiness to assume risks*.
   The negative effects on the production caused by breakdowns are meant to be strongly felt in order to build up sufficient elimination pressure. The *feeling of security* resulting from high inventories and long-term forecasts only serves to camouflage deficits and waste.

3.5 Conception and Implementation

Conception. The first step in the redesign of any production is the structuring of the factory into production segments based on product families and business types (Sect. 3.1.1). The choice of a segment to be redesigned is followed by the development of the future state conception of the value stream based on the eight design guidelines applied in their numerical order. In combination with external logistics and dimensioning, this results in a *ten-step approach* to value stream design:

1. Determination of the *customer takt time* and development of the *capacity profile* for the entire value stream.
2. Grouping together of production processes with the aid of *technical integration* and/or *flow production* wherever possible.
3. Coupling of production processes in an upstream direction, starting from the shipping end by way of *FIFO logic* wherever possible.
   Note: This is compulsory in customer-specific production; in serial production it is generally feasible whenever changeover times are roughly identical.
4. Linking of production processes for multiple usage parts requiring lot production by way of supermarket pull system.
   Note: If applicable, two or more subsequent processes within a *kanban control cycle* may be coupled by FIFO logic.
5. Determination of the *shipping principle*, i.e. direct shipping or shipping from a finished goods supermarket store, depending on the respective

products and delivery times, and creation of *a supplier connection*, i.e. just in sequence or just in time.
6. Definition of suitable kanban quantities and container sizes as well as *lot sizes* for production processes of high changeover effort by way of the EPEI value.
7. Determination of the *pacemaker* process and, should there be no finished goods storage, of the resulting customer decoupling point.
8. Definition of the *release unit* for production smoothing by way of order release.
9. Definition of *sequencing* rules and und possible *campaign formation* as well as determination of the *release horizon* for levelling of the production mix.
10. Provision for capacitive and restrictive *bottlenecks* in control logic.

The value stream approach at hand provides a standardized way of redesigning any given piece production towards significantly improved production processes. For lasting success, however, *innovative* reorganization must be dared, leaving far behind traditional, set habits and observing the production from a totally impartial, external point of view. The conceptional phase must remain unimpeded by seemingly immovable restrictions. More often than not, what started out as an unrealistic workshop vision on production redesign has in the past turned into strategic corporate policy, once the positive results on the overall production procedure had become clear. Entire shop floors with numerous interim storages have shrunk to a few handy U shaped lines, giant furnaces for lengthy thermal processes and large lots were transformed into lean continuous furnaces etc. Many a sceptic has dropped all objections and was totally won over, once old and dusty habits had lost their appeal in the face of newly developed targets. Once a workshop on value stream oriented factory planning with production managers realizes this, not only new concepts will be achieved, but new consensus, too.

Redesigning a production with the aid of value stream design

Goal is the improvement of production to reach a lean factory.

Essential characteristics

1. Consistent and systematic application of clearly defined design guidelines.
2. Material flow-oriented production design.
3. Consistent alignment of production processes and customer takt time.
4. Reduction of waste in the logistical overall process.
5. Comprehensive standardization of all production processes.

Result is a transparent factory which promptly meets all customer demand. Clear information flows, low-inventory material flows and production processes in perfect tune with customer takt times characterize all production segments and their respective value streams.

Implementation planning. Once the future state conception for the value stream has been approved, an implementation plan for the achievement of the defined goals must be developed. The value stream drawing depicting the redesigned production includes all production processes and their logistic linkages as well as all future parameters for cycle times, changeover times, lot sizes, container sizes etc. These technological and organizational improvements require suitable improvements of the respective processes yet to be achieved. The required *improvement measures* are symbolized by flashes of lightning (Fig. 3.63). These symbols, called 'kaizen flash' after the Japanese name for continuous improvement processes in production, are drawn into the future state map wherever necessary. In a value stream perspective, the general value stream context is highly transparent, thus facilitating the evaluation of the overall effects of the improvement measures thus defined.

Fig. 3.63 Kaizen flash symbolizing improvement measures

Once all improvement measures have been defined, the value stream is split up into sections for incremental implementation. For each *value stream section*, the respective tasks as per kaizen flash are defined and the pertaining goals are quantitatively recorded. Subsequently, an action plan is devised defining clearly described milestones, completion dates, and, above all, responsibilities. The implementation of measures leading towards the future state should be conducted step by step during regular factory operations, because only short feedback cycles allow for solutions to be tried out, improved and channelled in the right direction.

This will be realized in a suitable *succession* of previously defined sections. As the pacemaker process and its inherent characteristics have the greatest influence on the overall value stream, the pacemaker process will be the first to be designed in accordance with the future state concept. Subsequently, all downstream processes will be redesigned. The material supply should initially be generously designed in order to be able to fully concentrate on the process itself. Once the pacemaker and its follow-up processes have been redesigned from a technical point of view, the new planning and control system may be implemented in the customer-related part of the value stream. When the pacemaker section has been redesigned, the conceptual requirements placed on the customer-anonymous upstream processes will have become reality. Based on the above, the redesigning process may then move upstream step by step; as a rule each kanban control cycle will constitute one value stream section to be redesigned. At the same time, the existing current production planning system is gradually phased out.

To speed up the implementation process, various value stream sections may be redesigned simultaneously. Consistent adjustments enable repeated aligning of the processes. The requirements of the pacemaker process remain superordinate to all other processes and procedures at all times, though, regardless of their time of implementation.

Typical improvement measures

- Drastic lead time reduction through flow production implementation and definition of a pacemaker process determining exactly one planning point as well as restrictions with a view to order release.
- Elimination of waste in the production process in order to undercut customer takt times.
- Improvement of uptimes by way of preventive maintenance and increased process reliability.
- Reduction of changeover times through changeover optimization, alternatively, elimination of changeover times through jigmaking.
- Introduction of multiple machine operation.
- Reduction of waste in production processes.
- Conduction of value stream design at the suppliers'.

Case study. At *Liquipur,* the changes are implemented in three stages (Fig. 3.64): First, the flow production and shipping kanban are implemented. The second step adjusts the semi-finished goods store to the kanban control, organizes changeover workshops at the milling machines, creates service schedules and prints out kanban. Thirdly and finally, the supplier connection is optimized and converted to kanban.

The specifications previously determined with the aid of the design guidelines define the following targets: Reduction of the changeover time per part from currently 522 seconds by 6 per cent to 487.5 seconds as well as a reduction in the lead time of currently 13 days by 65 per cent to 4.5 days (Fig. 3.64), thus outlining the lead time potential achieved.

The operator balance chart reflects the achieved improvements concerning the alignment of the individual production processes' performances with respect to each other. Solely the current state of the washing process remains unchanged. Thanks to the implementation of the supplier kanban (Sect. 3.3.2), no inspection of goods received is required any more, the remaining minimal time effort of 0.5 seconds per piece spent on the unloading of goods is allocated to the shipping for simplification. The assembly process was redesigned as a flow process (Sect. 3.2.2) with a cycle time of 90 seconds. Milling continues to incur time losses through changeovers, breakdowns and scrap, though no changeovers are required on the machine used for variant no. 1. Thus, equation 2.22 shows different gross cycle times for the two machines:

$$CT_{CO} = 82\,\text{sec.} \times \frac{\left(\frac{58.000}{540} + \frac{30.000}{300} + \frac{8.000}{60}\right) \times 0{,}5\,\text{h}}{96.000 \times 164\,\text{sec.}} = 82\,\text{sec.} \times 3{,}9\% = 3{,}2\,\text{sec.}$$

$$CT_{UP} = 94{,}5\,\text{sec.} \times (1 - 92\%) = 7{,}56\,\text{sec.} \tag{3.22}$$

$$CT_{Q} = 82\,\text{sec.} \times (1 - 98\%) = 1{,}64\,\text{sec.}$$

$$CT_{gross} = 82\text{sec.} + 3{,}2\text{sec.} + 7{,}56\text{sec.} + 1{,}64\text{sec.} = 91{,}2\text{sec.} + 3{,}2\text{sec.} = 94{,}4\text{sec.}$$

According to this both machines are depicted separately in the operator balance chart.

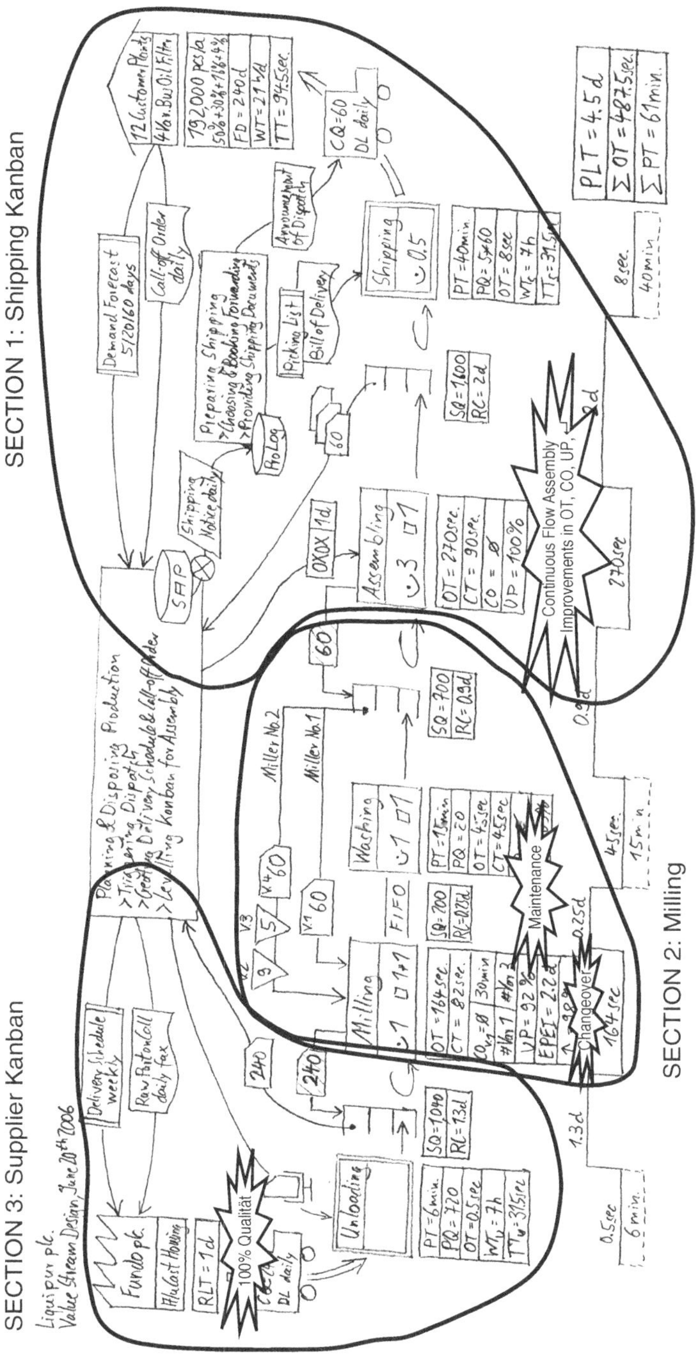

Fig. 3.64 Value stream design at *Liquipur* (6): Kaizen and implementation steps

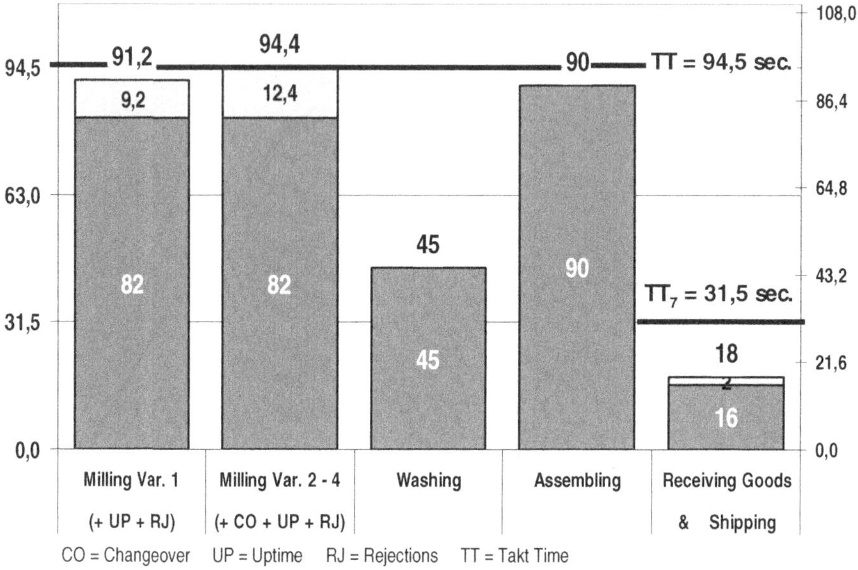

Fig. 3.65 *Liquipur* future state operator balance chart

3.6 Value Stream Management

The foregoing comments on the eight design guidelines for value stream design describe the course of action for a systematic redesign of a production procedure up to its implementation planning. Once the defined measures have been realized and the respective goals have been achieved, the *planning project* of value stream-related reorganization will have been completed, but day-to-day production will continue on. The implemented value stream must now keep up the achieved performance level of *factory operation*, ideally even gradually improve the same over time. Besides, there will always be minute alterations in product design, the composition of product families, the product mix required by the customers, the available resources and many other changes, not necessarily requiring complete re-conception but value stream adjustments. These tasks will be fulfilled by value stream management (Sect. 3.6.2).

Value stream-oriented production development requires suitable *organizational anchoring* within the company. Both project-related design functions and routinely conducted planning and control tasks must be allocated to responsible persons. If one project manager with staff function is deemed sufficient for a re-conception project, a value stream manager with line responsibility must be appointed for day-to-day factory operations. As a value stream will generally flow across the company's technical and technological core competences, the respective requirements are best met organizationally with the aid of value stream-oriented matrix organization (Sect. 3.6.1).

3.6.1 Value Stream Manager

The measures derived from the conceptual design and subsequently scheduled by implementation planning need to be allocated to the respective responsible parties. Process owners such as master craftsmen and/or department heads may be put in charge of improvements effected to individual production processes, but beyond that it is recommended to appoint one responsible person for the project management of the respective value stream. The main task of this *person responsible for value stream* consists of coordinating the various individual measures to enable the achievement of the goals aspired in the overall concept. In particular, this includes the balancing of the interests of all production departments involved in the value stream in question.

Another vital prerequisite for the sustainable application of value stream design beyond a one-time re-conception is the mandatory application of value stream-oriented planning as a basic precondition for all *investment decisions* taken in the respective departments. Decisions as to how well the respective alternatives would match the overall value stream performance may be taken by the value stream manager on a case-by-case basis. This will thus include tasks from the areas of work preparation, resource allocation und factory planning and may be integrated in the general factory organization in the form of project organization.

One possible method of consistent integration of the value stream method is the establishment of a correspondent *staff function*, often realized in the form of 'in-house consulting' to support sustainable, standardized implementation and providing the employees of the respective departments with the opportunity to systematically develop theoretical and practical competencies as well as company-specific approaches. The combined responsibility for all product families enables the principles of lean production to be similarly implemented in all value streams and to determine cross corporate priorities. It also facilitates the introduction of a uniform performance measurement system for all value streams and provides an institutional framework for regular production audits. This requires top level in-house advisors, though, to prevent understandable acceptance problems on the part of production supervisors. Besides, any additional indirect manpower will also need to be financed internally by the respective departments.

The most important task of the value stream management, though, is not the actual implementation of measures resulting from the re-conception of the value stream, but consistent value stream adjustment and improvement within the factory operation. This includes the monitoring and aligning of planning and control parameters such as the number of kanban (Sect. 3.6.2). In this context, the *value stream manager* assumes classical production planning tasks, though the methodological task profile has changed considerably and indicates the manner in which the value stream manager may be entrenched within the corporate organization.

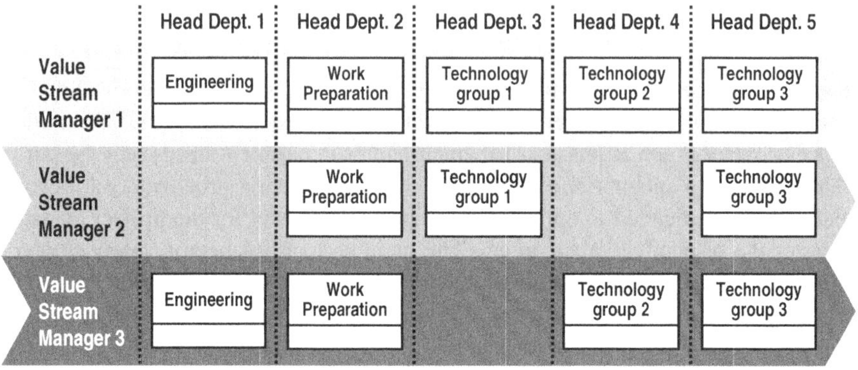

Fig. 3.66 Schematic and exemplary illustration of a matrix organization for three value streams

As structural organization for the value stream management suits a form based on *matrix organization* (Fig. 3.66), which enables the organisational linking in flow direction of corporate functions structured in departments across the value stream. This is the only way to achieve two diametrical goals within a uniform organization. On the one hand, the value stream manager ensures the product family-related customer orientation of the value stream in line with the creation of value and all corporate functions involved. This value stream perspective allows the consistent pursuit of a flexible, laminar, fast and lean production flow. Secondly, technological and organizational standardization of corporate competencies across the value stream is assured by way of technical leadership within the various departments. Without this, diverging developments could split the organization and virtually create two individual companies.

In a matrix organization, disciplinary leadership is also up to the technical supervisors of the respective departments across the value stream. This will sooner or later result in *conflicts* between technical supervisors and value stream managers, as the latter do not really have authority over the personnel in the various departments but are responsible for the entire value stream performance. Such conflicts, however, are equally well-known in other organizational structures and traditional production procedures. The term 'progress chaser' for instance illustrates the conflict between the overall production performance 'being chased' and the local 'chasing' optimization of individual departments. Order control departments for instance, such as in mechanical engineering, are responsible for the coordination between the urgencies and highly diverse special requests of the different sales departments and the contingencies and goals of the producing departments.

3.6.2 Value Stream Management Tasks

Planning levels. In factory operation value stream management is responsible for the operational implementation of the planning and control conceptions developed by way of value stream design. This is done on three different planning levels with different time horizons concerning long-, medium- and short-term planning tasks. The production planning and control level model is generally depicted as a hierarchically structured triangle (Fig. 3.67). Each planning level comprises the planning and control decisions to be taken within a similar time frame and is thus allocated a suitable planning horizon.

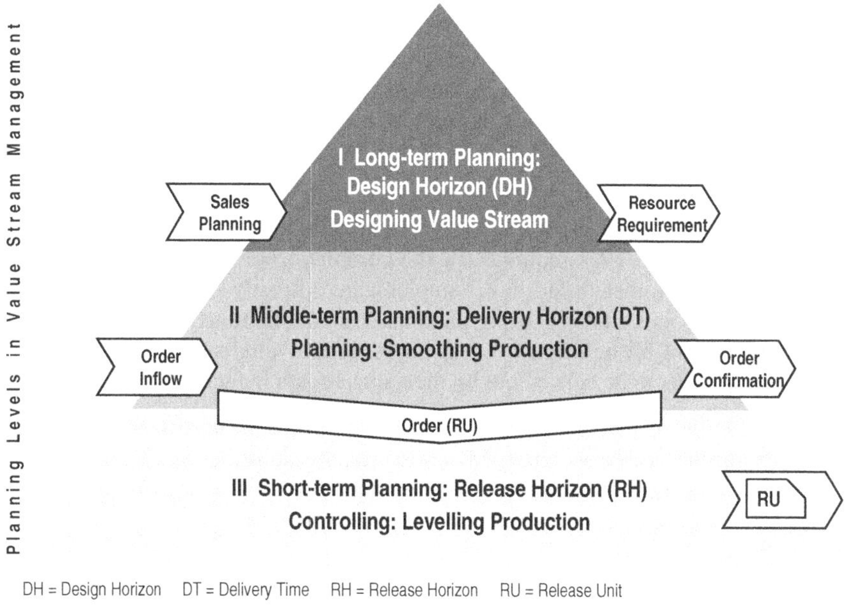

Fig. 3.67 The three value stream management planning levels

The uppermost level, i.e. the top of the triangle, corresponds to the long-term perspective generally assumed in planning projects. The *design* of the value stream may be regularly tested during factory operations. This includes re-dimensioning of the resources to match changed customer takt times through capacitive adjustment of resources and changes in personnel numbers. On the other hand, the basic logistical configurations may be re-adjusted by changing material flow connections from kanban to FIFO or maybe from signal kanban to container kanban or vice versa. This is required whenever the number of variants in a product family increases or additional customer-specific product variants are being produced. Likewise, procurement-related disposition methods may need to be adjusted. The long-term horizon corresponds to the design horizon (DH) of the value stream and may for instance be based on the business year to allow dimensioning in tune with budget planning and sales planning.

The medium- and short-term time horizons directly influence factory operations. The *planning* of the value stream by way of production smoothing is done on a medium-term level. Incoming customer orders are combined in release units (RU), which are then referred to the lower level for detailed planning and/or control. Order confirmations stating the respective delivery dates are generated. In customer-specific production, the medium-term horizon roughly equals the delivery time (DT). In make-to-stock production, the horizon depends on storage dimensions and production lead times, beginning at the pacemaker process. In both cases, the horizon should equal a multiple of the short-term level's time horizon and at the same time comprise the entire order queue.

Detailed planning on the subordinate, short-term level is done by way of sequencing for production mix levelling of the release units scheduled by the superordinate level as well as the value stream *control* by order release in said sequence. Order release will be in accordance with the bottleneck, if applicable. The horizon of short-term planning corresponds to the release horizon (RH) of the load levelling box, which in case of a bottleneck will correspond to the production lead time.

Planning and control The core task of value stream management lies in the planning and control activities required for the completion of orders. This is always triggered by a customer order, either implemented directly or initiated indirectly by the finished goods store via order picking. It should be noted, though, that each order must first of all be *allocated* to an appropriate value stream, giving rise to the following tasks to be carried out by the value stream manager (cf. Fig. 3.69):

1. *Generation of production orders* – Customer orders need to be converted into uniform release units, which may require splitting up or grouping together of orders. Suitable order types may need to be determined, for instance in case of urgent orders permitted to 'jump the queue'. This task also includes the balancing of release units in cases of changed or cancelled customer orders, which is easily done before order release, but should be prohibited afterwards, as this would considerably disrupt the production process.
2. *Confirmation of delivery dates* – As long as the queue is shorter than its defined maximum length, delivery dates in accordance with standard delivery periods may be confirmed to the customers. When the maximum length is exceeded, though, capacity adjustment measures must be considered (cf. below). Urgent orders may be allocated a confirmed shorter delivery period roughly equalling the production lead time, but should be kept at a minimum, for instance by way of price scaling, swapping of orders or partial deliveries. There is no 'traditional' determination of delivery dates.
3. *Sequencing* – In value stream management, the order sequence is determined by sequencing of the release units in accordance with clearly defined rules, generally with the aid of the load levelling box, possibly by way of a two-step approach with campaign formation or sequencing limited by a restrictive bottleneck.
4. *Material availability check* – This task should really be obsolete, but is usually required nevertheless, as not all materials may be procured by either of the two value stream-compatible modes of just in sequence delivery or supplier kanban. Besides, certain exotic products may require special materials.

5. *Order release* – Release units are triggered in a predefined rhythm or in accordance with the bottleneck's release signal at the pacemaker process. Reliable synchronization of parallel value streams to be initiated via golf ball signal must be duly monitored.
6. *Order control* – As the value stream management knows no process supervision, no step-by-step feedback on work progress is required. Only completed production orders, i.e. release units, are reported once they have left the last production process and are ready for dispatch and are subsequently booked into storage accordingly. In customer-specific manufacturing, scheduled shipping as scheduled may already have been confirmed and the shipping documents printed at this point.

With the aid of the six above core tasks of planning and control, value stream management also implements the queue principle (cf. Sect. 3.4.1) in the production procedure. This is of particular importance with products requiring customer-specific adaptor construction, as is typically the case in manufacturing systems engineering.

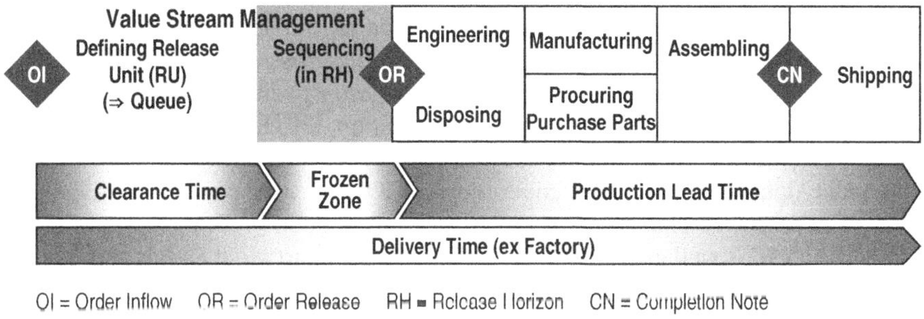

Fig. 3.68 Integration of value stream management in the production procedure

The two primary steps of value stream management – the definition of release units following order placement by the customer (planning) and the determination of processing sequences (control) – are carried out during the *clearance time* (Fig. 3.68). The better a value stream optimization manages to reduce the production lead time to significantly less than the (unchanged) delivery time, there more time will be available for order clearance at low effort. Customers' change requests may be taken into account during the entire waiting period without disturbing the production. Once the processing sequence has been determined in the 'frozen zone' immediately before order release, though, time related changes would be disruptive, as opposed to technical ones, as the actual processing of the orders is commenced at the latest possible point in time. The maximum queue length results from the difference between production lead time and delivery time and is dimensioned to match demand fluctuations as well as the different possible flexible capacity adjustments.

Monitoring. Apart from the operational core tasks discussed in the foregoing, value stream management is also responsible for the consistent monitoring of the value stream to evaluate the respective logistical performance. This is done with the aid of the *key figures* of the value stream method. For said *monitoring* of the value stream, the key figures must be duly processed and graphically depicted in chronological sequence. Prompt provision of succinctly defined and comprehensibly presented key figures to the persons in charge of operations is recommended. This helps the production staff to immediately recognize the effects both of their own actions and of unusual occurrences, thus enabling fast and timely operational rerouting rather than 'blind flying', should the need arise. Besides, in cases of negative performance drift of a key figure, suitable operational trend reversal measures are developed in cooperation with the responsible parties. The key figures for value stream monitoring may be split up into three groups concerning dynamics, productivity and customer orientation of the value stream:

The *dynamics* of a value stream reflect the first type of improvement potential for the production procedure, i.e. the production lead time expressed in ranges of coverage, generally separately recorded as range of coverage of the finished goods storage (RC_{FG}), work in progress (WIP) coverage from the provision of the raw material at the first value-creating production process up to the inventory posting in the finished goods store, and finally, the range of coverage of raw materials and bought-in parts (RC_{RAW}) in the various storage places.

The *productivity* of a value stream is reflected by the second type of improvement potential for the production procedure, i.e. the gross cycle time (CT_{gross}) per production process, expressed as time units per part. Priority will usually be given to output-related employee productivity. In addition, all factors potentially affecting the value stream performance may be recorded, such as for example machine and plant uptimes (UP), changeover times (CO) per process (not partial ones, though, because the changeover portions in lean production always constitute actuating values) as well as quality key figures like first part yield and rework rate.

The *customer orientation* of a value stream is reflected by the percentaged delivery reliability (DR). The respective customer complaint quota may be recorded as well. Delivery reliability is generally measured by comparison of actual and confirmed delivery dates; the difference between the latter and the dates desired by the customers reflects the delivery capability. Delivery capability within a quoted standard delivery period will be at risk, if the length of the queue containing orders not yet released expressed by the respective range of coverage (RC_Q) – either customer orders or stock replenishment orders – exceeds a certain predefined value. Timely availability of raw materials, which should be recorded for the purpose of supplier evaluation anyway, may also greatly influence delivery times and delivery capability. Monitoring of the production mix in percentaged portions ensures the timely triggering of possibly required adjustments with a view to procurement and resource availability.

Flexibility control. The monitoring of key figures, however, only makes sense when suitable subsequent measures are taken. In value stream management, this typically leads to medium-term measures controlling value stream *flexibility* with a view to volumes and capacities, an approach differing considerably from the usual production requirements planning with secondary demands planning, time scheduling and capacity planning. Value stream management substitutes these by the following two measures for inventory balancing and queue monitoring with capacity control (cf. Fig. 3.69). In addition, deviations in value stream performance should be counteracted by measures to eliminate breakdowns and generally consistently improve the overall process.

Thanks to kanban logic, there is no need for material requirements planning; material availability is automatically assumed. In order to assure long-term availability, though, the respective inventories must be adjusted to medium-term trends and customer demand. The *inventory adjustments* required from time to time in the finished goods store and in the supermarket store do not differ that much from the traditional method used in demand-oriented storage with reorder level. In demand-oriented storage, however, the respective reorder level and/or volume-related parameters are adjusted, while the value stream manager monitors the number of circulating kanban per article code. This may be proactively triggered by changes in the predicted customer demand forecast, preparations for special promotional sales activities, or new requirements caused by logistical system alterations, such as increased delivery frequencies or changed inventory policies on the part of the sales partners. A typical reactive trigger would be inventories dropping as a result of increased consumption. Conversely, monitoring of 'residual' kanban, i.e. the minimum number of kanban remaining on stock over a lengthy period of time, indicates an inventory obviously exceeding the volume required to cushion demand fluctuations. In a similar manner, the FIFO lines' buffer stocks may be adjusted in case of repeated material flow breaks or hold-ups which have not been able to be rectified (yet) through improvement measures effected to the production processes.

The value stream does not require any time scheduling in the original sense of the word. Initially, only the earliest possible start date resulting from the delivery time and partial production lead time following the scheduling within the release horizon is known for the orders waiting in the queue. In order to be able to reliably meet delivery dates, *queue length smoothing* may be required before the pacemaker process. Such control intervention will be necessary whenever the length of the queue drifts out of a clearly defined corridor of permissible fluctuation range. Insufficient queue lengths lead to lack of customer and/or stock replenishment orders, i.e. current demands fall significantly below annual average. Unless suitable measures are taken, production standstill through lack of orders threatens. Overlong queues, however, result in fast growing order inventories due to strongly increasing demand exceeding annual mean values. In comparison, the production takt time will be too slow, and unless suitable measures are taken, the value stream will run the risk of delivery delays caused by excessive delivery times and/or restrictions in delivery capability from the finished goods store.

Appropriate *capacity adjustment* may rectify the queue length to match the permitted flexibility range. The required *capacity control* in the value stream primarily serves the capacity flexibility designed in the future state conception. The best options in this context are provided by flexible working hours in one or two shift models. By permitting employees to work up to one extra hour or up to one hour less per day at short notice if required – providing suitably scheduled shifts cases of two-shift models – the daily output may fluctuate by a good plus/minus 10 per cent. Another possibility would be extra or non-working shifts, which would require higher organizational effort and costs as well as longer notice, though.

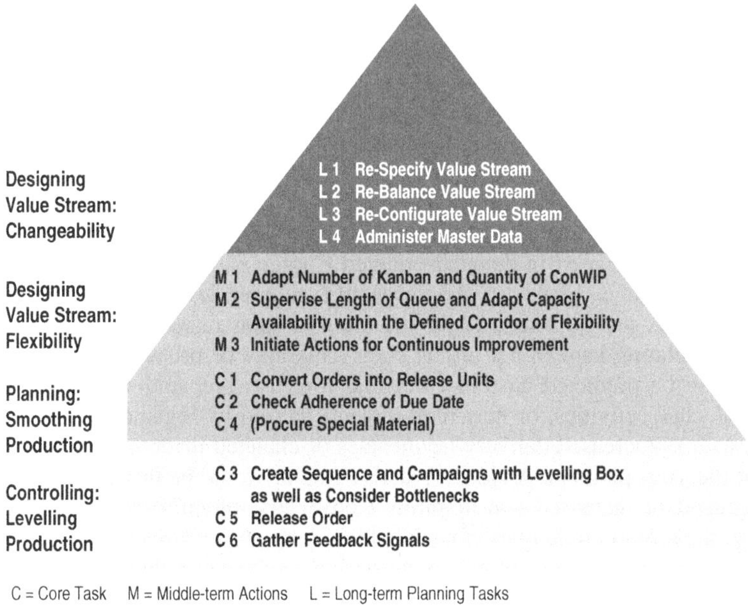

Fig. 3.69 Overview of the tasks of a value stream

The key figures watched in value stream monitoring will also indicate possibly deteriorating value stream performance. This mainly happens as a result of increased work in progress or decreased performance of employees and/or machines and should give rise to immediate *troubleshooting* on the shop floor. One possible solution could be to try and find the respective catalysts in cooperation with the employees involved by way of the traditional 5-W-method, i.e. the persistent child-like fivefold asking of the why-question, and subsequent definition of suitable measures for sustainable improvement in a CIP workshop. Periodic production audits may preventively secure and consistently improve the value stream performance through shorter production lead times, reduced inventories, smaller release units, increased employee productivity through multiple machine operation, minimized changeovers, improved uptimes etc. Accordingly, the medium-term level of value stream management uses methods already known from various production systems (cf. Sect. 4.2).

Changeability control. Long-term measures are required in cases of lasting changes in customer demand, modifications in product range and/or resources, necessitating a suitable reorganization of the value stream. The value stream uses certain planning tools to control the dimension-related and structure-specific *changeability* of the value stream. The capacitive reorganization of volumes and/or performances is effected via re-specification of the value stream on the one hand; in case of changes to the production mix as a result of demand shifts or product launches on the other hand, the value stream needs to be re-balanced. The third possibility, changed piece number characteristics per variant or application of new technologies, however, often necessitates a partial re-configuration of the value stream (cf. Fig. 3.69). A fourth option, namely master data maintenance, is another long-term planning task.

Consistent capacity adjustments required to keep the queue length within the given flexibility corridor indicate a lasting change in customer demand. Similarly, future developments of the customer demand may be predicted proactively by way of trend projection or demand planning. Conversely, consistent successful technological innovation and/or possible replacement investment may alter machine performance and uptimes to such an extent as to cause production takt times and customer takt times to deviate on a long-term basis. These changes may only be counteracted successfully by way of a *re-specification of the value stream.*

The operator balance chart is a suitable *planning tool* for dimensioning, where mean customer demand deviations, expressed in customer takt time and production process performance data, expressed in cycle time, are entered on a regular, maybe monthly, basis. The respective cycle time and/or customer takt time differences indicate need for action i.e. adjustments concerning the available capacities of the respective production processes. A factory's changeability is reflected by the extent to which necessary alterations may be realized through reallocation of resources and/or personnel without excessive effort.

Within a value stream's a product family, demands for the various variants may from time to time shift. As each value stream is specifically designed for a certain product mix, such changes may influence the general output or the position of a bottleneck, depending on whether the respective changes affect the lead time of one process or that of the entire value stream. As far as the new production mix concerned, *the respective value stream* will need to be *re-balanced.*

This will ordinarily occur whenever a new product variant is allocated to or removed from a product family. Such new product variant may have been taken from an existing product family or may constitute a newly launched product. As a rule, *product phase-ins* or (less frequently) product phase-outs affect the customer takt time and the specific demand in special appliances and parts. The respective master data are compiled and the pertaining production processes are technically and logistically laid out accordingly. The management and (new) determination of *master data* in cases of errors or product phase-ins is thus also up to the value stream manager. Product phase-ins may be planned in various advance steps to allow for the pertaining capacity adjustment measures to be implemented in the value stream in good time.

Changed piece numbers per variant may also lead to changed overall production characteristics. Accordingly, a serial product variant may become a small batch serial product or even turn into a customer-specific product variant. This may require changes to the logistical principle, for instance a change from kanban logic to FIFO lines. New technologies may integrate several production processes, significantly reduce lot sizes, make certain storages obsolete and thus also necessitate an adjustment of the production procedure. These cases require *re-configuration of the value stream*, which comes close to a complete re-conception of the value stream in an individual planning project.

Design criteria for a lean production

Value stream-oriented factory operation is planned and controlled in accordance with the six core tasks of value stream management. Top priority is given to the planning tasks controlling the flexibility and changeability of a lean factory. Six design criteria already known from order management determine the functional characteristics to be met by the respective planning and control systems (Wiendahl 2002). In a value stream-oriented production, the six criteria may be applied in *value stream management* as follows

1. **Production structuring** The production is subdivided in accordance with *product family-related segmentation* – not only logistically but as far as possible also spatially in order to create a transparent factory. This results in a continuous value stream from order receipt to the shipping end.
2. **Logistical overall concept** The objective is a *laminar production flow* with consistent production quotas, standardized order sizes and fixed order sequences with no changes in priority after order release.
3. **Market synchronization** The customer demand is implemented into the value stream at the *pacemaker* process in a fixed release rhythm. In customer order-related productions, this is also the customer decoupling point. In cases of delivery from the finished goods storage, the customer demand is indirectly passed on to the pacemaker process.
4. **Synchronization principle** Fluctuations in customer demand are *levelled with a view to volumes* and *smoothed regarding variance*. The strain on the production is exclusively levelled by volume, i.e. by way of the inventory buffer of the finished goods storage or the order buffer in the queue. There is no determination of utilization-related delivery dates; at best, conformation with the maximum delivery periods may be verified.
5. **Planning levels** In value stream management, three time horizons are differentiated. In *long-term planning*, the value stream is capacitively dimensioned with the aid of the operator balance chart. *Medium-term planning* converts customer orders into standardized release units and thus smoothes the production. In addition, the logistical value stream performance is monitored on this level to possibly derive suitable parameter adjustments. *Short-term planning* arranges the release units defined at superordinate level in sequence and releases the same for production in a fixed rhythm.

6. **Planning objects** All orders are scheduled as *release units* of uniform quantities, though not all deviations from the representatives are taken into account. Value stream-oriented production planning is greatly simplified thanks to the fact that only one production process and its capacities are controlled per value stream, usually the *pacemaker* process at which orders are triggered, or, alternatively, the bottleneck process. Accordingly, only orders for *finished products* need to be produced and scheduled, the production of which starts at the pacemaker process. Pre-products are automatically controlled via kanban and do not require any requirements explosion planning. This does not apply to value streams for complex products, though, which need to synchronize several starting points. Dates are only relevant with a view to sequencing, not with a view to control.

Chapter 4
Towards a Lean Factory

An ideal production manifests itself in a factory where each and every product is manufactured individually and according to customers‘ specifications by way of highly efficient industrial means of production, such as division of labour and mechanization – not only to minimize costs, but also to achieve the other essential goals of production (Sect. 1.2.2). Industrial means enable increased product quality and precision, shorter manufacturing processes and the realization of a highly diverse product range in spite of standardization. For continuous production improvement, the target of an ideal factory must be consistently focussed on.

With the aid of the eight value stream design guidelines discussed in detail herein (Chap. 3), any piece production may be guided towards an ideal target state, i.e. the goal of a *lean factory*. Besides, value stream design as discussed in the book at hand provides a well-proven sequence for the conceptual application of said design guidelines, with the aid of which an entire production procedure may be designed.

However, the path towards a lean factory will inevitably be subject to two types of *compromise* with respect to the application of the design guidelines:

1. One consists of the application of the respective 'weaker' of two design guidelines, e.g. line production instead of flow production, batch rather than piece production, observation of sequencing restrictions.
2. The other constitutes ‘imprecise’ implementation of design guidelines, such as low safety stocks in flow production, overtaking on the FIFO line, planning supervision of kanban control cycles by way of superordinate EDP systems etc.

According to experience, the ‘temptation’ of imprecise implementation is particularly high in day-to-day factory operations, as the effort required to strictly observe all the rules and regulations often seems to be disproportionately higher than the immediately resulting benefits. From time to time, violations of the rules may even have distinct advantages – but *exclusively* subject to certain specific circumstances pertaining to the respective production processes *and* products. Accordingly, highly fluctuating work content, such as in manufacturing systems engineering, may cause small plants to overtake larger ones with a higher number of components in the assembly line – an assembly system which, strictly speaking, disagrees with an ideal procedure aimed at evenly flowing production – not always achievable in a simple and cost-effective manner, though.

K. Erlach: Value Stream Design: The Way Towards a Lean Factory, LNL, pp. 231–250.
springerlink.com

Value stream-oriented factory planning. The value stream design approach provides a safe and promising *way towards a lean factory*. The eight design guidelines discussed above perfectly suffice for the design of a factory's production procedure. For value stream-oriented factory planning, however, two more aspects need to be supplemented:

1. The physical realization of a factory production procedure devised in accordance with the design guidelines requires a suitable spatial layout of resources allocated to the respective production processes. The possible *value stream-oriented layout planning* for a lean factory is going to be approached in more detail in the following (Sect. 4.1) with the introduction of two design guidelines for layout planning.
2. In value stream analyses as well as in value stream design, the field of observation embraces everything from factory gate to factory gate, thus methodically covering the planning levels of resources, segments, buildings, and the entire factory, leaving aside the level of the production network with its cross-locational correlations. Both the determination of *production division* in terms of the allocation of production tasks to various locations and the design of logistical correlations concerning *supplier networks* are subject to additional specific aspects which shall not be further elucidated within the framework of the book at hand. An appropriately expanded set of design guidelines would be required to cover the necessary planning tasks for a lean factory in a production network.

The remaining production design tasks refer to product design and various required production technologies. The design guidelines explicitly and implicitly define certain requirements and offer initial design advice which may not suffice for the technical design, though. The actual utilization of *innovation-driven technology development* as well as *production-appropriate product design* to achieve factory optimization should therefore not be considered part of the factory planning process but ought to be regarded as an independent scope of duties. According to the requirements placed on the technology development by value stream oriented factory planning, flow processes for example must be given general preference over batch processes. The pertaining technological principles to be applied and the respective technical design to be devised is then up to the respective technical specialist discipline.

Value stream-oriented factory operations. Since both the conceptual design and its subsequent realization in a factory hardly ever (euphemism for 'never') manage to reach perfection straight away, the production must be continuously improved during factory operations. This may be done on a small scale operationally, i.e. at the individual work places – planned and carried out by on-site staff – or conceptionally on a larger scale, i.e. by way of *continuous factory planning*. In addition, changes concerning available production technologies, products, and not least market-related goal prioritizations, need to be constantly integrated in the redesign of a factory. The basic concept of lean factory operations is therefore based on continuous improvement of all processes.

The organizational responsibility for the *value stream-oriented factory operations* implemented accordingly lies with the value stream manager. The day-to-day realization of the eight value stream design guidelines is done by way of the six core tasks of value stream management (Sect. 3.6.2). In addition to changes explicitly initiated externally, direct medium- and long-term adjustment requirements may be identified through monitoring of company-specifically defined value stream performance figures. These indicate the required planning tasks to be completed by the value stream manager with respect to *flexibility control* and *changeability planning* for a lean factory. With completed value stream management, the production process within the factory operations has been specified in its entirety.

However, factory operations are not only determined by value stream management, but in particular also by the way the different work steps within the production processes are organized. A company's work organization laid out in a structured manner is also referred to as a *production system* – not a technical system like a flexible machining centre, but a work organization-related framework providing directions for all employees, from production managers to unskilled workers, based on the Toyota production system. A production system includes all principles and methods required for a production organization with clear responsibilities, lean production procedures and supporting production controlling. The main components of a production system in the sense of an action system are briefly outlined below (Sect. 4.2). The subsequent development of design guidelines for the creation of production systems will enable a methodical approach for the implementation of clearly defined, company-specific production system components similar to the value stream design approach.

4.1 Value Stream-Oriented Factory Planning

Once the production procedures have been defined and the production logic has thus been determined, the physical realization is planned, beginning with the allocation and positioning of the already dimensioned resources. This conforms to the traditional field of factory planning, more often primarily associated with *factory layout* and *factory buildings* than with factory structures and production procedures. Value stream design covers the two latter aspects – structuring by way of product family formation and optimization of the production procedures with the aid of the eight design guidelines. The approach introduced herein efficiently fulfils and also perfectly visualizes these two tasks, thus the integration of value stream design into traditional factory planning has excellently proved itself. This is reflected in the most diverse planning tasks, so that even 'old hands' at factory planning wonder from time to time how they ever managed to plan any factories at all in the times before value stream design.

Largely unaffected by the value stream method, though, are those tasks which concern spatial design in the widest sense of the word. *Value stream-oriented factory planning* aims at applying the fundamental principles of value stream design to aspects of factory layout and building design. The general factory planning approach will be discussed in the following. Basic planning principles known from traditional factory planning may also be adapted for application in value stream oriented factory planning. In addition, the specific characteristics inherent in value stream-oriented planning will be explained in three steps, i.e. space planning for better space consolidation, ideal planning as an approach for material flow-oriented layout planning, and thirdly, real planning as a benefit analysis of different real layout alternatives based on the evaluation criteria of a lean factory. However, this approach will only briefly be outlined herein, as a substantiated analysis of the architectural and building-related aspects and the space-related planning content would rather require a lengthy independent treatise.

Planning Approach

In 2011, the German Association of Engineers (Verband deutscher Ingenieure, VDI) first brought out a detailed guideline on the subject of factory planning, the so-called '*VDI-Guideline 5200,* part 1': ‚Factory Planning – Planning Procedures'. Analogue to the German Fee Structure for Architects and Engineers (Honorarordnung der Architekten, HOAI), said guideline proposes a general planning procedure for product-related factory planning divided into different stages as a basis for methodological planning procedures describing pertinent planning services. The allocation of the nine different work phases as per HOAI §15 also reflects the connection with architectural building design and thus improves the temporal as well as the contentual coordination between production-related and architectural factory planning.

With what initially seems to be a slightly complicated definition of the *factory planning concept*, said VDI guideline comprehensively points out all the different aspects to be observed, converted into ten 'W-questions': **W**ho? is *planning* in **w**hat manner? Why? **W**hen? **W**hat with? **W**hat? For **w**hich reason? To **w**hat purpose? **W**here? In this context, an expanded planning concept was applied to incorporate not just the actual planning process but also the pertaining realization including factory start-up: Factory planning is a »systematic, objective-oriented process for planning a factory, structured into a sequence of phases, each of which is dependent on the preceding phase, and makes use of particular methods and tools, and extending from the setting of objectives to the start of production. Factory planning may also include a subsequent phase of adjustment while production is actually in progress. There may be different reasons for initiating the factory planning process and it may also include a number of different planning scenarios. Tasks are processed in the form of projects by teams and are controlled by project management methods« (VDI-Guideline 5200).

The *planning phase model* developed by said guideline divides into seven main functional phases plus the accompanying project management with an evaluation

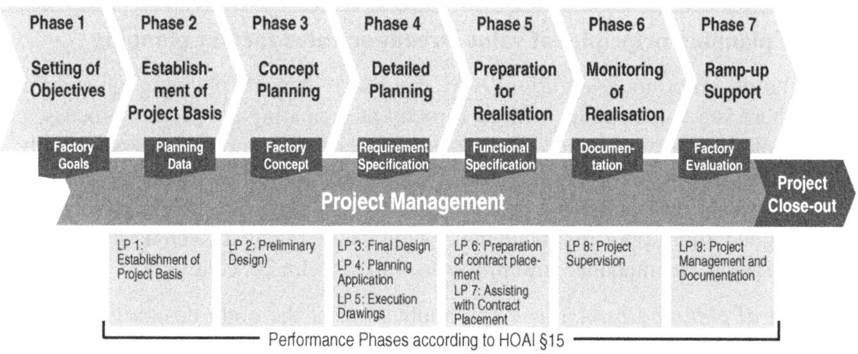

Fig. 4.1 Phase model of the factory planning process as per VDI 5200

of the quality of the entire planning process included with the project close-out (Fig. 4.1). Each phase concludes with the completion of an individual document to be jointly approved by the planning team and the principal, summarizing the results of the respective planning phase – a project corner stone, so to speak.

The actual *planning procedure* is conducted as follows: Factory planning is either triggered by internal changes (technology development, product changes) or changed external requirements (market situation, legal requirements). These must be converted into tangible factory goals with weighted evaluation criteria as well as into appropriate project targets and a suitable project plan (phase 1). The input parameters required in factory planning are the planning criteria to be determined (phase 2). Based on this, a viable factory concept is then developed by way of the four main steps of structuring, dimensioning, ideal planning and real planning (phase 3). The subsequent detailed planning includes the compilation of implementation plans, performance specifications, and permit applications (phase 4), followed by the contracting of external services and the implementation of internal activities (phase 5). The planning result pursued consists of a factory consistently producing at target performance level including evaluation (phase 7) and documentation (phase 6) of the planning services rendered.

The book at hand has so far dealt with several aspects of the planning procedure. In particular, space-related planning tasks shall be approached in the following, so that at the end of this section a comprehensive overview of the seven phases of value stream-oriented factory planning may be provided. The methodologically correct approach to factory planning is not only based on above seven planning phases, but also conforms with the following six *basic planning principles* adjusted for a value stream-oriented approach.

Basic planning principles of value stream-oriented factory planning

1. *Value creating orientation* – At the heart of each planning procedure, there is a value stream with its directly participating production processes, material flows and order processing-related business processes. Only when these procedures have been conceptionally devised and the pertaining resources and their space requirements have been conclusively designed, the supporting incidental functions are planned, such as supply and disposal, maintenance, training workshops, technical centres.
2. *Ideal planning principle* – The application of the eight design guidelines aims at the initial creation of ideal production processes. Similarly, initial resource layout planning always starts out as a restriction-free ideal layout, i.e. regardless of the existing building stock, 'out in the open', so to speak. The second then comprises the so-called real planning, taking into consideration restrictions mainly resulting from spatial layout and limited financial means
3. *Development of variants* – Due to restrictions, real planning will always be subject to compromise with a view to the desired ideal solutions. As a rule, various alternatives featuring different disadvantages will be possible. The conception of various different possible versions to explore the viable solution space enables the targeted choice of the best solution by way of comparative evaluations.
4. *Scenario planning* – In order to ensure sustainability of the developed factory concept, it should be safeguarded against different possible future scenarios, for instance by way of envisioning possible worst case/best case scenarios to test the concept's sensitivity. In particular, this includes ensuring low-effort updateability of the factory concept and at the same time enables the provisional planning of options of uncertain realization prospects.
5. *Pyramid principle* – The value stream perspective initially focuses on the overall production process, disregarding details and specificities, to develop the respective targets for the detailed planning of the production processes and their allocated resources and pertaining material flows. By way of exemplary detailed planning, the achievability of said targets may be verified and the basis for validated projections for the entire factory may be laid.
6. *Iteration principle* – During the application of the eight design guidelines and/or with advancing project status, previous definitions may turn out to be incomplete and/or incorrect and new insights may be gathered, requiring a rerun of previously completed planning phases, an iteration which ensures continuous planning improvement.

Space Planning

Space analysis. The value stream analysis does not specifically observe the utilization of space in the current state, instead this is done in connection with the value stream–oriented factory planning to establish the project basis. The individual elemental areas in the factory layout are broken down into space categories according to their respective utilization characteristics to enable the determination of partial spaces according to space types with the aid of space analyses. In general, the following seven *space categories* are differentiated:

1. *Production space* for production processes in manufacturing, assembly and quality control including production process-related supply areas;
2. Buffer and *storage space* for raw materials, bought-in parts, semi-finished products, finished goods, tools, operational and auxiliary materials, and pertaining facilities;
3. *Circulation space* for material and personnel flows;
4. *Functional areas* with production-related office space for production control, NC programming, dispatch handling, staff information and other information flow-related functions;
5. *Special purpose areas* indirectly related to the production but not directly allocated to any value stream, such as training workshops, technical centres, or tool making;
6. *Social* and *sanitary areas*;
7. *Free spaces* not included in the evaluation but observed in the later layout planning.

The partial spaces and space types may be graphically visualized by way of a pie chart. The ratios between the production space and the other three value stream-relevant space categories – i.e. storage, circulation and functional spaces – indicate the *space utilization degree* of the value stream. In addition to the degree of flow and the degree of utilization (cf. Sect. 2.4), this constitutes the third type of value stream improvement potential and indicates possible ways of space consolidation. For a better evaluation of the respective consolidation potential, the space utilization level may be compared to a respective – usually trade-specific – space ratio.

Overlaying of the machine layout in the factory layout with the *material flow linkages* defined in the value stream analysis results in an exemplary depiction of the material flow pertaining to the main components of a product family's representative. A grown factory structure's typical movements of material transport shuttling back and forth including transfer to and removal from stock visually illustrate the improvement potential possible changes to the existing machine layout; besides, the measured distances may then be evaluated as to costs.

Another analytical factory planning tasks consists of the analysis and evaluation of building structures and factory structures. Building structure-related analyses include an evaluation of *building stock* with a view to possible need for sanitation, but also a description of *space quality*, evaluating the available space as to usability with a view to the different resources. This takes into consideration floor loads, infrastructure, media supply, statutory constructional limitations, structural grid patterns and other structural restrictions.

In cross-factory structural planning, factory structure-related analyses are of higher significance than in the redesign of value streams, which only affect part of a factory. The former focus more on the correlation between all value streams of a factory. The respective analytical tasks include topological assessments of the respective areas and their *development*, transport connections, statutory and other restrictions concerning *site utilization* and/or expandability. The design is largely subject to urbanistic aspects, which shall be largely disregarded herein, though.

Space requirement. In value stream design, the dimensioning of a value stream's capacity is done as early as the first step. The resource requirement, i.e. the number of required resources is determined for each production process, using the operator balance chart as a capacity model. Based on said resource allocation to the different production processes, factory planning then determines the space requirement per resource, the summation of which will add up to the total production space. Resource space does not just specify the base area of the respective machine, but the space required for all pertaining facilities including operator and supply areas. The *determination of the space requirement* is then continued with the dimensioning of the buffer and storage space required for the respective FIFO coupling and kanban connections, the space requirement of which strongly depends on the type of storage used, which should be chosen to match the respective operating mode of a lean material flow. Thirdly, the means of transport and conveyance resulting from the respective type of logistical linkage and pertaining infrastructure requirements need to be considered, though the actually required circulation space will finally be determined in connection with the final layout planning.

The basic space planning concept for a lean factory observes a *strict spatial and personnel-related separation of production processes and material flows* on the shop floor. Producing employees are exclusively responsible for activities serving a value-creating alteration of the respective parts and products with the aid of production processes. Logistics employees on the other hand, are solely responsible for ensuring that the producing employees may concentrate on their value-adding tasks without ever running out of material or having to see to the removal of finished parts. The respective working areas should therefore be distinctly separated spatially. The respective design guideline is as follows:

> **Design Guideline 9:** *Separation of Production Process and Material Flow*
>
> Value-added operations in production processes should be separated from the supporting logistics operations both with a view to space and to personnel.

The merging of processing and logistical activities generally leads to seemingly necessary waste which thus goes unnoticed in day-to-day operations. The distinct separation of tasks pertaining to the very different fields of production and transport enables independent design and optimization of the respective work processes and facilitates the conformation with the respective requirements, thus aiding the optimization of efficiency (Fig. 4.2).

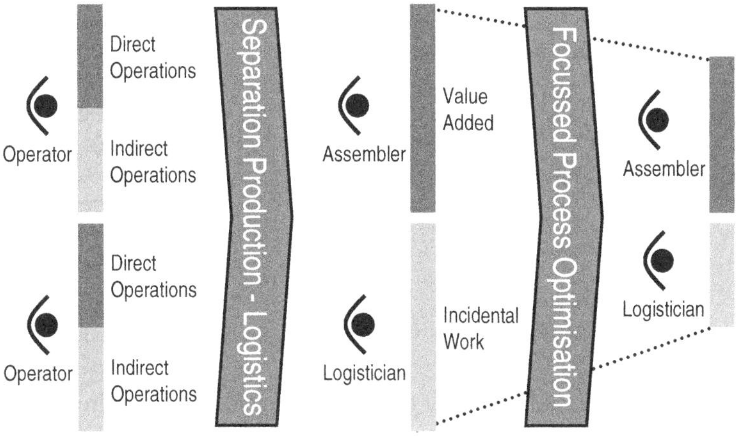

Fig. 4.2 Focussed process optimization through separation of activity types

Accordingly, operators carrying out direct as well as indirect operations, need to change over 'intellectually' more often. Besides, the respective work places will either need to compromise between the ergonomics of material processing and those of material supply, or perform complicated and time-consuming changes. The separation of tasks on the other hand eliminates time spent on changing activities. Besides, logisticians supply several operators on their routes – in contrast operators' direct self-sufficient supply. In factories, this *separation principle* is typically implemented in a factory with the aid of flow racks serving as interfaces. Besides, clearly defined supply and removal zones are created with the aid of distinctly marked staging areas, both of which will be described in further detail in the following.

In a flow production, flow racks constitute the interfaces between value-creating production processes and the staging logistics for material supply (Fig. 4.3). The buffers containing the supplied parts *decouple* the different time rhythms between processing and material flows, enabling operators to efficiently carry out their duties in accordance with the specific logic inherent in their respective tasks.

Logisticians supply assemblers with all required parts in retrieval-friendly containers placed on slightly inclined shelves for easier access. Assembly process and container position should be aligned to ensure each part being stored at its place of usage – ideally arranged in such a way as to enable assemblers to simultaneously use both hands without crossing over. Similarly, all mounting devices should be arranged to provide all tools at their respective places of use. Each container is equipped with clear identification markings concerning content and quantity. The assembler places empty containers on the return rail, thus signalling the logistician to supply a full replacement container according to the identification markings the next time round. The logistician follows a defined route on a fixed schedule and thus supplies several assembly places. In variant assembly of large parts unable to be arranged for container access, the logistician controls the assembly sequences by way of material supply via roller belt.

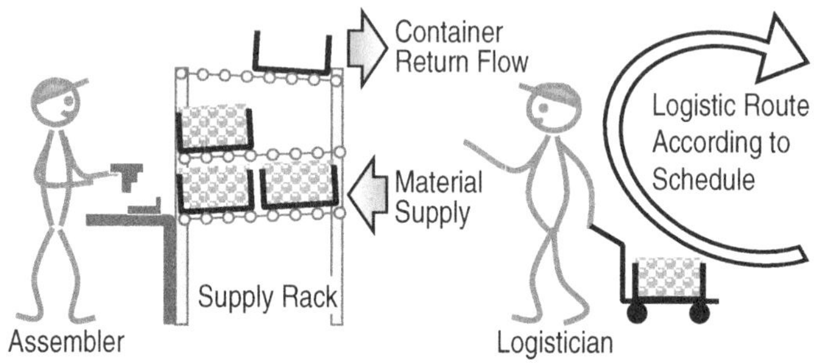

Fig. 4.3 Flow rack as interface between value-creating activities and logistics

A highly beneficial side effect of flow production is its high degree of *space efficiency* which automatically results from the design target of minimal distances for operators in flow production to eliminate waste due to long distances. Accordingly, value-creating work stations must be designed with minimal work areas and arranged as closely together as possible. As stock formation between work stations should generally be avoided to prevent batch creation, there is no space requirement for buffers. According to experience, it has proved very helpful methodically in the initial flow production layout design to pretend that the production procedure be carried out by one circulating operator only. This enables consistent adherence to the flow principle and at the same time ensures that there are no obstacles obstructing the paths.

Analogue to the above example of flow production, the same principle may also be applied to single machine production where operators should avoid wandering around the shop floor looking for required parts, then maybe finding a pallet truck or waiting for a crane to become available. Manufacturing will be much more efficient if the material required for the next following order is readily provided and finished parts are retrieved and taken to the next work station by another employee. Analogue to the clearly marked shelves in flow production, clearly defined and distinctly marked supply/removal areas aid the *visualization of the material flow*, providing each production process with separate areas for goods receipt and goods issue. This eliminates searching effort and makes the work in process easily recognizable. The respective reserved spaces must be aligned with the defined FIFO line buffer sizes, or kanban quantities, respectively. The logistics principle described above may also be applied to the provision of tools from an equipment vehicle for instance if unable to be stored where needed, or for the removal of metal cuttings etc.

As clearly shown above, the principle of strict separation of production processes and material flows makes production space more efficient. In addition, the distinct reduction of production lead times pursued in value stream design also leads to a decidedly lowers inventories on the shop floor and thus significantly reduces the required space. Now obsolete storage facilities may be dismantled and sold to Chinese competitors. The determination of a lean factory's space requirement is characterized by the target of *space consolidation*. The avoidance of waste through storage and transport significantly increases space productivity. Accordingly, the value stream method is also highly beneficial with superordinate planning tasks such as the consolidation and/or expansion of locations.

Layout Planning

Ideal layout. Once the space requirement has been determined, the spatial arrangement of all resources within the factory layout may be developed. According to the design rules of traditional factory planning, non-intersecting material flows must be created, i.e. a *material flow-appropriate* layout must be designed. Part of this task was already completed with the value stream depiction, the production process sequences were determined and the respective resources were clearly allocated. Simple arrangement of the resources in accordance with the material flow depicted in the value stream results in a *flow oriented ideal layout.* Accordingly, the respective design guideline is as follows:

> **Design Guideline 10:** *Flow-oriented ideal layout*
> The production equipment of a factory should be placed corresponding to their succession in the value stream and as closely together as possible.

Above design guideline replaces the triangle method for the arrangement of elemental areas already known from factory planning but rarely applied – for good reason. A value stream depiction directly translated into a layout would result in a *line structure* (Fig. 4.4, case A) – not necessarily always the best option. It is therefore recommended to incorporate the different possible basic material flow structures in the conceptional design. In cases of branched value streams, a parallel structure is recommended (Fig. 4.4, case B). Line structures with high-volume inflow of bought-in parts at several points lend themselves to a logistics area running parallel to the producing line (Fig. 4.4, case C). In cases of several merging partial value streams, the spine structure is usually the best option.

The various deviations from the line structure, i.e. U-shape, ring structure, angular structure, or loop structure, respectively, primarily result in changing positions of goods receipt and shipping end in relation to each other (Fig. 4.4, cases E to H). This enables value stream design in a building in accordance with the topological and/or development-related requirements without affecting the direction of the material flow. Besides, the changed layout of the production processes in relation to each other leads to changed *vicinities* which do not exist in line production. This is of particular significance when certain production processes need to be situated closely together or far from each other for technological or organizational reasons. Certain production processes could impair each other due to noise, dirt, heat, vibrations or other emissions, some critical media should only be supplied to production processes located close to the edge of the shop floor, certain technologies require special employee qualifications and are thus better arranged in a cluster.

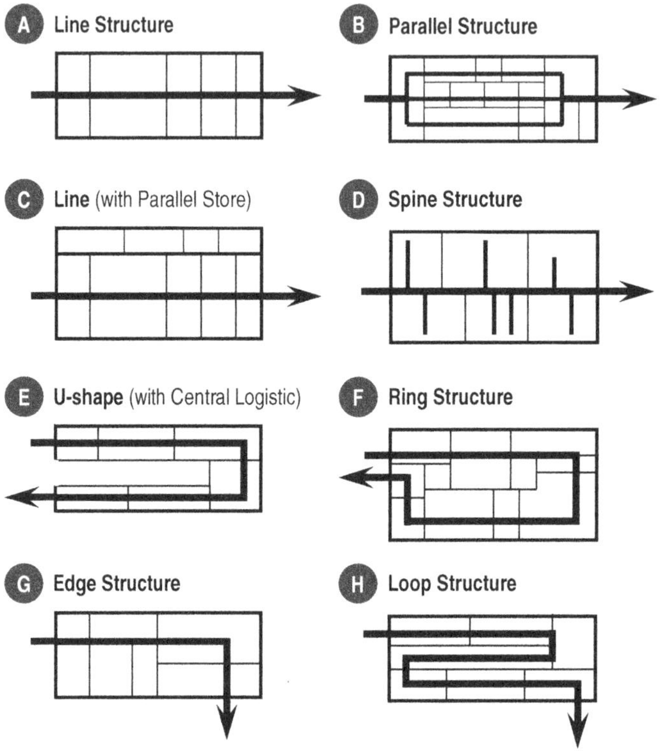

Fig. 4.4 Eight fundamental material flow structures for value stream-oriented factory layout

The basic structures illustrated in figure 4.4 are depicted in a highly simplified manner compared with the actually achievable material flow, as the space requirements of the various resources may differ significantly. In addition, the specific space geometries mainly resulting from the base areas of the respective resources will largely influence the ideally realizable material flow structures. An ideal layout will therefore always be characterized by the space geometry of the required resources and their specific requirements on the building-related infrastructure (media supply, air conditioning, soundproofing, cleanroom, special safety requirements with a view to fire protection, health protection, explosion control). As long as no building-specific requirements have been determined and no area-related restrictions have been taken into account, one single ideal factory layout solution will usually materialize.

As a rule, though, the mere draft layout modelling of each production process with only one elemental area will not be enough. *Machine layout plans true to scale* are needed to precisely illustrate the material flow developed by way of value stream design. Several identical machines allocated to the same production process and their respective supply areas should each be arranged and aligned to match the direction of the material flow. This is achieved through careful observation of all resources in a production process. Losses through interlacing, however, constitute a criterion somewhat diametrical to the flow-oriented single-machine layout (Fig. 4.5).

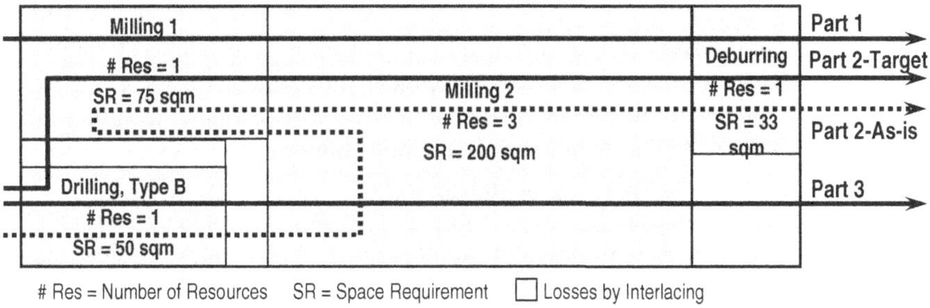

Fig. 4.5 Scaled machine layout plan in manufacturing

Another significant aspect is the depiction of the material flow for different parts within a product family, or parts family, respectively. As product families are characterized by similar (not identical) production procedure sequences, material flows will always differ slightly, especially in parts productions. In the illustrated example, parts 1, 2-target and 3 are processed by identical machines in different production sequences (Fig. 4.5), yet the material flows remain aligned. The value stream perspective, however, demands the conversion of the processing sequence of part 2-as is into part 2-target, possibly through clamp adjustment in milling, to prevent backflow (and in this case also duplicate milling 2).

Real layout. Building-related restrictions, such as grid dimensions, evenly shaped structures (no zigzag-shaped base), structural conditions, area loads, differing media availabilities, lighting, air conditioning, etc. are incorporated in *real layout planning*. In an ideal, lean production, the creation of building-related restrictions preventing a value stream-oriented layout of resources should be strictly prohibited. Day-to-day practice, however, will always be subject to compromises due to ever-present, inevitable building-related restrictions to be observed. The methodological tool for the systematic determination of the best possible solutions, i.e. the development of *planning variants* with subsequent *benefit analysis* as follows.

As the restrictions allow different deviations from the ideal solution, the conversion from ideal layout to real layout will result in different planning variants, both with a view to quantity as well as quality. In the first case of purely monetary observation, the various variants will be accompanied by different investment costs concerning modification and/or redevelopment of shop floors, extensions and annexes, complete new constructions as well as the moving of machines and other equipment. The respective costs per variant may be systematically determined and totalled with the aid of a suitable *cost schedule*. In order to realistically assess the various variants' investment effort, the calculation for an ideal solution with complete new construction should always be included as well as a minimal variant at the bottom end of the costs scale, possibly with no alteration costs at all. Regardless of the respective planning tasks, independent investment requirements may arise without any need for structural adjustments, though, for instance in the form of inevitable redevelopment, investment for the substitution of machines unable to be repaired economically or investment resulting from statutory provisions. In the face of such required minimum investment, however, some of the conversion costs may seem less grave.

The variant requiring the lowest investment, though, will not necessarily be the most economical one. Accordingly, the *running costs* to be expected following the realization of each variant must also be considered. A simple marginal analysis will do in this case, i.e. only the monetary differences will be compared. Variants entailing higher expenses both with a view to investment and to running costs may be disregarded. As a rule, variants with higher investment costs will entail lower running costs, so that the respective payback periods may be calculated and compared. However, the best possible variant may not be chosen on a monetary basis alone. Instead, apart from cost-effectiveness, the other three goal dimensions of variability, quality and speed (cf. Sect. 1.2.2) must be taken into account. For an effective evaluation of variants, suitable *qualitative evaluation criteria* must be determined for each of the goal dimensions to be then weighted in *one-on-one comparisons*. In order not to neglect or overrate any of the goal dimensions, possibly as a result of the sheer diversity of the evaluation criteria, the same number of criteria should be determined for each goal dimension, if possible. It may be helpful for the determination of the evaluation criteria to observe the following four aspects: product (material), technology (machine), time (method) and employees (man). The following exemplary criteria may be applied in the layout evaluation:

- **Variability** – Implementation of dynamics: In case of required modifications, the factory layout should be able to be adjusted with as little effort as possible thanks to suitable spatial, technical and work organization-related conditions.
  - Changeability in case of alterations to the product portfolio
  - Attractive working conditions
  - Expandability for future growth
  - Spatial conditions suitable to support flexible team work
- **Quality** – Reliability assurance: The factory layout should comprehensively support a stable production process by way of suitable spatial structures.
  - Physical proximity of supporting and producing processes
  - Clearly structured building utilization depending on the respective requirements
  - Target layout suitable for step-by-step implementation
  - Clustering of technological competencies

- **Speed** – Transparency assurance: With the aid of clearly structured material flows, the factory layout should enable and accelerate transparent production procedures.
  - Short, direct transport; suitable connections to other factory areas
  - Material flow-oriented layout
  - Short implementation periods of factory planning projects
  - Ergonomics-friendly and multi machine operation-supporting spaces

The principal evaluates the planning by way of credit points with a view to above criteria, possibly in cooperation with the members of the planning team. The allocated credit points are then multiplied with weighting factors depending on criteria relevance – for instance in accordance with the ranking order determined in the one-on-one comparisons. In order to achieve significant results, the four weighting factors should be chosen as a geometric series (1 – 2 – 4 – 8). For better qualitative comparability of the planning results, the two borderline cases, i.e. the current state to be phased out and the ideal state to be realized by way of new construction, should be included in the evaluation.

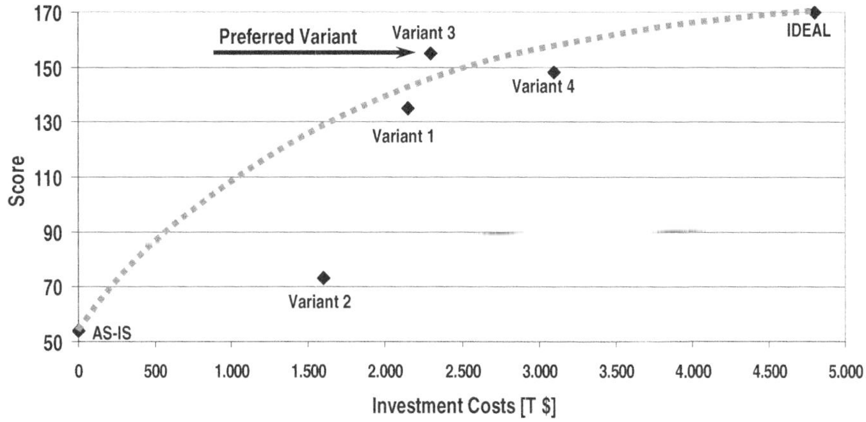

Fig. 4.6 Exemplary comparison of variants by way of benefit analysis

For the concluding choice of the *preferred variant*, both the qualitative and the quantitative evaluation of the benefit analysis are taken into account. For a descriptive evaluation, the credit points of the various variants are marked above the respective cost value (Fig. 4.6). As a rule, some variants will look better (here variant 2) than others (here variant 3). The illustration clearly shows the difference between the preferred variant and the ideal. With this summarized illustration of the evaluation by investment costs and credit points, the conceptional phase of a factory planning project has been completed.

Value stream-oriented factory planning

The systematic planning procedure is roughly structured by the seven *planning phases* built upon each other and the *milestones* placed in between (VDI 5200). Based on the components of the approach to value stream-oriented factory planning described and developed herein so far, the following applies to the planning phases of lean factory planning:

1. *Setting of objectives* – Numerous decisions have to be made within the framework of the design guidelines during the course of the planning stage, enabling and requiring differently specified and shaped details. In order to be able to make the right decisions based on comparative evaluations of the various alternatives, the factory goals and their respective interdependencies within the *logical square of goals* comprising variability, quality, speed and cost-effectiveness must be clearly defined.
2. *Establishment of the project basis* – The *value stream analysis* efficiently (though roughly) depicts the entire current state of a factory. Points identified as mission-critical from the value stream perspective may be supplemented by suitable detailed analyses. To complete the factory planning process, though, an additional *space analysis* and a building stock analysis are required. The result identifies three different types of improvement potential of the value stream analysis, i.e. flow degree, degree of utilization and space utilization degree.
3. *Concept planning* – The factory is structured into independently designed value streams by segmentation in accordance with the *deduction of product families*. The capacitive dimensioning of the resources results from alignment with the *customer takt time* and the development of the capacity profile for the entire value stream. The ideal planning is conducted by way of *value stream design* of the future production procedures with the aid of the eight value stream design guidelines as well as value stream-oriented *ideal layout planning* according to the tenth design guideline. The stringent application of the value stream-oriented factory planning approach concludes with a *benefit analysis* of alternative *real variants* in line with the goals of a factory with flexible, laminar, fast and lean production procedures. This results in a realizable future-state value stream with draft layout and preliminary building design in the form of the preferred variant.
4. *Detailed planning* – Upon completion of the conceptual future-state value stream and pertaining draft layout, the detailed planning of the improvement measures defined in the kaizen flashes starts – which may also define a complete new construction. More often than not, this will constitute the conception of a flow production to match the value stream. In this – and all other detailed planning procedures – the ninth design guideline concerning the *differentiation* between *value-creating* and *supporting activities* of guideline needs to be given particular attention.

5. *Preliminary implementation steps* – The value stream-oriented approach is typically characterized by the step-by-step realization of a structured sequence of *value stream sections*. As a rule, these sections consist of decoupled control cycles, implemented from the customer connection up to the supplier connection in an upstream direction. At the same time, externally procured resources and other services are contracted.
6. *Implementation supervision* – The responsible *value stream manager*, if any, will be responsible for the initiation and supervision of projects for the re-specification, re-balancing und re-configuration of the value stream.
7. *Ramp-up support* – The focal element is the continuous evaluation of the value stream performance by way of value stream *monitoring*. The performance ratios applied herein may be broken down into three groups for the evaluation of the value stream's dynamics, productivity and customer orientation. Depending on the definition of the respective performance figures, the factory goals are weighted differently in relation to each other. Dynamics usually evaluate speed and cost-effectiveness, productivity usually indicates cost effectiveness and quality, while the customer orientation defines variability and speed.

4.2 Production System

Thanks to the success of the Toyota production system (TPS), primarily developed by Ohno Taiichi, the systematic organization of productions by way of *production systems* has become widely used. Nowadays, almost all car manufacturers feature production systems of their own (Mercedes – MPS, Audi – APS, Valeo – VPS, to name but a few). Numerous comprehensive brochures and modified production procedures have been developed. Highly successful producers such as Porsche, Trumpf, Festtool, SEW Eurodrive, etc. now offer advisory services and pass on their own positive experiences to other industrial enterprises. While solution approaches and methods may be similar, experiences differ significantly from enterprise to enterprise – why?

Temple model. Not surprisingly in view of the protagonists, the Toyota production system as the main reference for all existing production methods is often depicted as a *temple* and referred to as the 'House of Lean'. Its threefold base consists of stable and standardized processes, levelled production, and visual management, its three pillars are called just in time, jidoka and continuous improvement (Liker 2004). The illustration chosen herein has been slightly modified in comparison with the traditional presentation, reflecting a more condensed, systemized arrangement of the elements of a production system (Fig. 4.7). The elements' various functions in the system are briefly described in the following:

- The *base* of a production system consists of absolute reliability of all production processes and production procedures and is ensured by way of standardization of all work processes.

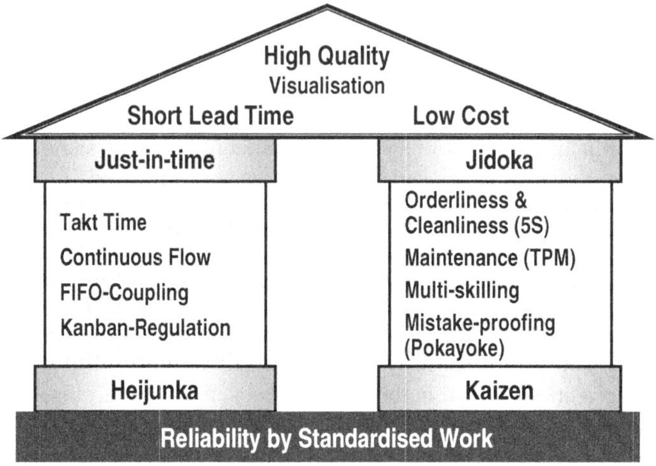

Fig. 4.7 Toyota Production System (modified illustration)

- The avoidance of waste is realized by way of methods represented by the two *pillars*.

  The *logistical* pillar on the left comprises methods for the avoidance of waste in production procedures, enabling even and balanced production (heijunka) and also timely production without overproduction (just in time). This conforms to the production procedure design methods defined by the value stream design guidelines.

  The *technological* pillar on the right includes methods for the avoidance of waste in production processes, mainly consistent improvement in cooperation with the operators at the work stations (kaizen) as well as intelligent automation (jidoka). In particular, the latter provides automatic functional monitoring, which frees operators from having to watch producing machines and enables multi machine handling. Other process-related methods include shop floor design according to criteria such as orderliness and cleanliness, consistent preventive maintenance, multi-skilled workers for flexible assignment and error prevention through fool-proof design of equipment and control elements (poka yoke).
- The *roof* consists of the three well-known essential goals of production: top quality, short lead times and minimal costs. The visualization of scheduled production procedures and production results by way of graphic illustrations depicting the production key figures adds transparency to the achievement of the goals.

Above mentioned and other company-specific production systems are generally characterized by a higher degree of detail. Additional methods and general principles are described in detail and arranged differently. The pillared model with roof and base is the most widely used illustration, though the number of pillars, their arrangement, the designation as production principle or method, the number of tools and methods observed, the allocation to base, pillar or roof differ considerably. This may seem somewhat arbitrary – and quite justifiably so: The convincing basic structure required for any production system to be founded on is missing.

Operational model. A suitable solution approach could be to conceive a production system as a description of an *action system* called 'factory' rather than as a random collection of methods. Accordingly, a suitable operational model serves to develop a precisely formulated production system. Technical actions may usually be broken down into five action phases: Preparation of the initial situation, initiation of an action, the event triggered, consequences caused by the actions triggered as well as post processing of the respective action (Erlach 2000). The operator for instance may (1) put into operation an available, readily tooled machine (2) which will process (3) the respective part (4) to give it the required shape. Finally, (5) the machine will be cleaned and retooled.

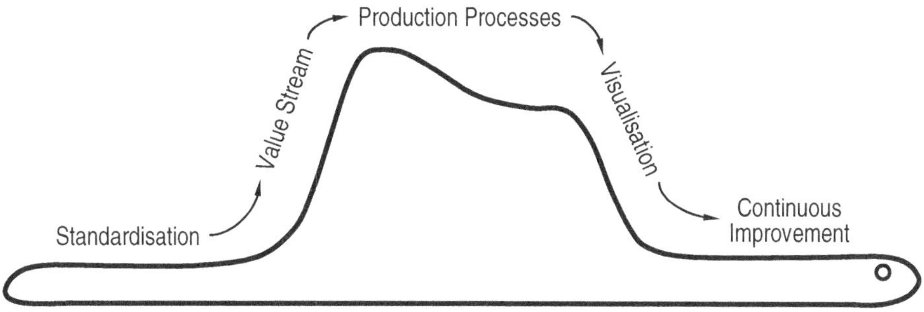

Fig. 4.8 Operational model with its five central fields of action, modelled as a snake that has swallowed an elephant (from. Le Petit Prince de Antoine de Saint-Exupéry "Mon dessin [...] représentait un serpent boa qui digérait un éléphant."; with a special tribute to Thomas T. Ballmer)

In line with the structure of the five action phases, the five provisionally parallelized pillars of the customary production systems may be substituted by a stringent sequence of *fields of action*. For the time being, we shall assume the respective fields of action for the systematic configuration of production systems to conform to the five 'pillars' of preparatory standardization, triggering information and material flows in the value stream, robust and operable production processes to be designed, visualization of the results to be measured by way of performance figures as well as post processing measures for the consistent improvement of future actions (Fig. 4.8). For the depiction of action phases, however, a pillar model is conceivably unsuitable, because it does not reflect the dynamics of factory operations at all. A detailed conception of the modules comprising a phase model-based production system could be well worth a comprehensive analysis elsewhere, though.

This new perspective clearly illustrates the fact that one of the essential features of a production system is its action-guiding function with a view to factory operations. As a result of the obligatory nature of the design guidelines as well as the action guidelines, these must be stringently enforced for a successful production system. This also explains the different degrees of success of the various production systems found in the German industry – for which strict adjustment of the corporate culture is a vital prerequisite. Accordingly, individual guidelines and approaches, such as kanban or improved orderliness and cleanliness may well be profitably implemented by the middle management; extensive implementation of an entire production system, however, will thus not be possible. It requires clear commitment at top management level as well as consistent patience in the realization of the respective sine qua non conditions for the successfully producing action in a production system.

Productions systems

A *production system* constitutes the classification scheme which intrinsically and logically compiles the permissible methodological standards of a production. A production system defines *how* to produce.

Production systems are characterized by the following criteria:

1. They are broken down into *fields of action.*
2. They contain descriptions of the *design guidelines* and subordinate design rules and methods, defining the same as *standards.*
3. They serve the realization of a lean factory, based on the four *goal dimensions* of variability, quality, speed and cost-effectiveness.
4. The consistent improvement of production procedures pursued with the aid of clearly defined standards is measured by way of suitable *performance figures.*
5. As *action systems* they are supported by employees assuming a high degree of *joint responsibility* and can only ever be successful if all participating parties commit themselves to the respective goals and methods.

Factory planning always leads to factory operations. With increasing pressure for the continuous improvement of a production in order to keep the same competitive, the time-related differences between planning and operations will diminish. It is often said that factory planning in itself is about to become a continuous tasks. A catchy point, maybe, but we should not forget that changes to overall procedures such as those devised with the aid of the value stream design guidelines introduced herein, generally cannot be planned in day-to-day factory operations. There is only one solution, namely to step back and observe the production from the bird's eye view of an externally supported value stream analysis, become a stranger, and without constraint wonder at seemingly necessary, yet inefficient, solutions. Finally, with all the might and main of active oblivion concerning the old restrictions deeply rooted in day-to-day operations we must develop (initially visionary) new solutions with the aid of the value stream design guidelines. From time to time we must dare think beyond the rigidified factories 'firmly inhabited' by their operators and think well ahead in order not to grow old with old habits but instead allow rejuvenating growth in order to survive.

Chapter 5
Case Studies

Only practice will tell if and how a method really works. The value stream approach and its various possible applications and operational modes are therefore going to be illustrated in detail by way of several project samples. For a better illustration of the wide range of possible applications of value stream design, different types of project samples from different industrial sectors were chosen. In contrast to traditional object lessons merely illustrating certain methods and applications, the examples included herein are based on industrial projects. These sample projects are rather complex and comprehensive in order to better illustrate the factual courses of action applied. The four examples discussed below are based on the following real companies: Beurer (located in Ulm), Brand (Anröchte), Saint Gobain Deutsche Glas (Aachen) and Dronco (Wunsiedel).

Though based on real-life models, the examples were slightly simplified and partially modified in order to better distinguish certain effects, experiences from other projects were integrated as well. The successful application of the value stream method depends on a clear depiction of the production-related specificities; special care must be taken not to accidentally 'simplify' these away. On the other hand, excessive in-depth details must be avoided in order not to smother the informative value with the diversity of too many different individual results.

The sample projects cover three essential manufacturing principles. A production of electric underblankets is converted into a single-piece flow production connected to further production processes via FIFO coupling (Sect. 5.1). A rather intransparent batch production of spring sets is greatly simplified through introduction of kanban control in the planning logic, which accordingly speeds up greatly smoothes the production flow (Sect. 5.2). The planning process of an entirely customer-specific production of safety glass is greatly improved following the identification of a bottleneck. Besides, suitable control is achieved through conformation with simple, clearly defined sequencing restrictions with a view to customer orders in the now decidedly faster flow of production (Sect. 5.3).

K. Erlach: Value Stream Design: The Way Towards a Lean Factory, LNL, pp. 251–304.
springerlink.com

5.1 Comfort Ltd. – Transparency by Continuous Flow

The Company

Product. Among other things, Comfort Ltd. of Oxford produce electric blankets for the European market at their Hungarian factory. In the winter, these electric underblankets warm the beds of the elderly, or of those in Southern European houses without heating. An electric heating cord is placed between two layers of foam-lined fabric and directly connected to the electricity network by a power plug. Product variants feature different shapes as well as different requirements concerning fabric and/or electrical equipment.

As seasonal demand fluctuations are immense (approx. 80 per cent are sold in the three months of autumn), a large part of the demand must be produced to stock well in advance. A distinctive feature of this special case is the differentiation of three market sub-segments with a view to the different types of demand behaviour, or customer groups, respectively, as follows:

1. There is a group of European discount shops, who generally order few variants in large piece numbers. Their entire annual demand is delivered at specific dates in September or October.
2. The high-quality own brand of Comfort Ltd. (larger formats, electronic circuits, higher heating power, different fabric qualities and designs with trimming, different foam thicknesses) is sold to British retailers in numerous variants. The demand is spread over the entire year, but peaks in winter.
3. External brands are produced to order for other European countries in numerous country-specific variants according to the respective local standards (low voltage, safety regulations) and brand-specific packaging variants.

The product family of electric blankets under observation includes blankets for single as well as for double beds and some with special foot warmers. The value stream analysis concentrates on the most prevalent standard format of 150 x 80 cm for the determination of the operation time.

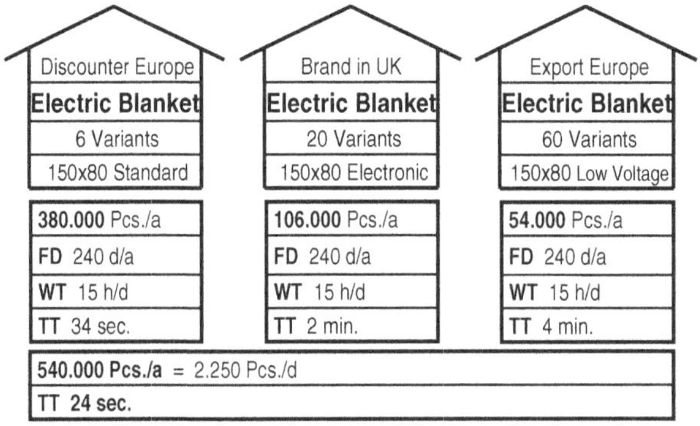

Fig. 5.1 Customer demand at Comfort Ltd

Customer demand. The annual total piece number of 540,000 is divided between the three customer groups of discount stores, own brand and external brands at a ratio of approx. 7:2:1. The customer takt time is individually determined for each customer group. The Customer takt time for the entire production, calculated from the reciprocal value of the total of all reciprocal values, amounts to 24 seconds (Fig. 5.1).

In cases of purely seasonal business, the mean sales rate generally says very little about the actual sales rhythm. As the production is going to be conducted all year round, though, the mean value-related customer takt time may be used in the production design.

Value Stream Analysis

Production processes. The first impression of the shop floor is that of a storage area with masses of fabric piled up high – there seem to be no employees at all. Dotted around the work stations are not quite one metre high shop floor creepers, each piled with 40 electric blankets in process folded in the middle. Depending on the materials used, pile heights are up to 2 metres – one daily demand alone is almost 60 piles. Hidden in between the piles are individual seated work positions with sewing machines and other equipment. At one end of the shop floor, there is an aged lining plant and a new automatic ultrasonic welding plant. The process cycle, observed in an upstream direction, is as follows (Fig. 5.2):

- **Packaging.** At each of six work stations, one worker positions a cardboard box, folds up a blanket and, together with an instruction leaflet, places it in the box, which is then sealed by a special appliance. Subsequently, one worker correctly sorts the electric blankets (now in their individual cardboard boxes) on pallets, prepares them for dispatch and transfers them to the staging area.
- **Testing.** The blankets must be tested for high voltage. Available are two testing devices each with two test beds for alternate testing of the blankets. Each testing device is operated by one worker. One testing process takes 12 seconds, the feeding in and removing of one blanket takes 20 seconds in all. Approx. 1 per cent of the blankets fail the high voltage test. The testing devices have an uptime of 95 per cent, changeover takes 5 minutes.
- **Bordering.** Bias tape and labels are sewn on at five stand-alone workplaces. Changeover takes 20 minutes.
- **Interlocking.** The power connections are sewn to the blankets in 33 seconds at three work stations.
- **Overmolding.** The connections are moisture proofed, i.e. overmolded with resin in 38 seconds. The resin-squirting and heating device has an uptime of 90 per cent, changeover time is 30 minutes.
- **Connecting.** The heating cords are electrically connected with the power cables in 36 seconds. The power cables are preassembled in another area disregarded herein. Changeover takes 5 minutes.
- **Skinning.** The loose ends of the heating cords are brought outside and skinned to enable external power connection. This takes 44 seconds.

- **Ultrasonic welding.** This new, highly automated plant welds together the upper and lower layers of the blankets. At the same time, the heating cord is laid out in loops between the two layers. The takt time is 16 seconds. As a total of six workers are responsible for the change of material and the piling of the sealed blankets, the operation time is six times the takt time, i.e. 96 seconds. A total of 78 variants (which due to partial dual use result in 86 finished products only) are produced in lot sizes of 1,000 to 20,000, the average lot size is 16,800. From the skinning process onwards, the mean order lot size for the 86 final product variants is 15,500.
- **Lining.** This aged – and with an uptime of 60 per cent also rather unreliable – plant lines the fabric with 2,500 metres of foamed material per shift, or 10.8 seconds per metre. This already includes the effort required for the changing of raw material rolls. Two lined panels of a length of one metre each (including scrap) are needed per electric blanket, resulting in a cycle time of 21.6 seconds per end product. The plant is run in three shifts. Accordingly, the process-specific customer takt time is 1.5 times that of the customer takt time of the remaining value stream, i.e. 36 seconds. Once lined, the fabric rolls need to dry out for one day.

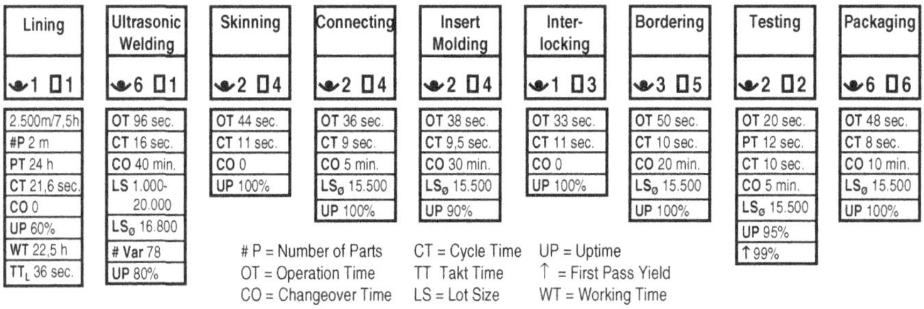

Fig. 5.2 Production processes at Comfort Ltd.

Operation times, changeover times, uptimes and first part yield are recorded for all production processes and entered in the value stream depiction (Fig. 5.2). There are three to five identical alternative workplaces per manual production process. Not all of these were manned during the mapping of the current state, accordingly lower employee numbers are shown in some cases. The calculated cycle times refer to maximum capacities, i.e. the number of work stations, not the number of workers. Depending on the respective demand, workers may be flexibly assigned to different workplaces, so that the number of workers involved in the various production processes may vary during the course of the day.

Material flow. 500 finished goods were counted but not included in the calculation of the production lead time, as they had already passed into the responsibility of the finished goods store. A total of 11,640 electric blankets in different stages of production were counted in between processing steps and on carts. At a daily customer demand of 2,250 pieces, this results in a range of coverage of 5.4 days for the inventories on the shop floor. In addition, there are stocks of 10,800 metres of lined raw material. At a demand of 2 metres per piece, this conforms to a range of coverage of 2.4 days, though the drying out time of 24 hours needs to be deducted from the inventory as process time. In addition, an inventory of 117,000 metres of foam material on rolls was recorded, equalling a range of coverage of 26 days. As due to the seasonal nature of the business, the finished goods store is subject to different criteria compared to those underlying the production, these inventories are disregarded here.

External logistics. The area observed by the value stream ends with the provision of the pallets for loading or storage. Distribution logistics with buffer stocks close by and/or the central Oxford warehouse holding all Comfort Ltd. products are disregarded in the observation.

The delivery of the foam material to be used as lining is depicted in the value stream map as an example for supply logistics. The foam comes in three different thicknesses and two different widths – i.e. in five variants – and is supplied on call by Megafoam AG in Scotland on the basis of annual contracts. The fabrics are procured in large minimum volumes from numerous different, suppliers, located mostly overseas. The electrical components are supplied by the UK headquarters.

Business processes. The logistics departments at the Hungarian factory carries out the production planning based on orders from the central warehouse as well as sales forecasts and large customer orders submitted directly from sales. With the aid of 'ProPlan' software, production orders for the calculated lot sizes including deadlines are generated on a calendar weekly basis, which serve to control the ultrasonic welding process. Over and above these lot size-related production orders, welding is given a production data sheet and time ticket with each 40 blankets. Both production documents accompany the shop floor creepers on which the respective blankets are piled, thus the operators at downstream processes receive all production documents together with the respective material. Production orders are placed on the last cart of each lot and thus trigger a ready release at the packaging station.

Daily capacitive planning of the manual processes downstream from ultrasonic welding is based on 'Excel'-scheduled production orders, resulting in order and capacity overviews in tabular form, according to which the respective operators are allocated to the various work stations, taking into consideration current worker and resource availabilities. Lining is controlled in conformance with order-specific material requirements for each roll of fabric by way of the accompanying time tickets – the pertaining production orders are not implemented in the software.

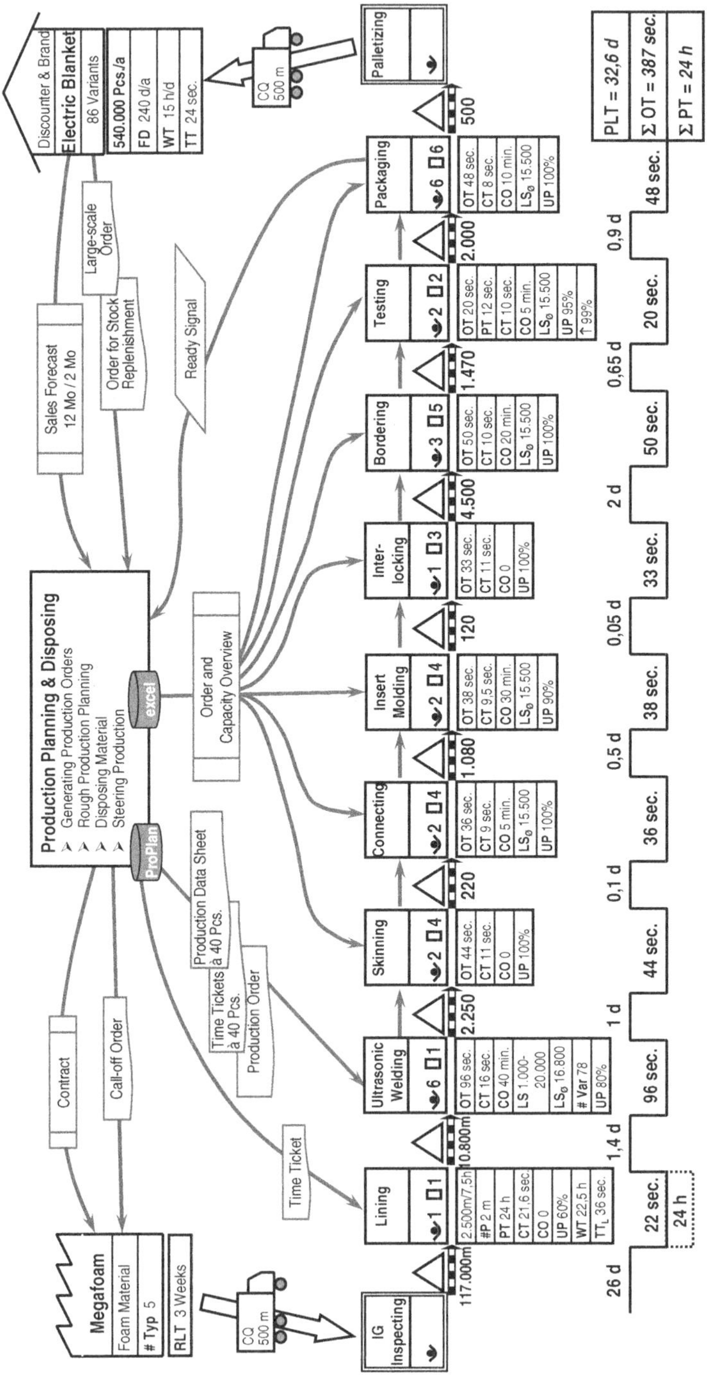

Fig. 5.3 Result of the Comfort Ltd. value stream analysis

Potential for Improvement

Lead time. As shown by the timeline, the raw material stock with its 26 days of coverage accounts for the largest part of the lead time, i.e. 80 per cent (Fig. 5.3). Work in progress with its 6.6 days stands in contrast to a total operation time of 6.5 minutes – accordingly, one production minute requires one day's lead time – a value not unusual for this type of production control with independent capacity planning for all production processes. In the case of nine production processes, the lead times on the shop floor amount to 0.75 days per process step. With each of these processes integrated in a line production, their entire respective lead time portions would be eliminated. For seven manual production processes, the potential lead time reduction is six times the time portion quoted, i.e. 4.5 days.

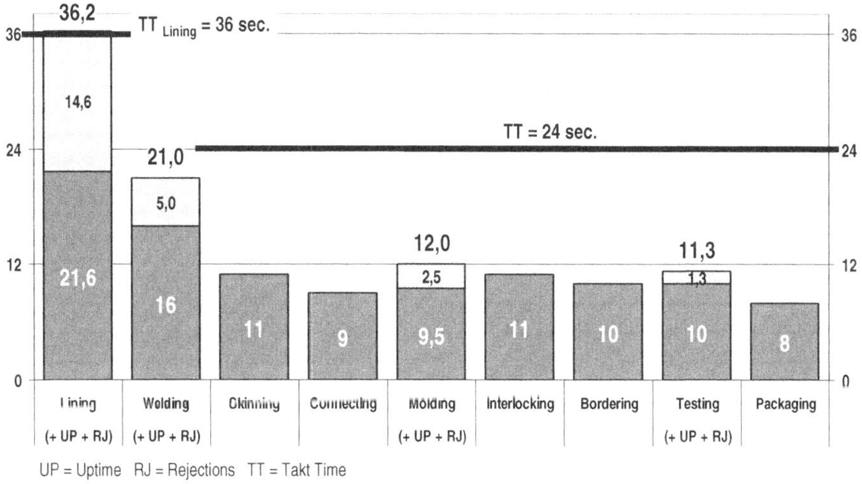

Fig. 5.4 Operator balance chart at Comfort Ltd

Line balancing. For clarity purposes, the operator balance chart only shows the gross cycle times of production processes with restricted uptimes (Fig. 5.4). The rejections passed on to all upstream processes account for 1 per cent of the cycle time and are thus disregarded in the illustration, as they only account for mark ups of 0.08 to 0.11 seconds, depending on the respective process. Changeover effects are equally negligible. With a lot size of 16,800 pieces, the 40 minutes of changeover time in welding result in changeover losses of 0.9 per cent acc. to equation 2.20, i.e. even less than the scrap losses. All other changeover times are even shorter.

What catches the eye, though, is the fact that the cycle times of the manual processes amount to less than half the customer takt time, showing that there are more than twice as many of these low-cost workplaces as required to meet the mean customer demand. Also, it is not possible to pre-produce enough lined and welded raw material to utilize all these workplaces to capacity. In accordance with equation 2.21, ultrasonic welding on the newly acquired plant increases the uptime by 5 seconds, which leads to a capacity reserve of 12.5 per cent. The lining process needs to run in three shifts because of its poor uptime. It still fails to meet the customer takt time – but only just, so that the occasional break worked through and/or extra Saturday shift suffice to meet the demand.

Value Stream Design

Goal. The redesign of the production mainly aims at increased transparency on the shop floor and significant cuts in shop floor inventories. In view of the substantial finished goods storage needed for the highly seasonal business, this will not significantly affect the liquidity, but will save production space and direct the attention towards further potential process optimizations.

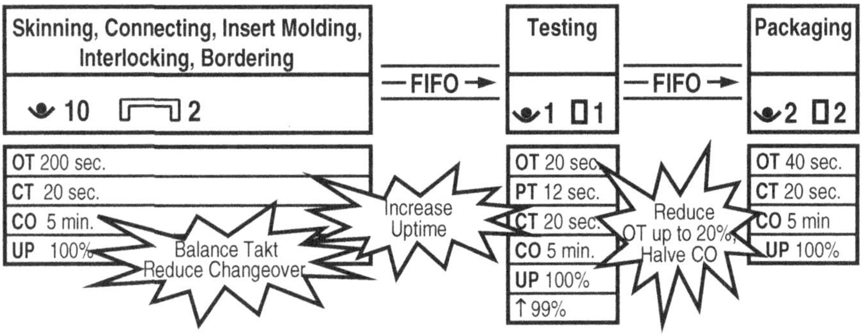

Fig. 5.5 Redesigned production processes at Comfort Ltd

Production processes. The five production processes of bordering, interlocking, overmolding, connecting and skinning may be integrated in a flow production (Fig. 5.5 – design guideline 2). This eliminates the four interim storages with their high piles of stock between the individual workplaces. The blankets are not piled up and removed from piles in between process steps anymore but are directly passed to the next operator. This improves the production flow and results in better ergonomics, too. As a first step, the existing resources are simply rearranged in line and connected by ordinary tables. The operators' first impression was a decided decrease of work effort as a result of the elimination of piling and unpiling (Fig. 5.6).

Fig. 5.6 From 'stockpiling' production to flow production

The requirements to be met for a successful redesign of the workplaces as a line production constitute of an uptime increase to 100 per cent and a changeover reduction to 5 minutes in insert moulding. The respective requirements placed on the flow line are entered in kaizen flashes (Fig. 5.5). The introduction of a flow production will greatly increase shop floor transparency and clarity and thus considerably improve the efficiency of the manual work stations, which will be a great help with a view to the most difficult task, the line balancing of all five process steps. For this purpose, an even cycle time of 20 seconds must be achieved across the entire flow line at almost constant operation time (200 seconds instead of 201 seconds in the current state).

Once all cycle times have been entered in the operator balance chart, the capacitive bottlenecks are clearly identifiable. In the flow production at hand, these are the processes of bordering and skinning, which with 25 seconds, or 22 seconds, respectively, exceed the desired cycle time of 20 seconds (Fig. 5.7). A customer takt time of 24 seconds (with the implementation of two flow lines) still results in excessive additional time of more than 16 per cent, which is best compensated by non-working shifts. As in some cases there are twice as many workplaces as required, standard products can be produced in the two flow lines, while the remaining individual production facilities can take care of special formats, e.g. electric blankets with two heating cords each, or small order batches – for instance when one of the flow lines isn't needed. Such focussing would considerably facilitate the flow lines' process optimization and at the same time support a factory structure better suited for optimal adjustment of resources and products.

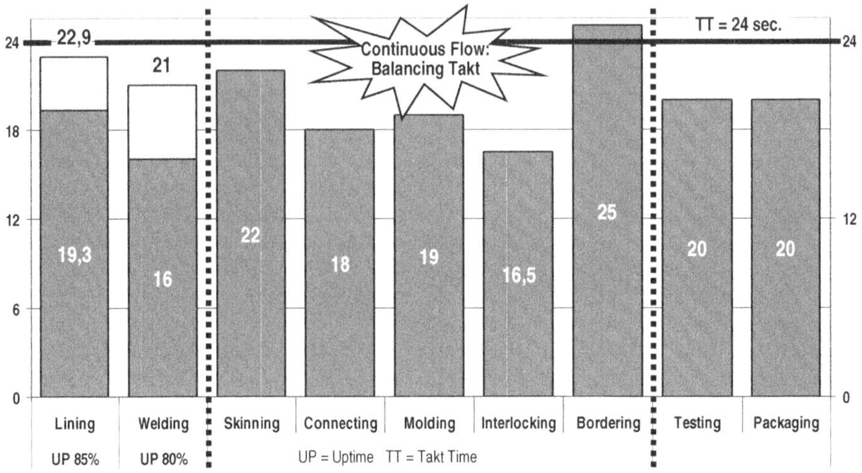

Fig. 5.7 Balancing of capacities at Comfort Ltd

Material flow. The testing process is approx. twice as fast as the flow production, which easily enables the coupling of one testing device with two flow lines via FIFO coupling (design guideline 3). Depending on the size of the FIFO buffer, the blankets may be provided on carts, or preferably be passed on directly. Each flow line could then supply one of the two alternately used testing beds. In this case, the uptime of the testing device should be increased to 100 per cent. The changeover time of 5 minutes conforms to the default value determined for the flow lines (Fig. 5.5).

In order to be able to also connect the packaging workplaces vie FIFO line, their changeover times should be halved to 5 minutes for better synchronization. (Fig. 5.5). The packaging time of currently 48 seconds for the two packaging places exactly conforms to the customer takt time of 24 seconds. In order to stay in tune with the flow lines, though, and to gain a little additional time, a decrease in packaging time to 20 seconds per unit is desired (Fig. 5.7).

The upstream production processes may also be connected via FIFO coupling. Taking into consideration the limited uptime of 80 per cent in ultrasonic welding, the achievable cycle time is not 16 seconds, but 21 seconds, though (Fig. 5.7).

The lining process is easiest connected to the remaining value stream via FIFO, if run in two shifts as well. A respective supplementary investment cut down on one worker currently working night shifts. Assuming a performance increase by 10 per cent for the new plant, the cycle time will drop to 19.3 seconds. Together with an increased uptime of 85 per cent, this will results in 22.9 seconds, thus keeping the plant well below customer takt time (Fig. 5.7).

Supplier connection. In contrast to the many different types of fabric which will continue to be procured in the traditional manner, there are only five foam variants, therefore a supplier kanban connection the foam supplier Megafoam suggests itself (design guideline 4). One kanban conforms to one roll of 300 metres. A successful reduction in delivery time to a few days only, would enable a significant inventory reduced from one month to 6 days (Fig. 5.8).

Production control. Production orders generated by the planning program are triggered into the production at the lining process (design guideline 5). The smallest suitable release unit is one lined roll of fabric of 300 m, i.e. 150 electric blankets (design guideline 6). At a customer takt time of 24 seconds, this equals one hour's customer demand. Such small release units, however, are opposed by the high changeover effort in ultrasonic welding. The respective changeover time of 40 minutes is required for changes of formats, cord types, welding patterns and in connection with release units of one hour would lead to an unjustifiably high changeover portion of 40 per cent. For simple changes of fabric, though, the changeover effort drops to 5 minutes. Accordingly, changeover restrictions could be helpful. As the production is mainly done to stock for the next coming season, short-term demand for variants not on stock is not to be expected. The most simple and pragmatic solution of release units of 1,200 pieces is therefore chosen. This conforms to eight rolls of fabric, i.e. the work content of slightly more than one shift and thus also fulfils the requirement of completing one order per shift.

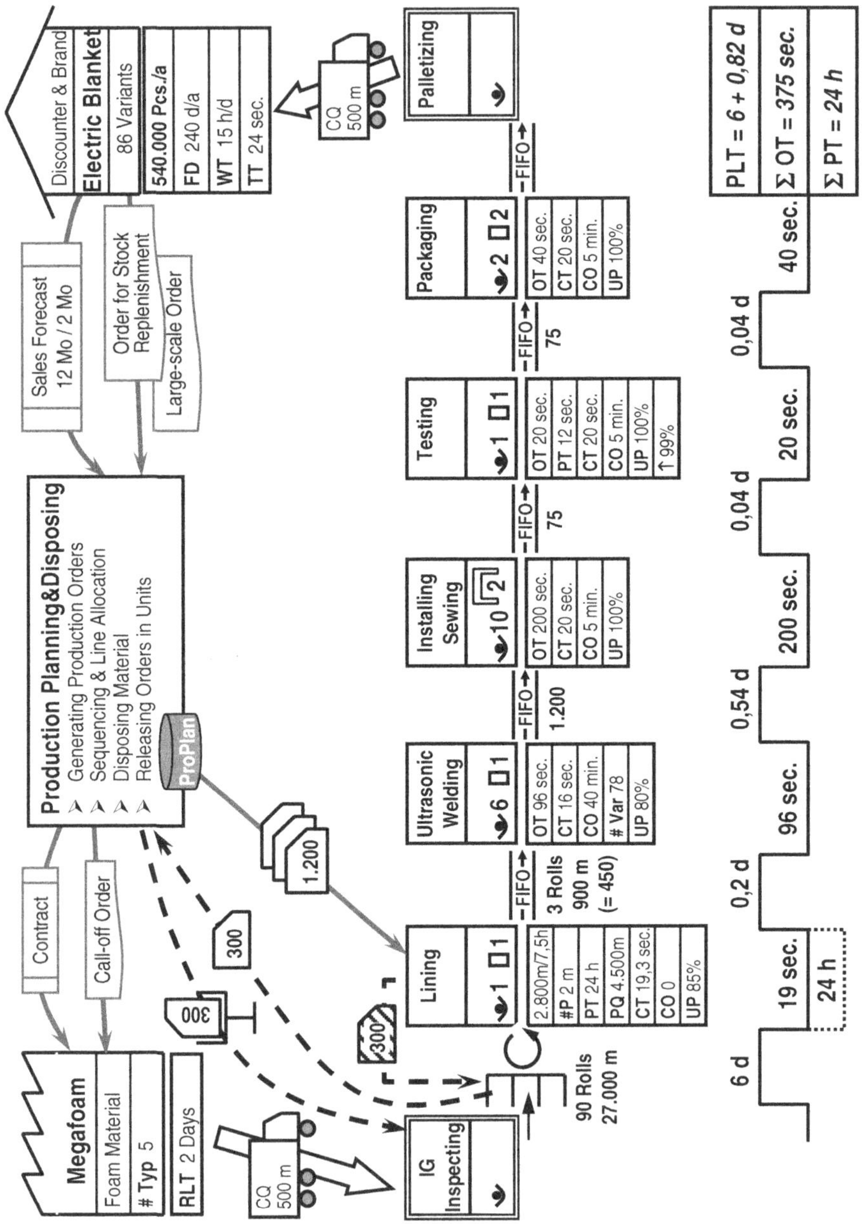

Fig. 5.8 Future-state map for the production of underblankets at Comfort Ltd

Conclusion The value stream analysis confirms and specifies the initial impression of the production. The inventory is particularly disadvantageous in this case because it keeps interrupting the flow of production in several places. The implementation of a flow production and several FIFO couplings decreases the production lead time of originally 6.6 days by almost 90 per cent to 0.82 days. Taking into account the supplier kanban as well, the total lead time is decreased by 80 per cent (Fig. 5.8).

Far more significant, though, is the fact that it was the transparency achieved on the shop floor that laid the foundation for the production optimization in the first place. Initially, a work content reduction by 10 per cent is pursued. Considerably simplified production control makes most production documents obsolete. Operators do not need to complete time tickets anymore, as the line output may now easily be determined by shifts.

5.2 Spring LLC – Lot Sizing in Variant Production

Company

Product. Spring LLC of Boston are in the market for different types of springs, such as compression springs, tension springs, valve springs, yoke springs, torsional springs, and spring assemblies for use with doors, gates and strike plates, in vehicle construction as well as in plant and engine construction.

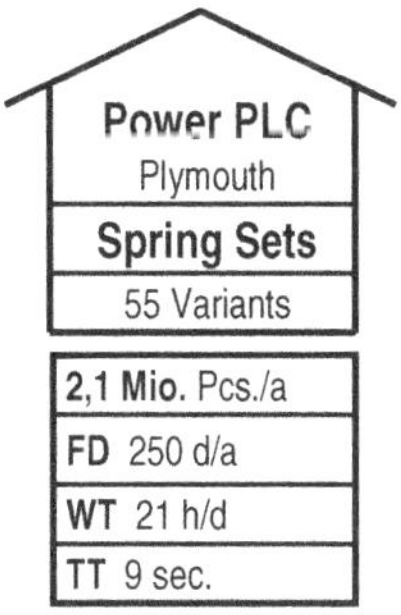

Fig. 5.9 Customer demand at Spring LLC

Customer demand. We shall take a closer look at the product family of straight coupling buffer springs for drive sections in vehicle construction. The buffer springs under observation are made in 55 variants for a major customer, Power PLC in Plymouth (Michigan), at an annual piece number of 2.1 million. The entire production is run in three shifts with one hour's break per shift on 250 days per year. This results in a customer takt time of 9 seconds (Fig. 5.9).

Value Stream Analysis

Production processes. The buffer springs under observation are supplied in the form of spring sets, each comprising an external spring (ES) inserted in an internal spring (IS), equipped with end caps on both ends. The production procedure is as follows (Fig. 5.10):

- **Shipping.** The spring sets are shipped as bulk cargo in skeleton boxes. On average, the provision and loading of one pallet containing 2,000 springs including printing and allocating of the respective delivery documents takes approx. 20 minutes. Receipt and storage of the respective goods supplied by production takes another 10 minutes. The shipping end is also responsible for other product families and works one shift per day with two workers.
- **Testing.** All spring sets are tested for the correct position of the end caps – fitted either to the external or the internal spring, depending on the variant. This takes one worker handling two testing positions 10 seconds. The assembling process supplies the spring sets to be tested on a conveyor belt. For a change of variant, the measuring device must be adjusted, which takes approx. 10 minutes. The testing procedure identifies about 5.5 per cent of rejections.
- **Assembling & jacking.** The internal spring is fitted inside the external spring and the end caps are fitted. Springs and caps are manually supplied to two manual jacking devices. This process manufactures all 55 variants of the respective product family.
- **Annealing & warm jigging.** One worker per continuous furnace places the springs in receptacles on a conveyor belt. One conveyor belt holds 240 springs and takes 30 minutes per run. After annealing, the springs are directly transferred to automatic jigging machines. One line's changeover time is 1.5 hours, the jigging machines' uptimes are 96 per cent. Process volumes and lead times result in a machine takt time of 7.5 seconds, accordingly the operation time for one internal plus one external spring is 15 seconds, i.e. in the case of two lines, the cycle time amounts to 7.5 seconds.
- **Ball peening.** The springs are ball peened in 25 minutes in tubs of 200 external, or 300 internal springs each, respectively. There are two plants with an average uptime of approx. 84 per cent. They are run by one worker also responsible for internal transport. A change of containers including material provision and change of program takes roughly 5 minutes. The partial volumes from the tubs as well as the time needed for ball peening and feeding result in a machine cycle time of 9 seconds for external springs and 6 seconds for internal springs, amounting to a mean value of 7.5 seconds for each plant. As each spring set requires two individual springs and as two plants are being used, this also conforms to the process cycle time.
- **Grinding & deburring.** The narrow ends of the springs are grinded and immediately afterwards deburred by a device directly linked to the grinding machine. The time required for grinding depends on the spring diameter and with 6 seconds is significantly longer for external springs than for internal springs (4 seconds).

- **Annealing.** The springs pass through a continuous furnace on a conveyor belt in 20 minutes. The two continuous furnaces each hold up to 300 springs at a time.
- **Winding.** The production begins with the winding and cutting to length of wire for the springs. Differences as to different wire material and gauge as well as spring diameter and length enable numerous variants. The required winding time per spring mainly depends on the diameter of the wire and is between 5.2 and 12.8 seconds. The average time weighted by piece numbers is 11.5 seconds for external springs and 9.5 seconds for internal springs, resulting in an operation time per spring set of 21 seconds. Changeovers on the grounds of changed spring geometries take 2 hours Depending on the demand for the various spring variants, lot sizes are between 2,000 and 20,000 pieces. Each winding machine is allocated one operator for changeovers, material changes, troubleshooting and removal of finished parts.
- **Unloading.** Wire coils are needed as raw material for the springs; the respective unloading and stockpiling of the coils is handled by the shipping employees. Due to their negligibility, goods receipt-related activities are not specifically recorded herein.

Operation times, changeover times, uptimes and first pass yield are recorded for all production processes (Fig. 5.10). Because of the high changeover times, springs and spring sets are manufactured in large lot sizes. For each article number, an economic lot size is calculated based on the lot size formula by Harris (cf. Eq. 2.4). Minimum volumes are determined for rare variants. This usually results in different lot sizes for each production order. Value stream mapping therefore uses the previous year's lot size ranges and mean values.

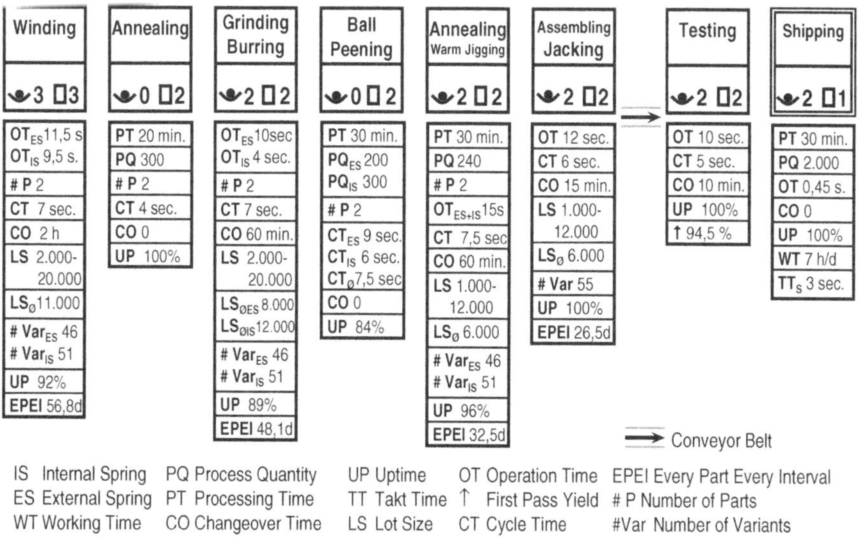

Fig. 5.10 Production processes at Spring LLC

Based on the mean lot sizes, the EPEI value is then calculated for the production processes incl. changeover effort. In assembly, lot sizes of between 1,000 and 12,000 pieces and an operation time of 12 seconds result in a machine loading time of 3.3 to 3.6 hours At a mean lot size of 6,000 pieces, the EPEI value for the 55 variants is calculated in accordance with equation 2.6 as follows:

$$EPEI = \frac{\#Var \times ((LS_{\varnothing} \times OT) + CO)}{\#Res \times UP \times WT} = \frac{55 \times ((6.000 \times 12\,\text{sec.}) + 15\text{min.})}{2 \times 100\% \times 21\,\text{h}} = 26{,}5\,\text{d}$$

with:
- # Var number of variants [Pcs.]
- # Res number of similar resources [Pcs.]
- $LS_{\varnothing}$ average lot size [Pcs.] (5.1)
- OT operation time per piece [time unit]
- CO changeover time [time unit]
- UP uptime [%]
- WT working time [time unit/d]

The processes of warm jigging, grinding and winding handle the springs in lots. As each spring set consists of one external and one internal spring, the arithmetical total number of variants is 110, though multiple use reduces that to a total of 97 pieces, i.e. 46 external and 51 internal springs. Due to the different numbers of variants and the different operation times, some of the EPEI values need to be calculated individually and then added for external and internal springs; accordingly, the EPEI values for the three processes are calculated as follows:

$$\text{EPEI}_{\text{Warm Jigging}} = \frac{97 \times ((6.000 \times 7{,}5\,\text{sec.}) + 1\text{h})}{2 \times 0{,}96 \times 21\,\text{h}} = 32{,}5\,\text{d}$$

$$\text{EPEI}_{\text{Grinding}} = \frac{51 \times ((12.000 \times 4\,\text{sec.}) + 1\text{h})}{2 \times 0{,}89 \times 21\,\text{h}} + \frac{46 \times ((8.000 \times 10\,\text{sec.}) + 1\text{h})}{2 \times 0{,}89 \times 21\,\text{h}} = 48{,}1\,\text{d} \qquad (5.2)$$

$$\text{EPEI}_{\text{Winding}} = \frac{51 \times ((11.000 \times 9{,}5\,\text{sec.}) + 2\text{h})}{3 \times 0{,}92 \times 21\,\text{h}} + \frac{46 \times ((11.000 \times 11{,}5\,\text{sec.}) + 2\text{h})}{3 \times 0{,}92 \times 21\,\text{h}} = 56{,}8\,\text{d}$$

Material flow. For the purpose of value stream mapping, inventories are counted at five locations and depicted by way of storage triangles (Fig. 5.11). These are converted into ranges of coverage on the basis of a customer demand of 8,400 spring sets per day, equalling a daily requirement of 16,800 springs.

- **Finished goods storage.** This contains 20 skeleton boxes complete with 2,000 spring sets each plus 5 skeleton boxes containing 2,000 spring sets ready for dispatch each. Besides, 14 partially filled skeleton boxes are on stock, containing mostly variants of low customer demand. According to the markings on the boxes, these contain a total of 8,800 spring sets. The respective total inventory amounts to 58,000 sets, divided by 8,400 this results in a range of coverage of slightly less than seven days.

- **WIP assembling, jacking & testing.** Due to the large lot sizes, the work in progress also significantly influences the lead time. These three processes are firmly connected by conveyor technique and at the time of value stream mapping were processing a production order lot size of 4,000 spring sets as well as another lot of 6,000 sets. These 10,000 sets equal a range of coverage of 1.2 days.
- **Interim storage.** Behind the annealing process, there are tubs containing a total of 33,600 internal and external springs including work in progress. The various possible end product combinations are irrelevant for the determination of the logistically evaluated coverages. Accordingly, with two springs per product, this results in a range of coverage of two days.
- **Semi-finished goods storage.** The interim storage for ball peened springs contains 400 tubs of 200 external springs each and 226 tubs with 300 internal springs each. The total of 147,800 springs conforms to a coverage of 8.8 days.
- **WIP grinding, deburring & ball peening.** Once grinded, the springs are immediately ball peened in the second process step. In the current-state depiction, one production lot of 8,000 external springs and one of 4,000 internal springs were being processed, equalling a range of coverage of 0.7 days.
- **Interim buffer.** As a rule, the annealed springs are further processed after a short storage period. The respective 23,500 springs in the interim buffer conform to a range of coverage of 1.4 days.
- **WIP winding & annealing.** For reasons of quality assurance, springs may not be stored for long, therefore all winding production orders include the immediately subsequent annealing process as well. The current-state mapping determined one production order of 6,000 external springs and one of 9,000 internal springs being processed, equalling a range of coverage of 0.9 days.
- **Raw material storage of wire coils.** Depending on the respective wire sizes, one coil will weigh between 500 and 1,500 kg. Currently, 74 coils weighing a total of 72.600 kg are on stock. The mean weight of 1,000 spring sets is 500kg. A mark-up of 10 per cent must be assumed for rejections and material loss from change of coils. This results in an inventory equalling 132,000 spring sets conforming to 15 days' coverage.

External logistics. The spring sets are shipped as bulk cargo in skeleton boxes on a daily basis in pallets of 2,000 spring sets each (Fig. 5.11).

The wire coils are provided by two suppliers. Filacier Inc. of Grenville, Québec, Canada annually supply 355 tons in three different qualities in five different measurements; Iron PLC in Pittsburgh, Pennsylvania provide an annual 800 tons in five qualities and eleven measurements. Deliveries are effected Tuesdays and/or Thursdays via carrier. The deliver periods are six weeks, however, three of the five most popular sizes are delivered on-call within a week from Grenville (Fig. 5.11).

Business processes. Spring LLC prognostically plan on the basis of customers' demand forecasts and generate production plans for several production processes accordingly. Semi-finished and finished goods are made to stock according to plan. Order processing is divided into three business processes as follows (Fig. 5.11):

- **Dispatch handling.** Customers submit their orders to Spring LLC via EDI on a weekly basis. Having factually checked correctness and deliverability of the orders automatically created in SAP, the shipping department releases the same and advises the carriers of the respective delivery volumes. Shipping lists and matching delivery documents are generated daily. Goods supplied by production are posted into the finished goods storage; goods provided for shipping are posted out of the finished goods storage. All tasks related to this business process are handled by the shipping staff.
- **Production planning.** Based on the demand forecasts for six weeks, daily updated by the customers via EDI, individual production plans of one week's horizon are generated for the processes of winding, grinding, warm jigging and assembling and also updated on a daily basis. The four production plans are created as Excel spreadsheets with production lot sizes determined manually to be able to take into account planned additions to and removals from storage. The production plans for grinding conform to the plans for winding, but at one day's difference – though depending on the actual availability of parts specific deviations are possible. Annealing and ball peening are not issued with production plans; they are conducted in overlapping lot production with their pertaining upstream processes and are directly controlled by the operators accordingly. They are marked with 'go see' symbols in the value stream depiction.

  In addition, production planning creates withdrawal documents for the coils needed in winding. In winding, the accompanying parts documents and time tickets are also attached to the orders. The released production orders are created in SAP and marked as completed after testing. Production planning is carried out by one employee.
- **Disposing & ordering.** Based on the customers' demand forecasts and Spring LLC's projections, the released production orders and the raw material inventories, SAP creates fortnightly and bimonthly demand forecasts for the suppliers. In addition, the system generates order suggestions which are checked and printed by the material planning staff and forwarded to the suppliers via fax.

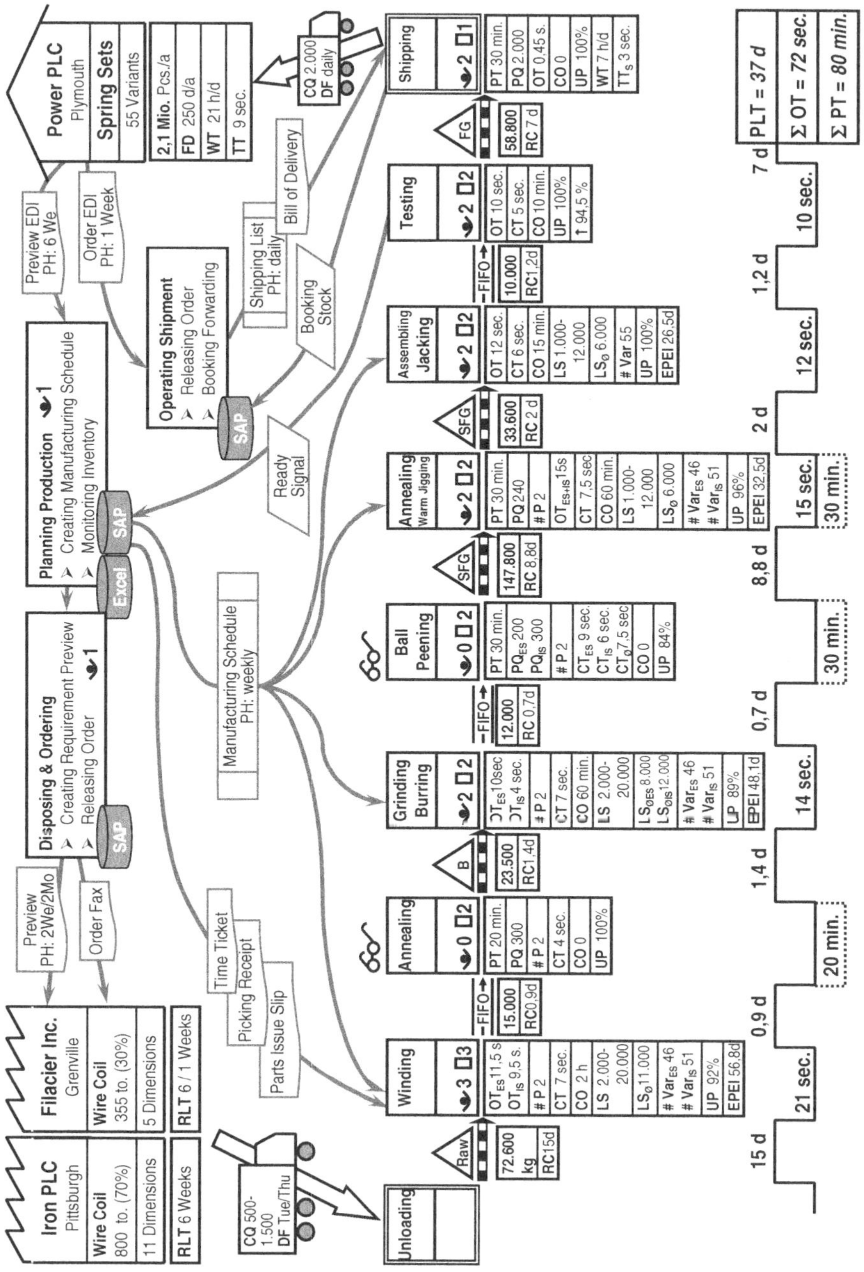

Fig. 5.11 Result of the Spring LLC value stream analysis

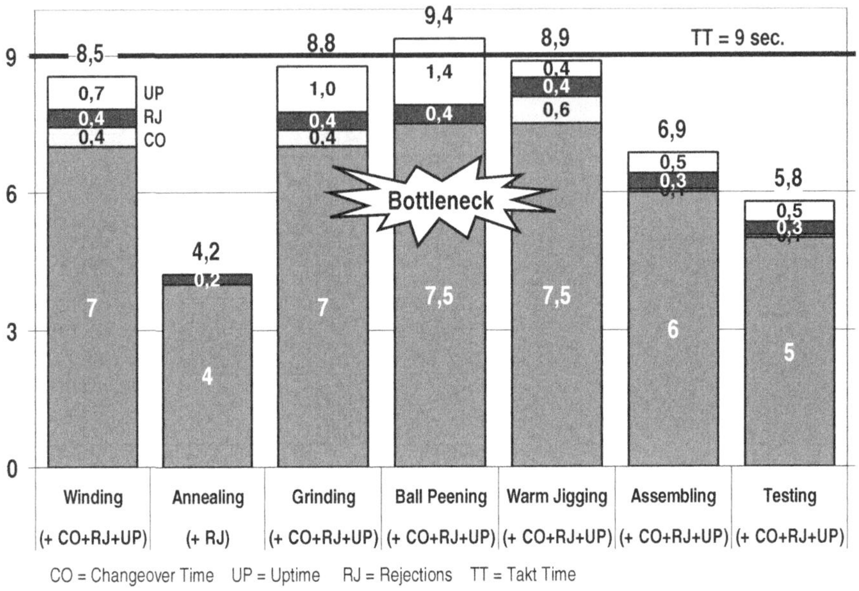

Fig. 5.12 Current-state Spring LLC operator balance chart

Potential for Improvement

Lead time. The total lead time of 37 days is very long, even after deduction of raw material and finished goods the inventory within production (WIP) still amounts to15 days' coverage, i.e. 40 per cent (Fig. 5.11). In comparison, the operation time per spring set of 72 seconds is negligibly small. As all well-established variants are bundled in units of 2,000 each, though, the work content of the smallest reasonable lot size amounts to 40 hours Sequential processing of such minimum lot in an otherwise order-free factory in view of a process time of 80 minutes would complete the lot in just under 42 hours, i.e. two working days – even less in cases of overlapping production. This is the theoretical lead time potential.

Line balancing. The operator balance chart identifies the value stream's bottleneck at an unexpected location: the ball peening process, which is not even a very costly one with need for tight planning. This critical point is not immediately recognizable, as the ball peening cycle time of 7.5 seconds is still clearly lower than the customer takt time of 9 seconds (Fig. 5.12, dark grey section of bar graph). As shown in equation 2.21, though, the limited uptime of merely 84 per cent increases the cycle time by 16 per cent of the customer takt time, i.e. 1.4 seconds (Fig. 5.12, white section of bar graph). Accordingly, the plant takt time almost equals the customer takt time. According to equation 2.7, however, a 5.5 per cent mark-up, or between 0.2 and 0.4 seconds, respectively, must be added (Fig. 5.12, black section of bar graph), slowing down the ball peening process even further by another 0.4 seconds to 9.4 seconds – obviously still not enough to meet the customer demand. How is that possible? Inquiries reveal that a large part of the break times are usually worked through, changing the available working time from a calculated 21 to an actual 24 hours, resulting in a factual customer takt time of 10.3 seconds (cf. Eq. 2.9).

Warm jigging and grinding also come fairly close to the customer takt time. Apart from rejections and breakdowns, the required changeovers also cause time losses (Fig. 5.12, light grey section of bar graph). According to equation 2.20, the respective mark-ups are calculated from the changeover time, the mean lot size and the operation time of one spring or spring set of the respective lot as follows:

$$CT_{CO,\,Warm\,Jigging} = 7{,}5\,\text{sec.} \times \frac{60\,\text{min.}}{6.000 \times 7{,}5\,\text{sec.}} = 7{,}5\,\text{sec.} \times 8\,\% = 0{,}6\,\text{sec.}$$

$$CT_{CO,\,Grinding} = 7\,\text{sec.} \times \frac{2{,}1\,\text{Mio.} \times \left(\frac{1}{8.000} + \frac{1}{12.000}\right) \times 60\,\text{min.}}{2{,}1\,\text{Mio.} \times (10\,\text{sec.} + 4\,\text{sec.})} = 7\,\text{sec.} \times 5{,}36\,\% \approx 0{,}4\,\text{sec.} \tag{5.3}$$

Solely the winding process with its difference of 5.5 per cent between its gross cycle time and the customer takt time shows a viable value which leaves sufficient additional time. Warm jigging could be somewhat defused through changeover reduction; improved uptimes in grinding and ball peening could improve the bottleneck. In the calculation of downtime-related time losses, the limited uptimes of the testing facilities must be taken into consideration even for the fairly fail-safe assembly processes, as both are linked by conveyor technique. As the testing requires less time than assembling, the assembly workers will always incur waiting times. Besides, both processes are too fast for the customer takt time by 25, or 35 per cent, respectively, which results in free shifts and/or uneven distribution of the workload. The annealing furnace is also clearly too big (by factor 2). As the annealing process runs automatically, this does not waste any operator time, but may possibly constitute considerable waste of energy.

Value Stream Design

Goal. The intended redesign of the spring production mainly aims at a significant inventory reduction through better adjusted lot production closer to the customer demand, taking into account weekly demand fluctuations of up to 30 per cent above average. Besides, a higher degree of automation is going to be introduced to cut personnel costs.

Production processes. The manual assembly process of jacking and subsequent testing may be automated as one (design guideline 2). A cycle time of 4 seconds is realistic, i.e. one plant will be sufficient. Special attention must be given to changeover efforts to avoid excessive lot sizes being enforced. A changeover time of 20 minutes, i.e. only slightly longer than the current state, is desired.

With a view to product quality, fast assembly of the ball peened springs is recommended to bring up the first pass yield to 97 per cent. A successful increase in ball peening uptime to 90 per cent would lower the required downtime-related mark-up to 0.9 seconds. In addition to a rejection-related mark-up reduced to 0.2 seconds, the gross cycle time of 8.6 seconds is well within the customer takt time.

Material flow. The value stream is divided into two sections – the production of springs as a semi-finished goods production and the production of the spring sets (Fig. 5.14). The *spring set production* begins with the ball peening process and runs downstream to the shipping end. For better process synchronization, changeover times in warm jigging must be reduced to 20 minutes to match the assembly plant and thus enable a connection of the processes via FIFO coupling (design guideline 3). The two external and internal springs to be assembled must be annealed at the same time. Buffer quantities must be adjusted to enable shuttle transport in tubs between ball peening and warm jigging as well as the subsequent transport to the sorters in assembly. The connection between warm jigging and assembly could equally be done by conveyor link, in which case two continuous furnaces, two automatic jigging machines and one assembling machine would form a unit, thus further reducing the need for interim buffers. In this case, though, downtimes would accumulate and variant-specific fluctuations in operation times could not be buffered anymore, therefore this solution shall be disregarded for now. The FIFO buffer is designed for 0.5 days in this section.

In order to make up for customer demand fluctuations, goods are produced in advance for up to three days, 1.5 days on average. Thanks to the customers' demand forecasts and weekly orders, this is realized with very little deviance – and to be on the safe side, the best selling articles are pre-produced.

The upstream *production of semi-finished goods* comprises the process steps of winding, annealing and grinding as well as deburring of the springs, which are all connected via FIFO coupling. As there are three winding machines, but only two furnaces and two grinding machines, continuous production in overlapping lots is not possible. Instead, at least one complete lot, preferably a small one, must be accumulated behind the winding process before it can be transferred to annealing. With skilful shop floor control, large lots may be continuously produced and small lots slipped in in between. The respective buffer of two days is larger in this section than in the final production.

The *linkage* of the two value stream sections is done via kanban control (design guideline 4). Decoupling of the two sections allows for relatively speedy customer-specific assembly of the spring sets. A large part of the production lead time results from the first, customer-anonymous, section of the value stream. Decoupling via kanban enables larger lot sizes in winding and automatically controls the multiple use of individual spring types in different set variants. Besides, depending on the respective material-specific available coil lengths, deviating lot sizes may be produced in winding without directly affecting any customers' orders. Thanks to kanban logic, the respective surplus and/or short quantities are easily buffered.

Supplier connection. Due to long delivery times and large distances, a supplier kanban does not seem possible at this point in time. Customers' forecasts are particularly suited for low-inventory raw material disposition, which considering the high degree of variance, has already been realized to a large extent for raw material in the current state.

Production control. The trigger point for customers' orders is located at the ball peening process (design guideline 5). The delivery unit of one pallet of 2,000 spring sets is the perfect size for a release unit (design guideline 6). This conforms to a customer demand of 5 hours The weekly customer orders may be evenly intermixed with the aid of a load levelling box within the 5-day release horizon (design guideline 7). There are 10 exotic variants with annual piece numbers between 160 and 2,200 pieces, totalling 0.7 per cent. In these rare cases, it makes sense to ignore uniform release units and conduct the production in a customer-specific manner – beginning with the winding of the individual exotic springs.

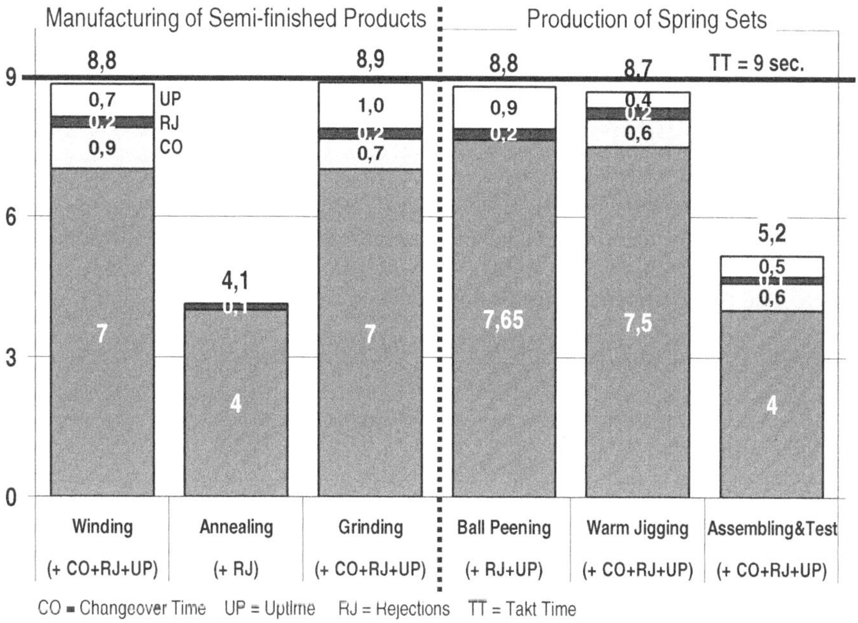

Fig. 5.13 Future-state Spring LLC operator balance chart

The uniform lot size of 2,000 pieces in ball peening does not match the internal springs' container size of 300 pieces. The internal springs required per lot will thus be provided in 7 containers of 286 pieces each in future. This increases the cycle time per internal spring from 6 to 6.3 seconds, the mean cycle time changes to 7.65 seconds. Taking into consideration the mark-ups for downtimes and rejections, the gross cycle time for ball peening thus rises to 8.8 seconds (Fig. 5.13).

The individual springs are re-produced by way of signal kanban, sometimes in larger lot sizes. An ABC analysis determines 12 A springs, 19 B springs and 66 C springs. Closer inspection reveals 34 of the C-type springs to amount to a mere 2 per cent of the total piece number. With an annual demand of between 160 and 5,400 pieces per variant, they form a D-class of exotic parts of their own, which are best produced according to customers' specifications in the required numbers. Due to their small numbers, their deviation from the uniform release units and triggering at the levelling box hardly affect the production flow at all.

Depending on the respective class, kanban quantities are allocated the factors 4, 2, or 1 and are thus calculated at 8,000; 4,000; or 2,000 pieces. To assess the mean supermarket store inventories, the total number of circulating kanban is assumed to conform to exactly one signal quantity per variant. This applies if signal quantities per variant are larger than the maximum stock, depending of the maximum daily demand and the replenishment lead time as per equation 3.9. At a production lead time of 2 days starting from the winding process, the maximum demand for 2 days of the A variant with the highest piece numbers must be less than 8,000 pieces. According to equation 3.7, the supermarket store's range of coverage is then calculated as follows:

$$RC = \frac{1}{2} \times \frac{(12 \times 4 + 19 \times 2 + 31) \times CQ}{DD} = \frac{1}{2} \times \frac{117 \times 2.000}{16.800} = 7\,\mathrm{d} \tag{5.4}$$

with: RC supermarkets range of coverage [d]

CQ container quantity [Pcs.]

DD daily demand [Pcs./d]

With the aid of the determined kanban quantities, the EPEI values for winding and grinding (disregarding exotic springs) may be calculated as per equation 2.8. For 31 external plus 31 internal springs this results in:

$$EPEI_{Winding} = \frac{(117 \times 2.000 \times 10{,}5\,\mathrm{sec.}) + 62 \times 120\,\mathrm{min.}}{3 \times 92\% \times 21\,\mathrm{h}} = 13{,}9\,\mathrm{d}$$

$$EPEI_{Grinding} = \frac{(117 \times 2.000 \times 7\,\mathrm{sec.}) + 62 \times 60\,\mathrm{min.}}{2 \times 89\% \times 21\,\mathrm{h}} = 13{,}8\,\mathrm{d} \tag{5.5}$$

For the determination of changeover mark-ups in winding and grinding, the different lot sizes as per ABC class need to be taken into account and must be weighted in accordance with the partial piece numbers of the respective classes:

$$\begin{aligned} CT_{CO,\,Grinding} &= CT_{net} \times \frac{CO}{LS_{ABC-class} \times \frac{OT_{IS} + OT_{ES}}{2}} \\ &= 7\,\mathrm{sec.} \times \left(\frac{60\,\mathrm{min.} \times 70\%}{8.000 \times 7\,\mathrm{sec.}} + \frac{60\,\mathrm{min.} \times 20\%}{4.000 \times 7\,\mathrm{sec.}} + \frac{60\,\mathrm{min.} \times 10\%}{2.000 \times 7\,\mathrm{sec.}} \right) \\ &= 7\,\mathrm{sec.} \times 9{,}64\,\% \approx 0{,}7\,\mathrm{sec.} \end{aligned}$$

$$\begin{aligned} CT_{CO,\,Winding} &= 7\,\mathrm{sec.} \times \left(\frac{120\,\mathrm{min.} \times 70\%}{8.000 \times 10{,}5\,\mathrm{sec.}} + \frac{120\,\mathrm{min.} \times 20\%}{4.000 \times 10{,}5\,\mathrm{sec.}} + \frac{120\,\mathrm{min.} \times 10\%}{2.000 \times 10{,}5\,\mathrm{sec.}} \right) \\ &= 7\,\mathrm{sec.} \times 12{,}86\,\% = 0{,}9\,\mathrm{sec.} \end{aligned} \tag{5.6}$$

with: CT_{net} cycle time [time unit]

CT_{CO} time losses by changeover [time unit]

CO changeover time [time unit]

OT operation time (internal and external spring) [time unit]

LS lot size [time unit]

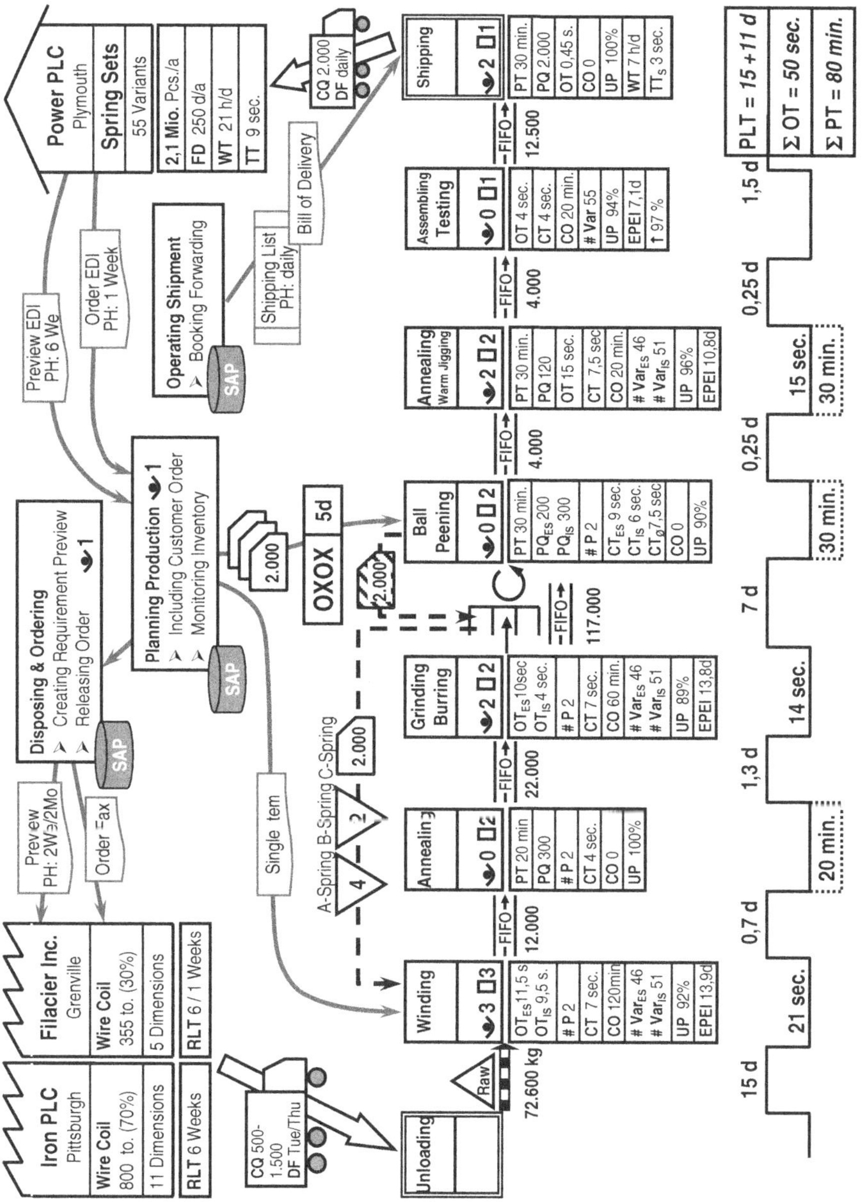

Fig. 5.14 Future-state map for the production of coupling buffer springs at Spring LLC

Conclusion Disregarding raw material, the overall process has a production lead time of 11 days (Fig. 5.14). The work in progress has accordingly been cut in half in comparison with the current state. More than half the inventory is now in a defined semi-finished goods store, while the remaining work in progress results mainly from lots currently being processed. Thanks to the separation of the value stream into two sections, the production process was not only made faster, but is now also more standardized and more transparent.

However, further potential lead time reduction and flexibility increase is hindered mainly by the winding process. Its large changeover portion enforces excessive lot sizes and comparatively long operation times per spring, thus preventing continuous, synchronized lines. Instead, three production lots from winding always need to be harmonized with two of the production.

5.3 Glass PLC – Short Delivery Times in Single Piece Production

The Company

Product. Besides the business segments of glass production and car glass, Glass PLC are also in the market for construction glass. In this field, three product groups may be differentiated, namely, toughened safety glass (TSG), laminated safety glass (LSG) and insulating glass (ISG) for windows. This business field of Glass PLC includes 11 factories in the United Kingdom with partially overlapping product portfolios and a highly region-specific customer structure. Among other things, the Bridgend factory under observation herein produce individual safety glass panels with customer-specific measurements (thickness, geometry, bore holes) for utilization in construction (exterior glazing of buildings, doors, partition walls). These TSG panels also differ with a view to the type of glass used (tint, ornaments).

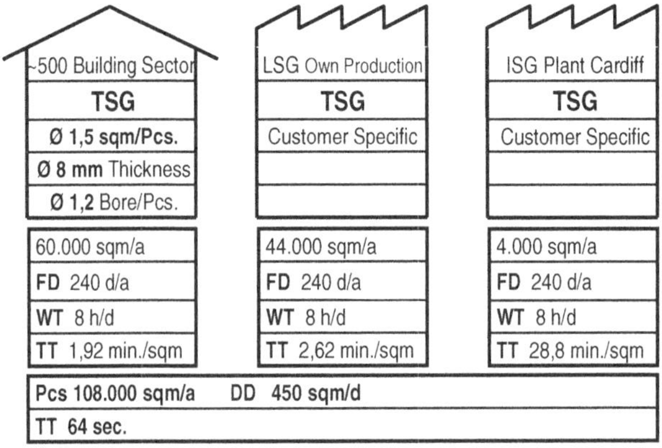

Fig. 5.15 Customer demand at Glass PLC, Bridgend

Customer demand. About 55 per cent of the TSG glass produced at the Bridgend works is delivered to approx. 500, mainly regional, customers and construction sites. Roughly 40 per cent of the production volume are used for the production of laminated glass at the same factory, the rest is needed for the production of isolated glass at the Cardiff works. The annual piece numbers comprise 72,000 glass panels of different types of glass in customer-specific sizes of between 0.1 and 10 sqm. The different sizes required clearly illustrate how in plate glass production the number of glass panels is irrelevant for the determination of the customer demand – instead, annual production volumes are stated in square metres (sqm). The annual output at the Bridgend factory amounts to 108,000 sqm, this corresponds to an average panel size of 1.5 sqm (Fig. 5.15). In single-shift operation, this results in a customer takt time of 64 sec./sqm.

Value Stream Analysis

Production processes. Glass machining is done in job shop production. The panels are cut to size and processed in various steps on alternative machines and finally tempered in a furnace. A few isolated panels are equipped with screen-print before tempering, which shall be disregarded herein, though. The detailed process is as follows (Fig. 5.16):

- **Picking & packaging.** The TSG panels supplied on carts by production are transferred onto holding devices by two workers in commission-appropriate manner and keeping in mind the sequence in which the customers will be visited along the shipping route. The panels are divided by spacers, partially shrink-wrapped, or protected by cardboard. As the panels intended for internal LSG production do not need to be dispatched, only 59 per cent of the total production need to be packed. Division by the customer takt of 64 seconds results in a process-specific customer takt of just over 108 seconds.
- **Tempering.** Heating and subsequent cooling by fan turns glass into safety glass. The glass panels supplied on roll carts are arranged on a roller conveyor by two workers and furnished with the respective stamps. The furnace automatically pulls in the panels and releases them on the other side after completion of the heating process, the length of which depends on the thickness of the glass in process. The same workers visually control and stack the panels. When changing to thinner glass panels, the controlled temperature decrease requires idle runs, visual controls sometimes reveal scratches, from time to time glass breaks – the resulting first pass yield is only 95 per cent. There are no statistical results on uptimes. The tempering furnace runs in two shifts; accordingly, the process-specific customer takt time changes from 64 seconds to 128 seconds.

  The furnace processing time depends on the thickness of the panels in process and amounts to 54 sec/mm. The determination of the cycle time therefore requires a detailed analysis of the previous business year's data. The analysis indicates a weighted mean thickness of 8 mm; glass thicknesses varied between

4 and 16 mm. Accordingly, the average process time of the furnace is 7.2 minutes. Eight square metres of furnace space are available per run. The mean furnace capacity utilization is 45 per cent, i.e. 3.6 sqm per run. Accordingly, 3.6 sqm are processed in 7.2 minutes, which results in a cycle time of two minutes. With two workers, this conforms to an operation time of 240 sec./sqm.

- **Washing.** The glass panels covered with glass dust are fed into the continuous washer by the upstream processing plants on roll conveyors. Once clean and dry, the panels are stacked and placed before the furnace. One worker needs 33 seconds per glass panel, which at an average size of 1.5 sqm results in an operation time of 22 seconds per panel. It takes approx. two minutes to adjust the washer to the glass thickness.
- **Drilling / milling.** Bore holes and cut-outs in the glass panels are made according to drawings. Available are one manual drilling machine with marking-out table and one NC-controlled milling and drilling machine to be programmed by the worker. Manual drilling takes 120 seconds per bore hole, automatic drilling takes 0 seconds. The panels are equipped with up to 16 counter-sunk bore holes. Due to the different numbers of required bore holes and the cut-outs to be milled from time to time, operation times fluctuate immensely. A precise evaluation of data results in 1.2 bore hole per delivered glass panel, i.e. 0.8 bore holes per sqm. Multiplied with the operation time per bore hole this results in cycle times of 24, or 96 seconds, respectively – ignoring the time spent on milling. Changing of drills takes 2 minutes.
- **Grinding.** Edge processing of medium-sized, rectangular panels is done on a double-sided continuous grinding machine and takes approx. 100 seconds per sqm. For complete processing, the panels are returned, causing minimal changeover times for the change of width. Larger and/or irregular glass panels are processes vertically on a single-sided grinder. This one-sided process is decidedly slower, but as it is used for larger panels, the resulting operation time of 150 seconds per sqm is only minimally longer. As there are two one-sided grinders, the actual cycle time is even shorter.
- **Cutting.** Raw glass sheets of 19.26 sqm are automatically scribed on a cutting table and then broken and stacked by two workers. The raw glass sheets are supplied by a storage worker. The machine is controlled by the foreman, who, similar to other managers, is not included in the value stream depiction. Scrap optimization enables a first pass yield of 80 per cent per raw glass sheet, 7.5 minutes are required, i.e. 23 seconds per sqm; allowing for scrap, the cycle time amounts to 29 seconds. The minimal portion of patterned glass and/or offcuts may be processed on a manual cutting table. The cut panels are equipped with adhesive labels containing all customer order data.

The employee numbers shown refer to the number of workers required to operate the respective machines. However, the production comprises more workplaces than operators, as these are allocated to different workplaces in accordance with the respective daily fluctuating product range requirements.

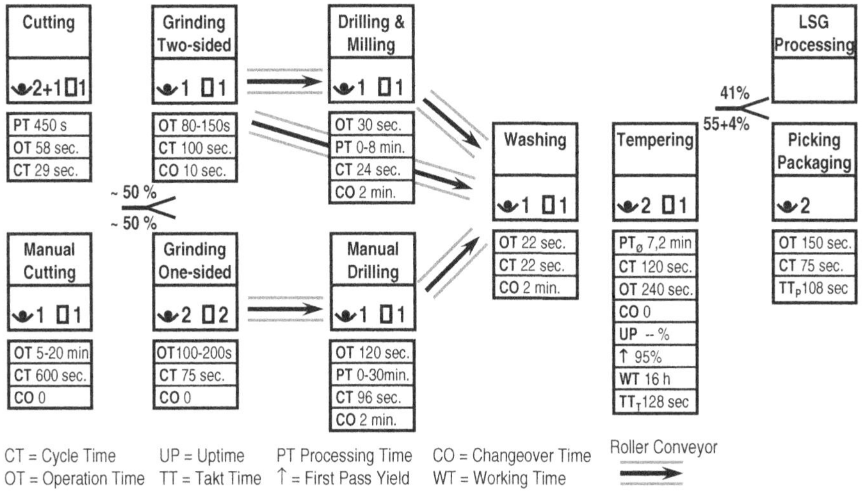

Fig. 5.16 Production processes at Glass PLC

Material flow. Some machines are directly connected by way of roll conveyors for easier movement of the glass panels (cf. Fig. 5.16). All other transport is conducted on internally used panel holders on wheels which can hold a large number of glass panels, but need to be pushed by two workers due to size and weight (Fig. 5.19, left). The counting of inventories seems impossible at first, because not the amounts of the differently sized and shaped panels on the holders must be determined, but their total surface areas. The adhesive labels, however, already state the respective surface areas, so that simple addition determines the inventory. The results are depicted in figure 5.17.

The inventory of goods already packed for shipping within the next few days was determined as 790 sqm; this also includes the supplies for the ISG factory. In addition, 425 sqm are made available for subsequent LSG production. At a daily demand of 450 sqm, this results in a range of coverage of finished goods of 2.7 days. Another 135 sqm are stacked behind the furnace, corresponding to a range of coverage of 0.3 days.

An expected 5 per cent of coverage-reducing scrap need to be subtracted from the 377 sqm before the furnace, though, resulting in 358 sqm of the final product, i.e. a range of coverage of 0.8 days. Dotted around the production and behind the cutting process are roll carts with another 811 sqm; assuming a 5 per cent scrap rate, this corresponds to 1.7 days' coverage (Fig. 5.16).

External logistics. 770 raw glass sheets are counted in the glass store, equalling 14,830 sqm; at a first pass yield of 80 per cent and a 5 per cent scrap rate, this corresponds to 25 days' coverage. Raw glass sheets can be supplied daily from the company-owned Liverpool plant. However, individual sheets are subject to price mark-ups of 10 per cent, while discounts are granted for purchases of several parcels of five tons each. According to an ABC analysis, 75 per cent of the total glass are used for nine variants in two glass types and six thicknesses. 15 per cent comprise few special glass types for major construction projects. Standard glass types of different thicknesses as well as approx. 40 variants of coloured and coated glass account for another 8 per cent, while 2 per cent consist of embossed cast glass supplied in half the normal size. 78 variants are currently on stock in small volumes.

Delivery is effected by own trucks via different fixed daily routes. The shipping department generates delivery documents and delivery lists, listing the respective orders sorted by routes, drivers and customer addresses. The shipping list specifies the order of placement of the panels on the panel holders. As many customers order individual panels, there cannot be an individual panel holder for each customer; instead, they are stacked and unloaded by individual customers' orders. The shipping list determines in which order the driver is to call at the customers' locations.

Business processes. The sales department receives the customers' orders and creates order files with the customers' requirements. Order processing records the orders with the aid of the trade-specific corporate 'GLAP' software. Following this technical clearance, the availability of special types of glass is verified, the order is confirmed with delivery date and finally the drawing is approved by the customer. Once the order has been dispatched, operations scheduling may include the order in the planning. As the software permits capacitive planning at machine level, all processing steps may be allocated to the respective machines. The orders are planned for the process steps of cutting, processing, tempering and picking with transition times of one day each. The planning result generates to-do-lists for each machine containing the customer orders to be processed every day.

Besides the to-do-lists, production receives work orders and adhesive labels with customer data; the latter are attached to the glass panels for identification. The cutting process receives the so-called 'jobs' from the G/Opt program, scrap-optimized cutting plans for one daily demand each. The tempering furnace workers release ready signals and thus trigger dispatch planning. For panels rejected in the quality control, re-cutting orders are immediately submitted to the cutting process by the foreman, where these are either manually slotted into the automatically generated cutting plans or cut from cutoffs. The recuts are supplied on yellow trolleys and are thus clearly marked for urgent processing.

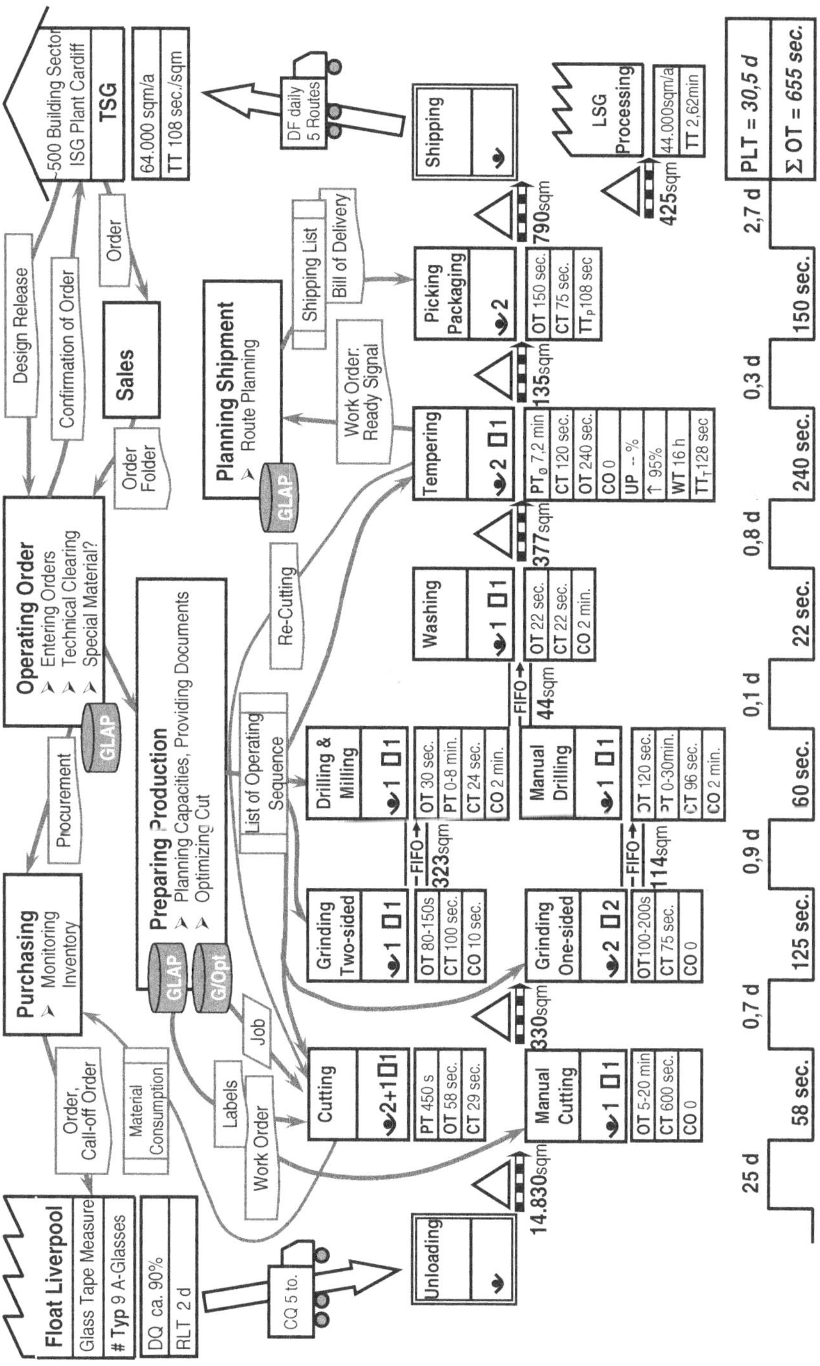

Fig. 5.17 Result of the Glass PLC value stream analysis

Potential for Improvement

Lead time. Production planning with transition times causes long product lead times and long delivery periods. It also results in high material volumes in the production, which in turn may cause extra effort in searching for certain orders. To-do-lists determine the daily workloads, but no processing sequences. The furnace workers sort the glass panels by thickness; while in processing the panels need to be sorted by size – as a result of the repeated sorting; accordingly, many orders remain in the production much longer than originally planned, while others are brought forward. All in all, the value stream analysis identified a lead time of 5.5 days plus a very high raw material inventory of 25 days – at an accumulated operation time of only 11 minutes (Fig. 5.17). However, the defined goal of a delivery time of only one day even for customer-specific glass panels made of standard glass seems by all means realistic in spite of the currently prevailing circumstances.

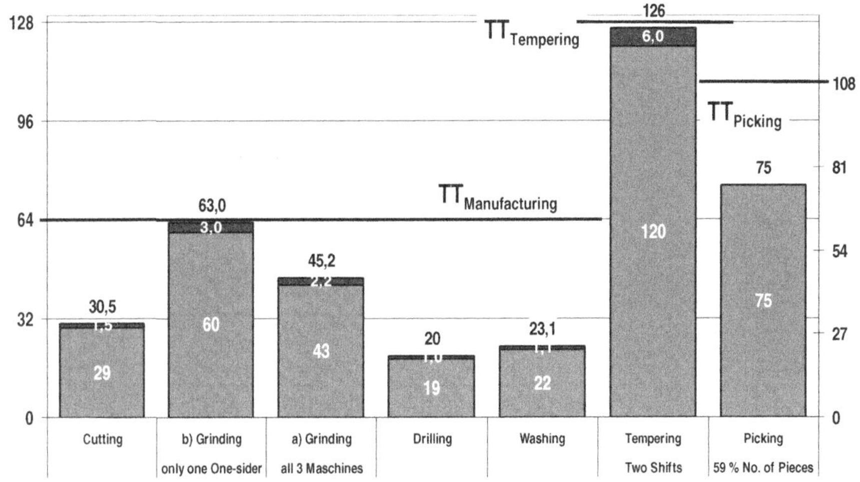

Fig. 5.18 Glass PLC operator balance chart

Line balancing. The operator balance chart clearly identifies the furnace as a bottleneck – in spite of two-shift operation and disregarding its limited uptime (Fig. 5.18). With a cycle time of 120 seconds and a scrap-related time loss of 6 seconds, the difference to the customer takt time is a mere 2 seconds – not enough to cushion breakdown-related downtimes.

With a cycle time of 19 seconds – calculated by adding the reciprocal values of both plants' cycle times of 24, or 96 seconds, respectively – drilling shows significant capacity reserves – which are needed for the very time-consuming milling processes required for some panels.

With a total cycle time of 43 seconds, the grinding process shows the considerable capacitive flexibility required for such customer-specific productions of highly fluctuating operation times of the respective order mix. Shutting down of one of the two one-sided grinders would result in 63 seconds incl. scrap, almost exactly conforming to the customer takt time, but then the order mix would always have to be specially aligned with the respective capacitive grinding requirements. Accordingly, higher machine capacities with flexibly assignable workers are preferable.

With its 29 seconds of cycle time, the cutting process (the manual cutting table may be disregarded herein) is almost twice as fast as the customer takt time of 64 seconds. The release of one day's order volume regularly leads to large inventories before processing – accompanied by lack of space and an unnecessarily clogged-up shop floor.

Value Stream Design

Goal. The redesign of the production primarily aims at radical cuts in production lead time by more than 80 per cent to one day. The realization of *delivery periods of one day only* for part of the customer orders is the defined goal.

Production processes. Different process steps place different requirements on their respective optimal order sequences.

- Only glass panels of identical thickness may be tempered together in the furnace. Besides, it is not recommendable to change glass thicknesses too often. Due to the different temperatures required, different thicknesses cause waiting times, in particular with large differences and/or when changing to extremely thin panels.
- The time-consuming milling portions in drilling should be spread evenly across the day.
- In edge processing, glass panels of similar geometries should be processed together in lots to minimize width adjustments.
- In the cutting process, identical glass types must be processed together. Customer orders comprising different types of glass therefore need to be split up.

These different order sequence requirements clearly oppose an integration of the production processes into a flow production (cf. design guideline 2), as consistent line balancing across variants seems impossible.

Material flow. In order to achieve an evenly balanced material flow, though, a fixed order sequence is required, which may not be changed with every other production process. For this purpose, the three process steps of cutting, processing and tempering are largely synchronized; the production is thus changed to a two-shift schedule in line with the furnace. In processing, this reduces the number of workers per shift; the remaining workers flexibly operate the machines as required. The cutting output, however, needs to be drastically decreased to approx. one quarter of the current yield. This can only be achieved by one cutting worker manually supplying additional glass sheets from the storage, while another one takes over machine control based on scrap optimization. In addition, the LSG plant must be directly supplied with TSG glass as well; so far this was always done at the end of the shift – instead, both areas will in future be alternately supplied by the same workers.

Fig. 5.19 Robust and manageable glass panel holders on wheels

The roughly synchronized production processes are coupled via FIFO (design guideline 3). Processing simply uses the existing roll conveyors, partially slightly modified for better material flow. Between the three process steps, roll carts are used small enough though to hold only cut panels and thus be able to be moved by one worker alone (Fig. 5.19, right hand picture).

Supplier connection. An analysis of the raw material inventories showed a large part of the raw material demand comprising nine variants, highly suited for implementation of a supplier kanban (design guideline 4).

Production control. As the production is entirely carried out in accordance to specifications, the trigger point is perforce located at the first process, i.e. the cutting process (design guideline 5). The default raw glass sheet size suggests itself as the perfect release unit (design guideline 6). In cutting, this always corresponds to identical work content, in tempering, however, the work content is highly dependent on the respective glass thickness. A buffer before the furnace serves to balance output fluctuations between cutting and tempering not regarded in the determination of the release units.

The raw material is the decisive factor for the determination of the optimal order sequence for a balanced variant mix in a customer-specific production. Accordingly, the different raw glass sheets must be spread as evenly as possible (design guideline 7). However, in view of the numerous process-specific downstream restrictions, this would not be wise after all. As the furnace constitutes the capacitive bottleneck, the output of which is further reduced as a result of idle runs required in connection with frequent changes of glass types, its restrictions have utmost priority with a view to the sequence determination (design guideline 8). Accordingly, the panels should be produced according to thickness, ideally in daily cycles starting with 4 mm and gradually increasing to 19 mm, thus eliminating all idle runs. Customer orders comprising different thicknesses are consolidated in commissioning.

Above described sequencing rule needs to be adjusted to the specific requirements pertaining to the customer groups to be supplied. The following five business types are differentiated:

- '1-day TSG' consist of short-term deliveries of small amounts of glass panels for regional craftsmen and construction sites.
- 'Scheduled TSG' comprise regular orders from customers generally supplied on a weekly basis, delivery periods depend on the respective shipping plans and are between four and eight working days.
- 'Project TSG' include orders from major construction sites with respective long delivery times as well as numerous deliveries effected on fixed dates per order.
- 'ISG TSG' concern the daily deliveries to the Cardiff works, currently at an agreed delivery period of three days.
- 'LSG TSG' is scheduled for internal further processing to be supplied to the LSG production with a transition time of two days.

All business types are subject to a production lead time of one day – a decided speedup, also with a view to LSG production. With respect to the '1-day TSG', however, there is now an additional requirement of having to complete not only the production within one day but also all order processing, packaging, and express delivery to the customers. Orders for 1-day TSG are permitted until 09.00 hrs. at the latest, allowing order processing, work preparation and cutting optimization until 10.00 hrs. For optimized cutting, orders for panels from other order types are supplemented to best utilize the default raw glass sheets. Cutting of the '1-day TSG' panels begins at 10.00 hrs., with one raw glass sheet being processes roughly every 15 minutes due to the meanwhile slowed-down cutting process. The subsequent production procedure largely depends on the required operation times; besides, the buffer before the furnace must be taken into consideration. The first 'fast' panels should leave the furnace between 14.00 and 15.00 hrs.. The remaining panels as well as commissioning and packaging are completed by the end of the second shift, i.e. 22.00 hrs.. As of 06.00 hrs. the following morning, the panels are loaded and delivered.

The '1-day TSG', however, requires a minimum of two daily cycles: one for the production of 'fast panels', another one for all other business types. The customer demand for '1-day TSG' will most likely fluctuate greatly. These demand fluctuations may be smoothed with the aid of the other business types, for instance by bringing forward of 'project TSG', by utilizing the VSG transition time to make up for delays, or by postponing some of the ISG glass within its allocated three-day timeframe where necessary. The GLAP planning software is very helpful in this respect. Impossible to plan and only marginally controllable, however, remains the required re-cutting. For 1-day TSG, this may be done during the second cycle. Rework cut from thick glass, though, will only just be finished before the end of the second shift, possibly requiring overtime for packaging.

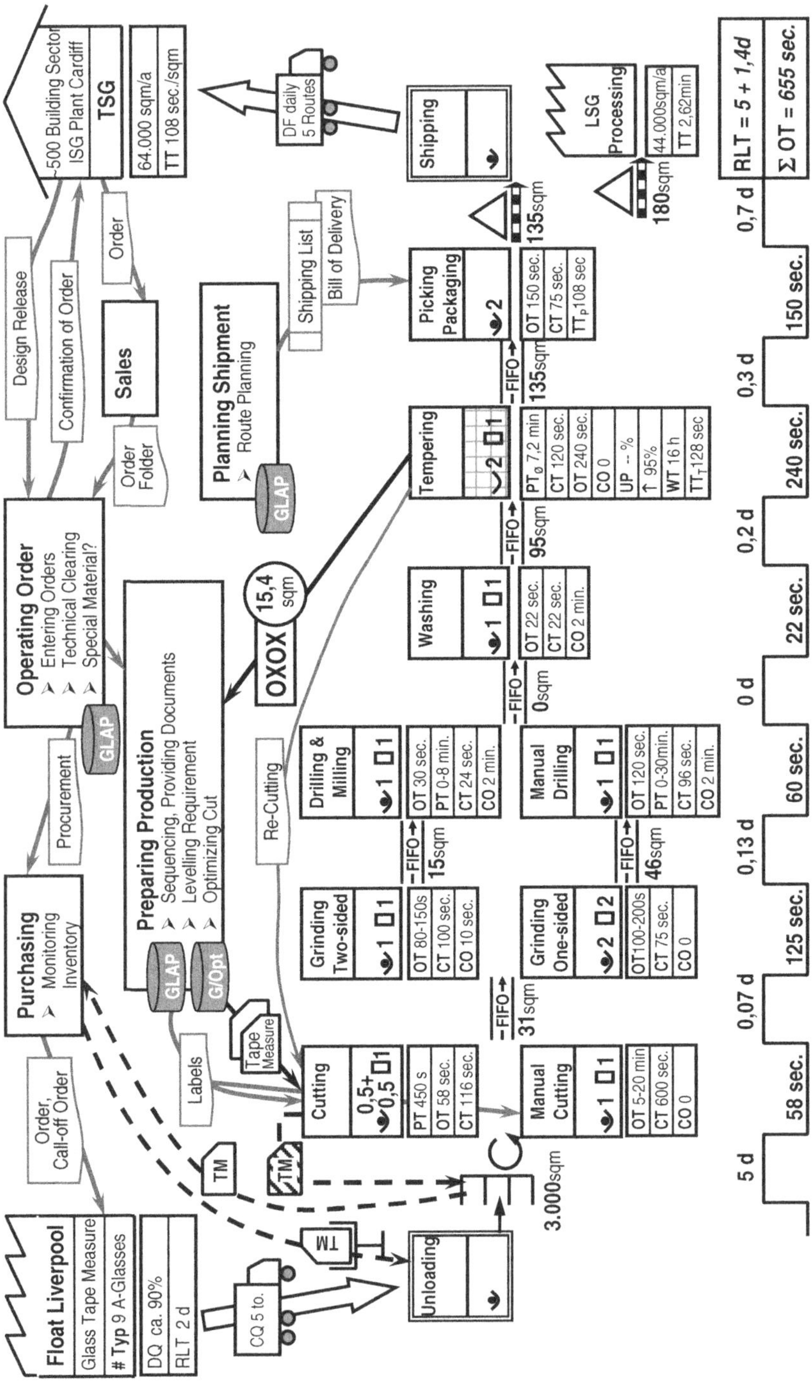

Fig. 5.20 Future-state map for TSG production at Glass PLC

Conclusion The value stream analysis illustrates the essential setting parameters and restrictions for order dispatch and production control, in particular with a view to the highly complex correlations inherent in a customer-specific production. At the Bridgend works, the furnace turned out to be a capacitive and restrictive bottleneck, the output of which may be increased by expedient sequencing in order dispatch.

Changes in production control resulted in a decrease in production lead time by 75 per cent from 5.5 days to 1.4 days (Fig. 5.20). Disregarding the finished goods provided for imminent dispatch according to plan, the production lead time amounts to only 0.7 days, so that with a two-shift schedule, one day's lead time will be feasible in future. The pertaining time decrease also results in a shortened lead time by almost one week with respect to the internal VSG glass production.

5.4 The Seco ULC – Flow Orientation by Technology

The Company

Product. Seco ULC, located in Toronto (Canada), produce and distribute a large product range of bonded abrasive products with a focus on disc-shaped products such as cutting discs, grinding discs, flap discs and diamond cutting discs. Just over one third of the products are sold as own brand via specialty shops in Canada and United States. The discs are also supplied to hardware stores under their respective names. Besides, many manufacturers of grinding machines like angle grinders have their initial equipment produced under their own product names. These so called 'private labels' account for 25 per cent of the production, one third of the production is exported – half of this to the USA. All in all, more than 2,000 customers are supplied.

Customer demand. We shall take a closer look at the product family of cutting and grinding discs. The discs differ with a view to geometrical dimensions and abrasives. 45 different formulas are in use for the latter, mainly consisting of abrasive grains as active media, resin-based bonding materials, and fillers. Disc sizes are characterized by nine different outer diameters and eleven different thicknesses. In addition, there are two different standard bore diameters and five rarer ones. Besides, the discs may have straight or depressed centre bore holes. Further variance is created by different brand names and different packaging formats, different combinations of which would allow thousands of variants, though only comparatively few are in practical use.

The actually manufactured number of variants is shown by the number of adhesive labels used for 300 items each, indicating product denomination, decisive characteristics (in accordance with respective compositions) and instructions of use for each variant. In addition, there are various forms of possible packaging – such as for instance customer-specific cardboard packaging containing between 5 and 20 discs in airtight cans or shrink-wrapping.

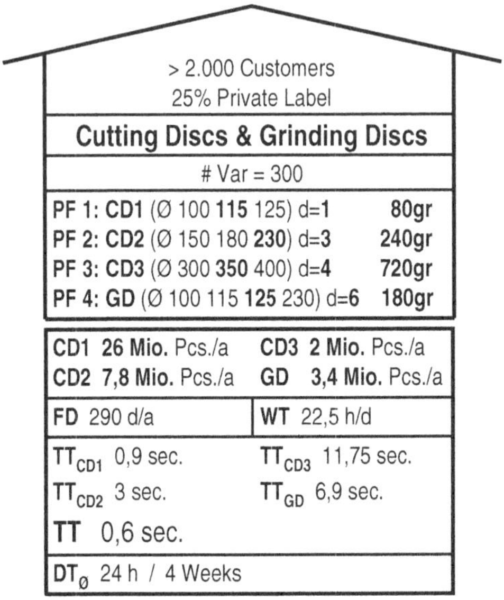

Fig. 5.21 Customer demand at Seco ULC

As far as the production is concerned, the essential differences between the different types of cutting and grinding discs are size-related, as most operation times grow considerably with increasing diameters and thicknesses. Cutting discs (CD) are therefore divided into three product families by diameter. As thicknesses correlate with diameters, they may be disregarded in the creation of product families. Grinding discs (GD) are in a product family of their own, as they are significantly thicker than cutting discs. Each product family has one particular thickness/diameter combination, comprising the largest part of the total piece number and is thus used as the representative. The first product family CD1 (diameter 115 mm, thickness 1 mm) is produced at an annual piece number of 26 million; the average weight per discs is 80 g. The annual piece number of product family CD2 (diameter 230 mm, thickness 3 mm) is 7.8 million at an average weight of 240 g each. Product family CD3 (diameter 350 mm, thickness 4 mm) has an annual output of 2 million pieces and an average weight of 720 g per disc, while 3.4 million grinding wheels (diameter 125 mm, thickness 6 mm) of 180 g each are produced per year (Fig. 5.21).

The production runs three shifts per day, six days per week, with half an hour's break per shift. Considering public holidays and shutdowns, there are 290 working days in a year, resulting in a total daily demand of 135,172 discs and a customer takt time for all discs of 0.6 seconds (Fig. 5.21). The products are available from stock, though bulk orders are usually subject to a delivery period of four weeks.

Value Stream Analysis

Production processes. The production of cutting and grinding discs is spread over two locations. One produces the bonding matrix which is later pressed into discs at the main factory. These raw discs are then cured in a thermal process including several additional tasks, and finally packed. The detailed production process is as follows (Fig. 5.22):

- **Shipping.** The finished goods are stored on pallets in high rack storage, from where they are removed in accordance with picking lists, commissioned at four workplaces and equipped with delivery documents. About 70 per cent are directly provided for shipping via carrier, the rest includes small orders of a total weight of less than 100 kg. These are transferred to three packaging stations, where they are parcel-packed and franked for postal delivery. As shipping will not be investigated in further detail herein, no times are recorded.
- **Packaging: Canning and/or shrink-wrapping.** Small and medium-sized discs are either packed in metal tin cans of 10 discs each or shrink-wrapped in stacks of 25. Approximately 25 per cent are canned.

  Two plants with one operator each are available for the canning process. Every 20 seconds a can is closed. The division by packaging size and number of plants results in a cycle time of 1 second. Changeover to a different diameter or thickness (can height) takes 30 minutes. The plants' availability is 80 per cent.

  Shrinking is done at a plant with a handling robot operated by one worker. Every 12.5 seconds a stack is completed, which conforms to a cycle time of 0.5 seconds. Changeover time is only one minute, providing the required material is provided parallel to the primary operating time. The uptime is 84 per cent.

  Large cutting discs and grinding discs are shrink-wrapped in stacks of 10 each. The shrinking is done manually at three work stations and takes 120 seconds, resulting in a cycle time of 4 seconds. Each change of variant takes 10 minutes for the provision of material.
- **Final Inspection.** During the final inspection, the discs are visually checked for defective edges, which takes 10 minutes per production order. In addition, two destructive tests are conducted per order for quality assurance, investigating stress rupture times and fracture behaviour. Both tests are carried out on a different disc each and take 15 minutes in all. 3.2 per cent rejections are determined in all.
- **Dumping.** During the pressing process, the discs are stacked on metal mandrils separated by dividers. Once cured, the stacks are placed upside down in one of six machines operated by one worker each, which requires considerable physical effort. Shaking motions are performed to loosen and separate the discs now firmly stuck to the mandrils – a very loud and dirty task which may easily damage both discs and mandril. The times required to remove all discs from a mandril are 120, 200, 160 and 100 seconds for the four product families. Changes of diameter require adjustments of the guide rails, which takes 5 minutes. The plants' uptime is 93 per cent. Operators fetch their respective orders from the cooling area. As they are not issued with fixed sequencing instructions, they randomly choose those orders which seem to have been there longest.
- **Curing & cooling.** Subsequent to the pressing process, the raw discs undergo a lengthy high-energy thermal curing process to achieve the required strength. Exposure time in the furnace including loading and unloading by way of a

forklift is 20 hours There are 44 furnaces, each holding two standard 800 x 1200 skeleton boxes.

Once removed from the furnace, the cured discs need to cool down for 6 hours in an anteroom – even longer in hot weather.

- **Clamping.** To prevent the pressed raw discs from distorting during thermal treatment, they are pinned down with weights or clamped to the mandril. One worker needs 40 seconds for one mandril at a workplace with an uptime of 98 per cent.
- **Pressing.** A circular cloth blank is placed in the pressing mould, then the correct amount of matrix is added, another blank is placed on top, and the label is added before the compacting process starts. The process is subject to 4 per cent of material loss.

  There are six automatic revolving transfer machines for cutting discs from product families 1 and 2 with a takt time of 6 seconds at an uptime of 85 per cent and a changeover time of 30 minutes. The cycle time is 1 second. In addition, three rather aged manual pressing moulds are used, where work steps are standardized, documented and clearly visualized on site. At eleven small plants for small cutting discs and grinding wheels, one cycle of 48 seconds generates 4 discs. Two medium-sized plants for product family 2 take 72 seconds to produce four discs. Six large plants for product family 3 produce two discs in 80 seconds. Division of the respective machine takt times by the number of discs per cycle and the respective number of plants results in cycle times of 1.1 / 4.5 / 6.7 seconds. The plants have an uptime of 85 per cent and a changeover time of 15 per cent. 2.2 per cent of the cutting discs and 1 per cent of the grinding discs are rejects; weighted by piece numbers, this corresponds to a total of 2.1 per cent. The pressed discs are stacked on mandrils by the respective operators of the machines.
- **Conditioning Bulk.** The matrix is delivered three times per day and needs to be specially processed to remove clumping caused by humidity in transport or storage, or even during the original mixing process. The straining of 300 kg of mixture by way of one of three sieves takes 12 minutes, followed by six hours of setting. The same three operators are also responsible for stacking the pressing moulds. 10 minutes are estimated for transports of 100 kg each. The pressing mould operators order the required mixtures in good time before completion of the previous orders by way of fixing call-off cards to the so-called 'red board'.
- **Mixing.** First, the product-specific grains are mixed with liquid resin for 8 minutes. Then fillers and powdered resin are gradually added over 4 minutes. The mixture is then filled into three sacks via another drum, taking another 3 minutes. All formulas are designed for a volume of 300 kg. This is a very dusty process – which is one of the reasons why this process is conducted at a different location, four km from the actual production site. The quality of the matrix very much depends on the weather. Four operators handle two mixing plants on a two-shift roster on five days per week. At an annual 242 factory days of 15 working hours each, this results in a process-specific customer takt time of 0.33 seconds.

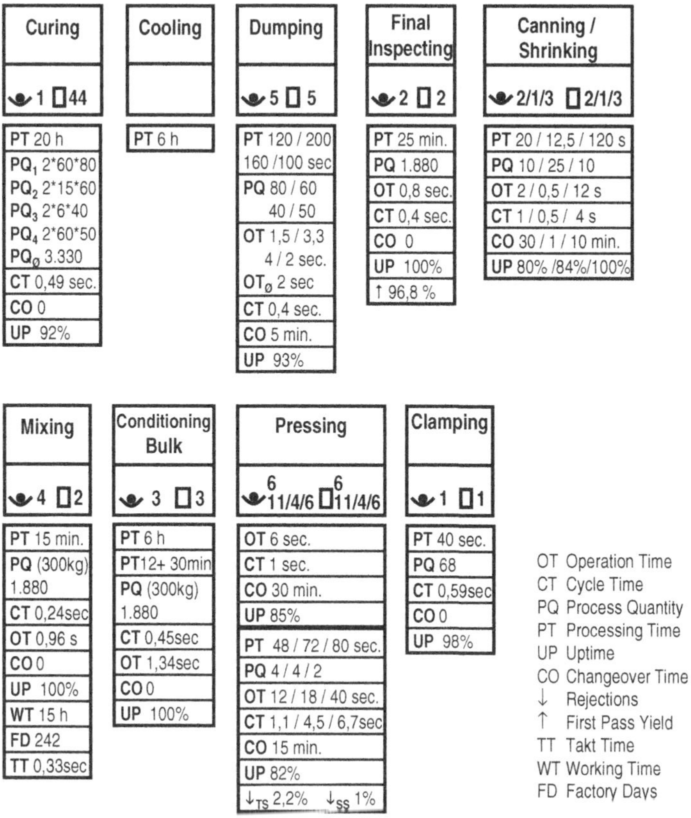

Fig. 5.22 Production processes at Seco ULC

As some of the processes work with batches, the value stream key figures, mainly the cycle times, still need to be calculated, taking into consideration the fact that though the production of four of the product families overlaps their specific process performance requirements differ considerably. Accordingly, lot sizes and production volumes of the batch processes are calculated for the four product families as well as their mean values weighted by annual piece numbers.

The order volume is determined at 300 kg by the mixing process. Based on this and the respective product family-specific unit weights, the lot sizes are calculated per product family (PF) and as weighted mean values – taking in to consideration, though, that only 288 kg of pressed discs remain upon deduction of 4 per cent material loss:

$$LS_{PF1}=\frac{PS}{W_{PF1}}=\frac{288\,\text{kg}}{80\,\text{g}}=3.600 \qquad LS_{PF2}=\frac{288\,\text{kg}}{240\,\text{g}}=1.200$$

$$LS_{PF3}=\frac{288\,\text{kg}}{720\,\text{g}}=400 \qquad LS_{PF4}=\frac{288\,\text{kg}}{180\,\text{g}}=1.600$$

$$\begin{aligned}LS_{\varnothing}&=\frac{PS}{\sum_{pf=1}^{4}\frac{W_{pf}\times Pcs_{pf}}{Pcs}}=\frac{288\,\text{kg}}{\frac{(80\,\text{g}\times 26+240\,\text{g}\times 7{,}8+720\,\text{g}\times 2+180\,\text{g}\times 3{,}4)\,\text{Mi(}}{39{,}2\text{ Mio.}}}\\&=\frac{288\,\text{kg}}{\frac{6.004\text{ to.}}{39{,}2\text{ Mio.}}}=\frac{288\,\text{kg}}{153\,\text{g}}\approx\mathbf{1.880\ Pieces}\end{aligned}\tag{5.7}$$

with: LS lot size [Pcs.]

W unit weight per disc [gram]

PS package size raw material [kilogram]

Pcs annual piece number [Pcs./a]

pf index of product family

The cycle times for mixing, straining and final testing are accordingly calculated from the batch sizes as per equation 2.3 (cf. corresponding caption for explanation of the abbreviations) as follows:

$$\begin{aligned}CT_{Mixing}&=\frac{PT}{PQ\times\#\,\text{Res}}=\frac{15\,\text{min.}}{1.880\times 2}=0{,}24\,\text{sec.}\\CT_{Conditioning}&=\frac{PT}{PQ\times\#\,\text{Res}}=\frac{42\,\text{min.}}{1.880\times 3}=0{,}45\,\text{sec.}\\CT_{Final\ Inspecting}&=\frac{PT}{PQ\times\#\,\text{Res}}=\frac{25\,\text{min.}}{1.880\times 2}=0{,}4\,\text{sec.}\end{aligned}\tag{5.8}$$

After pressing, the raw discs are stacked on mandrils of 60 cm each, separated by dividers of 3 mm. Depending on discs thickness, there will be between 40 and 80 discs on a mandril, though these are not specifically counted. The mean numbers per mandril for the four product families are estimated at 80, 60, 40, and 50 discs; compared with the respective piece numbers, the mean amount of discs per mandril is 68. Accordingly, the clamping-related cycle time is calculated as follows:

$$\begin{aligned}PQ_{Mandril\,\varnothing}&=\frac{Pcs}{\sum_{pf=1}^{4}\frac{Pcs_{pf}}{PQ_{Mandril\,pf}}}=\frac{39{,}2}{\frac{26}{80}+\frac{7{,}8}{60}+\frac{2}{40}+\frac{3{,}4}{50}}=\frac{39{,}2\text{ Mio.}}{373.000}\approx\mathbf{68\ Pieces}\\CT_{Clamping}&=\frac{PT}{PQ\times\#\text{Res}}=\frac{40\,\text{sec.}}{68\times 1}=0{,}59\,\text{sec.}\end{aligned}\tag{5.9}$$

For curing, each furnace holds two skeleton boxes with raw discs. In line with the respective diameters and stack heights, filling capacities differ with different product families. One skeleton box comprises 60 mandrils of product families 1 and 4, 15 mandrils of product family 2 or 6 mandrils of product family 3. Weighting by piece numbers results in a mean number of discs per furnace run of 3,300 pieces. The respective cycle time is calculated as follows:

$$PQ_{Oven\,\emptyset} = \frac{Pcs}{\sum_{pf=1}^{4} \frac{Pcs_{pf}}{PQ_{Oven\,pf}}} = \frac{39{,}2 \text{ Mio.}}{\frac{26 \text{ Mio.}}{2\times60\times80} + \frac{7{,}8 \text{ Mio.}}{2\times15\times60} + \frac{2 \text{ Mio.}}{2\times6\times40} + \frac{3{,}4 \text{ Mio}}{2\times60\times5}}$$
$$= \frac{39{,}2 \text{ Mio.}}{11.775} \approx \mathbf{3.330\ Pieces} \qquad (5.10)$$

$$CT_{Curing} = \frac{PT}{PQ \times \#\,\text{Res}} = \frac{20h}{3.330\times44} = 0{,}49\,\text{sec}.$$

Thy cycle times for dumping – taking into consideration the respective mandril quantities – are calculated from the individual times weighted by piece numbers as follows:

$$OT_{Dumping} = \sum_{pf=1}^{4} \frac{PT_{pf} \times Stck_{pf}}{PQ_{pf} \times Stck} = \frac{120\times26}{80\times39{,}2} + \frac{200\times7{,}8}{60\times39{,}2} + \frac{160\times2}{40\times39{,}2} + \frac{100\times3{,}4}{50\times39{,}2} = 2\,\text{sec}.$$
$$CT_{Dumping} = \frac{OT}{\#\,\text{Res}} = \frac{2\,\text{sec}.}{5} = 0{,}4\,\text{sec}. \qquad (5.11)$$

Material flow. The inventories are counted by pieces, or – in case of the matrix, by kg. The range of coverage is accordingly determined based on the daily demand of 135,172 discs. In the production, the first pass yield of 96.8 per cent as per final inspection needs to be taken into consideration; the resulting daily demand is 139,640 pieces. Based on disc weight and piece numbers per product family, the average piece weight is determined at 153 g. Considering a 4 per cent mark-up for material losses and deducting the rejections of 2 per cent incurred during pressing, the daily required volume is 22.64 tons. Volumes and ranges of coverage are entered beneath the respective triangular storage symbols (Fig. 5.23).

- **Finished goods storage.** According to a data request, the high rack storage contains 3.352 million cutting and grinding discs, which corresponds to a range of coverage of 24.8 days.
- **WIP – packaging.** 32 orders of a total volume of 62,400 discs are being handled at six work stations, conforming to a range of coverage of 0.45 days.
- **WIP – final inspection.** 28 orders, i.e. 49,960 discs are being tested in the final inspection; this corresponds to a range of coverage of 0.36 days.
- **WIP – dumping.** 38 orders comprising a total of just under 75,000 discs are ready for dumping or being dumped, equalling a range of coverage of 0.54 days.

- **WIP – cooling.** 26 orders of a total volume of 44,400 discs are currently undergoing a cooling process. This corresponds to a range of coverage of 0.32 days.
- **WIP – furnaces.** 98 orders are currently being cured or waiting to be cured. The total piece number as per order documents is 197.000, which equals a range of coverage of 1.41 days. It should be noted, though, that the processing time of the curing process of 20 hours results in a range of coverage of 0.9 days.
- **WIP – clamping.** Thanks to overlapping production, only discs from orders currently being pressed are waiting to be clamped, i.e. an estimated 4,200 discs, conforming to a range of coverage of 0.03 days.
- **WIP – pressing.** 23 orders of 300 kg each and an arithmetical coverage of 0.3 days are being processed. Deducting the respective inventory in clamping, the range of coverage amounts to 0.27 days.
- **Matrix inventory.** A total of 125 mixtures of 300 kg each are currently in interim storage, which corresponds to a range of coverage of 1.66 days.
- **Raw material storage.** The most important part of the raw material inventory is the grain. At the location under observation, 38 tons of 20 different grain sizes are being stored; another 688 tons are being stored at a provider's store. The total range of coverage amounts to 45.81 days.

External logistics. Shipping distinguishes small and large consignments. Small orders are shipped as parcels of up to 30 kg each via parcel service. Large orders of 100 kg and more are shipped on pallets via carriers (Fig. 5.23).

The raw materials required for the matrix are almost exclusively procured from China – grain is supplied by KeLi of Quingdao at delivery periods of 3 – 4 months. Execution and storage are handled by logistics service provider L&T, calling daily at the Toronto factory (Fig. 5.23).

Business Processes. Seco ULC fulfils customer orders either directly from stock or customer specific with an average delivery time of four weeks. Order processing is broken down into the following four business processes (Fig. 5.23).

- **Order processing.** Depending on the customer's technical facilities, orders are received via fax, telephone, email, or EDI and are entered into the system via AS 400. Following an inventory check, small orders are confirmed. For bulk orders, which account for approx. 70 per cent of the total volume, scheduled orders are generated accordingly. Taking into consideration the capacitive utilization of the pressing moulds, the system determines the delivery dates which are then confirmed to the customers.
- **Shipping preparation.** For small orders, picking documents and bill of ladings are directly generated from AS 400. The same applies to bulk orders – but only following the ready signal from production and notification of the carrier. In case of sea freight, loading dates have to be taken into consideration.
- **Production preparation.** The main task is the creation of the production orders The scheduled orders recorded in the system – either generated from entries for certain customers order or automatically triggered upon undercutting of reorder levels in the finished goods storage – need to be converted into

standardized formula volumes of 288 kg. Keeping in mind the confirmed delivery dates and an assumed production lead time of 5 days max., the production orders are allocated and scheduled for certain pressing moulds by way of Excel spreadsheets. This pressing mould utilization planning is conducted on a weekly basis; it controls the production and is in addition utilized for the volume planning of the mixing process by way of backward scheduling.

Also recorded is feedback from the pressing moulds concerning exact piece numbers and initial rejections, from the final inspection concerning scrap as well as ready signals from the furnaces, and from the dumping and packaging processes.

- **Procurement.** The raw materials are scheduled on the basis of usage reports from mixing and are ordered from Chinese, German and Italian suppliers once a week, allowing for at times very long delivery periods. In the following, we shall solely regard the daily call-off orders placed with the logistics service provider L&T who stores the raw material required for the matrix.

Potential for Improvement

Lead time. The very high lead time mainly results from the range of coverage of the finished goods inventory and raw material stock of a total of 70 days, or 6 weeks, based on weeks of six working days. The high raw material inventory is mainly due to the long replenishment times. The finished goods storage, on the other hand, seems unnecessarily high, even in view of the high number of variants. This is likely also due to the lengthy periods of time passing between order triggering on the grounds of undercut reorder levels and ready signals after scheduling and production of stock replenishment orders, which in turn is caused by the slow weekly planning rhythm and even slower planning horizon as well as the actual production lead time of 5 days (Fig. 5.23).

The work in progress inventory of just over 3 days is mainly caused by the preparing and processing of the matrix. The first includes noticeably high transport effort between locations, which does not do the product quality any good and which is responsible for the processing requirement of the matrix in the first place, while the very long process time in curing is disruptive to the smooth flow of the production. In addition, all remaining inventories between the various processes are decidedly too large. A medium-sized cutting disc from product family 2 for example has an accumulated operation time of 18 seconds, which at 5 days' work in progress results in a flow degree of an infinitesimal 0.045 per mill. In a medium-sized lot of 1,880 pieces, though, this still achieves a reasonable flow degree of 8.4 per cent.

Line balancing. The operator balance chart identifies a bottleneck at an unexpected location – clamping (Fig. 5.24). As times were only mapped by way of examples and then interpolated for the highly diverse product families, the actual clamping time requirement is likely slightly smaller, as otherwise the respective piece numbers could never have been produced. Nevertheless insufficient dimensioning of the most cost-effective production process with a view to resources

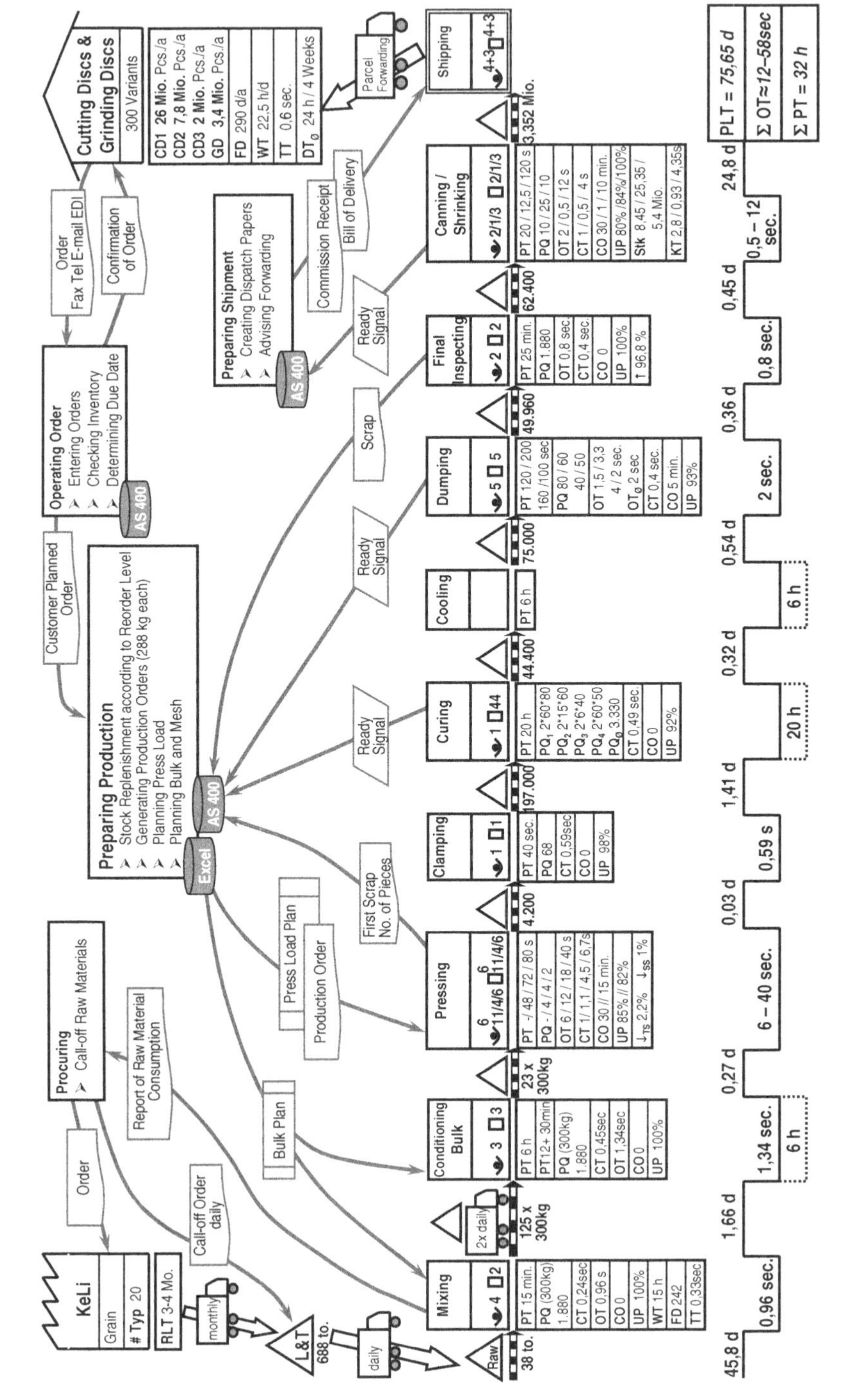

Fig. 5.23 Result of the Seco ULC value stream analysis

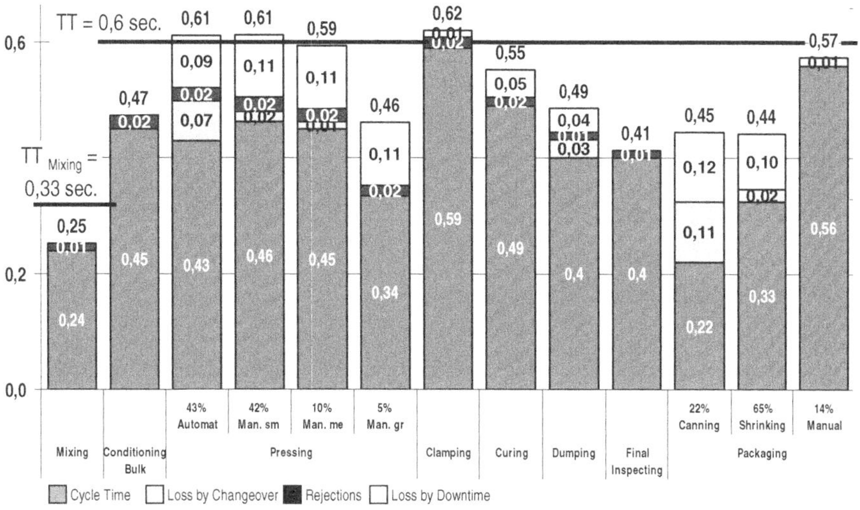

Fig. 5.24 Current-state ULC operator balance chart

should be carefully avoided to prevent material flow breaks – this applies especially to variant productions with constantly fluctuating capacity demands based on changing product mixes.

At the pressing and packaging processes, the value stream branches out into three branches, or four, respectively. In order to enable a consistent depiction of the operator balance chart, the respective cycle times are stated with their percentaged portions shown underneath the respective bar graphs (Fig. 5.24).

Taking into consideration the high changeover and uptime losses of the (partially) automated plants in packaging, their utilization is less than 75 per cent, though there are several additional manual packaging stations. The respective plants are optimized as to piece numbers, but only for certain diameters and with a view to canning processes with high changeover portions.

The allocation of volumes is somewhat more difficult with respect to the pressing process, as only product family no. 3 may be clearly allocated to one certain type of resource. Automatic pressing of product families 1 and 2 could generate a total piece number of 33.8 million discs, resulting in a process-specific customer takt time of 0.69 seconds. With a cycle time of 1 second, a maximum of 70 per cent of the demand may be processed on the automated pressing moulds. In view of uptime losses of 15 per cent as well as changeover losses of maybe 5 per cent, the splitting in half of product families 1 and 2 with a view to automated vs. manual pressing seems expedient for a first estimate. The resulting partial volumes for the four different types of pressing moulds are as follows:

$$\frac{Pcs_{Automatic\ Pressing}}{Pcs} = 50\% \times \frac{(Pcs_{pf1} + Pcs_{pf2})}{Pcs} = 50\% \times \frac{26 + 7,8}{39,2} = 43\%$$
$$\frac{Pcs_{Manual\ Pressing\ small}}{Pcs} = \frac{50\% \times Pcs_{pf1} + Pcs_{pf4}}{Pcs} = \frac{50\% \times 26 + 3,4}{39,2} = 42\% \tag{5.12}$$
$$\frac{Pcs_{Manual\ Pressing\ medium}}{Pcs} = \frac{50\% \times Pcs_{pf2}}{Pcs} = \frac{50\% \times 7,8}{39,2} = 10\%$$
$$\frac{Pcs_{Manual\ Pressing\ great}}{Pcs} = \frac{Pcs_{pf3}}{Pcs} = \frac{2}{39,2} = 5\%$$

The allocation of volumes enables relatively evenly balanced loads, with high to excessive loads on all pressing moulds. This could possibly be corrected by improved uptimes and lowered changeover times of the automated pressing moulds.

The exemplary calculation of cycle times and time losses in automated pressing, curing and canning is shown in equation 5.13:

Automatic Pressing:

$$CT = \frac{OT}{\#Res} \times \frac{Pcs_{Automatic\ Pressing}}{Pcs} = \frac{6\,sec.}{6} \times 43\% = 0,43\,sec.$$
$$CT_{CO} = CT \times \frac{CO}{LS \times OT} = 0,43\,sec. \times \frac{30\,min.}{1.880 \times 6\,sec.} = 0,43\,sec. \times 16\% = 0,07\,sec.$$
$$CT_Q = CT \times \sum(1 - \uparrow) = 0,43\,sec. \times [(1 - 96,8\%) + 2,1\%] = 0,023\,sec.$$
$$CT_{UP} = TT \times (1 - UP) = 0,6\,sec. \times (1 - 85\%) = 0,09\,sec.$$

Curing:

$$CT_{Curing} = 0,49\,sec. \quad \text{(cf. Eq. 5.10)}$$
$$CT_Q = CT \times (1 - \uparrow) = 0,49\,sec. \times (1 - 96,8\%) = 0,02\,sec. \tag{5.13}$$
$$CT_{UP} = TT \times (1 - UP) = 0,6\,sec. \times (1 - 92\%) = 0,05\,sec.$$

Canning:

$$\frac{Pcs_{Canning}}{Pcs} = 25\% \times \frac{(Pcs_{pf1} + Pcs_{pf2})}{Pcs} = 25\% \times \frac{26 + 7,8}{39,2} = 22\%$$
$$CT = \frac{PT}{PQ \times \#Res} \times \frac{Pcs_{Canning}}{Pcs} = \frac{20\,min.}{10 \times 2} \times 22\% = 0,22\,sec.$$
$$CT_{CO} = CT \times \frac{CO}{LS \times OT} = 0,22\,sec. \times \frac{30\,min.}{1.880 \times 2\,sec.} = 0,22\,sec. \times 48\% = 0,11\,sec.$$
$$CT_{UP} = TT \times (1 - UP) = 0,6\,sec. \times (1 - 80\%) = 0,12\,sec.$$

Value Stream Design

Goal. The entire production process is to be newly designed with the aid of investment in technologies partially yet to be developed in order to achieve consistent flow orientation and a significant reduction in personnel effort.

Production processes (1): Curing. The crucial point is the intended substitution of the current curing process conducted in batches by a continuous process. As no suitable continuous furnaces are being offered by the relevant plant manufacturers, a suitable prototype is going to be specially developed. This is expected to improve the production flow, save handling costs and production space as well as considerable energy as opposed to the old and poorly insulated furnaces.

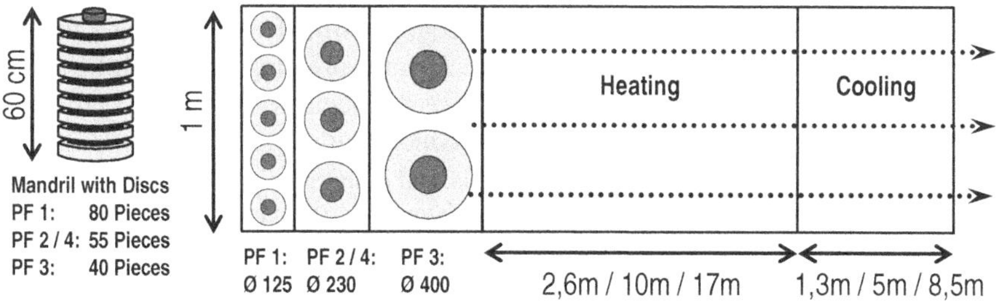

Fig. 5.25 Schematic diagram of a continuous furnace

For the preliminary design of the continuous furnace, an internal width of 1 m is assumed. Depending on the respective disc diameters and product families, five, three or two mandrils with discs will be able to be placed on a conveyor belt 5 cm apart to be processed in the continuous furnace (Fig. 5.25). The length of the furnace results from the goal of cutting the current process time in half, i.e. 10 hours, and to thus balance the furnace performance with the clock time (CLT) of an automatic pressing mould of 6 seconds (PF 1, 2, 4), or 12 seconds (PF 3), respectively. Taking into consideration the quantities per mandril (MQ) of 80, 55 or 40 discs, the following – highly differing – required furnace lengths are determined:

$$Lenght_{Oven,pf1} = \frac{PT \times (\emptyset + 50mm)}{CLT \times MQ \times \# D} = \frac{10h \times (125mm + 50mm)}{6\sec. \times 80 \times 5} = 2{,}63m$$

$$Lenght_{Oven,pf2/4} = \frac{10h \times (230mm + 50mm)}{6\sec. \times 55 \times 3} = 10{,}2m \quad (5.14)$$

$$Lenght_{Oven,pf3} = \frac{10h \times (400mm + 50mm)}{12\sec. \times 40 \times 2} = 16{,}9m$$

The cooling process is assumed to take 5 hours, accordingly half a furnace length needs to be added. As the required furnace for product family 1 is comparatively short, it is designed with twice the determined capacity and will thus be able to hold discs from two pressing moulds. In order to meet the respective customer takt times, the following numbers of furnaces and allocated pressing moulds will be required per product family (Fig. 5.27):

- **PF 1:** At a furnace takt time of 3 seconds at four furnaces with a cycle time of 0.75 seconds, the customer takt time of 0.9 seconds is undercut. As the pressing moulds' takt times are twice that, eight pressing moulds will be required.
- **PF 2:** The customer takt time of 3 seconds is undercut at a furnace takt time of 6 seconds and three furnaces and three pressing moulds of a cycle time of 2 seconds. Strictly speaking, two furnaces would suffice.
- **PF 3:** The customer takt time of 11.75 seconds is slightly exceeded at a furnace takt time of 12 seconds with one furnace and one pressing mould. However, as product family 2 features significant excess capacities, the smallest discs of product family 3 may be reallocated to product family 2.
- **PF 4:** At a furnace takt time of 6 seconds and a suitable pressing mould with a cycle time of 6 seconds, the customer takt time of 6.9 seconds is undercut.

In the development of the furnace, an increase in power efficiency in particular is highly encouraged. In this context, tube shapes turned out to be particularly suited as combustion chambers and were thus initially vertically used in the construction of the intended furnaces. However, this resulted in immense space requirement and adverse heat distribution. Splitting of the furnace space into several vertical tubes in an arrangement similar to that of a revolving transfer machine finally resulted in the development of an 'organ pipe-shaped' vertical continuous furnace (Fig. 5.26). It is sequentially fed from below, but due to the splitting up into several vertical tubes, the respective batch sizes are small and allow perfect integration in a line production with the pressing and packaging processes as described in the following steps.

Production processes (2): Dumping & packaging. The dumping process is going to be designed ergonomically, resulting in a clear reduction of noise and dirt and eliminating the manual lifting of heavy loads. Dividers have to be found which are easier to remove than the current ones. Matrix composition and process control in the furnace may need to be adjusted. Ideally, cured and cooled discs will be leaving the furnace on their mandrils at even speed and will be easily removed by hand for immediate quality inspection. Besides, integration of the canning and shrinking processes could prove expedient (design guideline 2).

With packaging materials (cans, cardboard boxes) being provided by a logistician, there would be no changeover times. A smaller degree of automation could actually increase uptime.

Fig. 5.26 Old heating furnace for skeleton boxes and the 'organ pipe-shaped' prototype of the newly developed furnace for vertical loading (here shown without insulation)

Large cutting discs (PF 3) and grinding wheels (PF 4) will continue to be shrink-wrapped manually in packaging units of 10 discs in 120 seconds. As the discs have to be manually picked up for packaging anyway, dumping and quality control may easily be integrated without extra time effort. This results in an operation time of 12 seconds per disc; accordingly, one packaging station will suffice to handle all cutting discs, while the grinding wheels will need two packaging stations in order to match the cycle times of the respective furnaces (Fig. 5.27).

In product families 1 and 2.25 per cent of the discs need to be canned in units of 10 discs each, the rest is shrink-wrapped in units of 25. This results in an average 21.25 discs per packaging unit. At an estimated operation time of 40 seconds, the required cycle times of 0.6 seconds, or 1.8 seconds, respectively, will be met (Fig. 5.27).

Production processes (3): Pressing. The pressing process is going to be completely automated, i.e. the current trend is consistently going to be developed further. The additional automation will result in savings of approx. 50 per cent of the personnel costs. Taking the existing plant as a measure for potential performance, a total of 13 automated plants will be required (Fig. 5.27) – with the seven additional ones replacing the current 18 manual pressing moulds.

Production processes (4): Mixing. The mixing process is to be reallocated to the production location, which will entail several advantages: Transport costs will be eliminated. Matrix production to order would also eliminate the need for straining conducted by three workers per shift (i.e. 12 workers allowing for leave times) (design guideline 2). Besides, more than 1.5 days' lead time and storage space at the matrix storage location could be saved. In addition, mixing in accordance with the actual demand would ensure constant high quality of the matrix, ideally in air conditioned premises.

To increase the quality, the mixing process time is doubled to 30 minutes. The newly defined processes will work with smaller lot sizes adjusted to the old straining processes of 100 kg each, thus cutting the mean lot size to one third, i.e. 627 discs. The daily matrix demand of 22.64 tons over three shifts results in a demand of just over one ton per hour. Accordingly, just over five, i.e. six mixing machines are required for 200 kg per hour. Ideally, one operator will handle two mixing machines while the pressing moulds will be supplied by a logistician. Changes of formula require 5 minutes' changeover (Fig. 5.27).

Material flow. The remaining four production processes will be closely connected via FIFO lines with minimal buffers (design guideline 3). Raw mixtures are to be prepared as closely to the time of use as possible and taken to the automatic pressing moulds. The pressing mould operators pick up the discs stacked on a mandril by the machine, clamp the same and place them on the furnace conveyor. After curing, a worker removes the discs from the conveyor belt, with the aid of a simple tool removes the non-sticking discs from the mandril, visually checks the quality and packs the discs.

The very close connection of the processes from pressing until canning virtually constitutes a line production; in the value stream depiction these processes are therefore additionally connected by a dashed line. In an extreme case, this would leave only two physically isolated processes, i.e. the 'mixing' and the 'shaping of discs' (design guideline 2). This requires consistently successful line balancing, though, as well as high availability of the components, which is why in our depiction the processes are slightly decoupled by FIFO buffers, just to make sure. The production lead time is significantly reduced from 3 days to just over one day, mainly caused by the furnace processing time of 15 hours, which for technical reasons in turn requires a minimum work in progress inventory of 0.67 days (Fig. 5.27, timeline). With processes protected by buffers, a flow degree of 60 per cent is achievable.

A supermarket store is envisioned for the finished goods (product type 1), with an initial range of coverage of two weeks (12 working days) (design guideline 4). This seems sufficient, as only standard products are going to be stored, while customer-specific products will be able to be produced at short notice, thanks to the short lead times. These customer-specific products (product type 2) will be directly provided for shipping by the production. Bulk orders spread over several production days, if any, will be accumulated at the shipping end.

The supplier connection shall not be specifically regarded herein, as the current solution with the logistics provider works very well and with its demand-oriented control is very similar to a supplier kanban. The unchanged current state is therefore marked in light grey in figure 5.27.

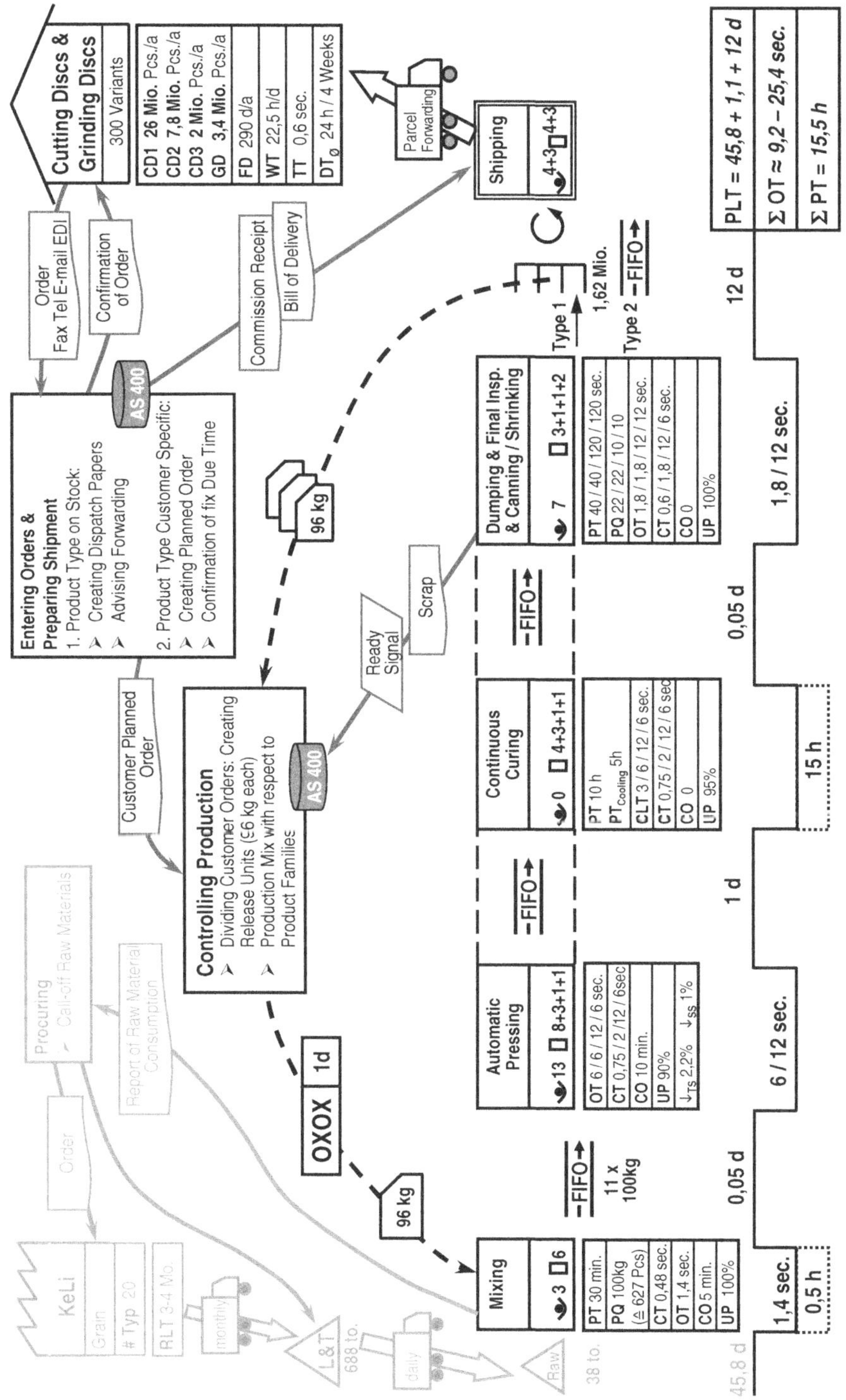

Fig. 5.27 Future-state map for the production of cutting discs at Seco ULC

Production control. The production begins with the mixing process and may be conducted either in a customer-anonymous or a customer specific manner (trigger point as per design guideline 5). A mixing machine's capacity determines the release units at a standard 100 kg, which corresponds to 96 kg after material loss (design guideline 6). The number of discs to be pressed therefrom depends on the respective disc geometries and on average amounts to an 627 pieces, minus 5 per cent of finished discs due to scrap losses.

The business process of 'production control' converts the scheduled orders for customer-specific products created in AS 400 into uniform release units; in addition, there are kanban from the finished goods storage. Both types of release units are accumulated in a queue. Within the release horizon of one day, an approximate daily 225 release units are released for production in the defined intermixture of the four product families. A load levelling box ensures an even utilization of the machines available for each product family (design guideline 7).

Conclusion Significant extensive technological changes were conceptually devised in the value stream design. The initial aim was the substitution of old furnaces by new continuous furnaces and an ensuing improvement of the production flow. Once this decision had been taken, the upstream pressing and the downstream dumping and packaging processes were easy to link. This was aided by product family creation in connection with the value stream analysis, based on which suitable performance classes were formed for the production processes. The production processes from pressing to packaging were suitably integrated, also with a view to operators' tasks. Accordingly, the achieved result is impressively close to a 'one process factory'. Only the mixing process will not be able to be integrated, mainly due to the inherent formation of dust.

Chapter 6
References

John, B., Matthias, H.: The Lean Toolbox. The Essential Guide to Lean Transformation, 4th edn. PICSIE Books, Buckingham (2009)

Duggan Kevin, J.: Creating Mixed Model Value Streams. Practical Lean Technique for Building to. Demand. Productivity Press, New York (2002)

Klaus, E.: Das Technotop. Die technologische Konstruktion der Wirklichkeit. LIT-Verlag, Münster (2000)

Klaus, E., Engelbert, W.: Energiewertstrom. Der Weg zur energieeffizienten Fabrik. Fraunhofer Verlag, Stuttgart (2009)

Klaus, E.: Energy Value Stream. Increasing Energy Efficiency in Production. In: Schuh, G., Neugebauer, R., Uhlmann, E. (eds.) Future Trends in Production Engineering. Springer, Heidelberg (2012)

Eliyahu, G.M., Jeff, C.: The Goal. A Process of Ongoing Improvement. North River, Cambridge (1995)

Liker, J.K.: The Toyota Way. 14 Management Principles from the World's Greatest Manufacturer. McGraw-Hill, New York (2004)

Hermann, L.: Handbook of Manufacturing Control: Fundamentals, description, configuration. Springer, Heidelberg (2012)

Nakajima, S.: Introduction to TPM – Total Productive Maintenance. Productivity Press, Cambridge (1988)

Taiichi, O.: Toyota Production System: Beyond Large-scale Production. Productivity Press, New York (1988)

REFA – Verband für Arbeitsstudien und Betriebsorganisation e. V, Methodenlehre des Arbeitsstudiums. Teil 2. Datenermittlung. Hanser, München (1992)

Günter, R.: Some Methodological Aspects of Modelling Socio-Technical Systems. In: Trappl, R., et al. (eds.) Progress in Cybernetics and Systems Research, vol. 10, pp. 525–536. Hemisphere, Washington DC (1982)

Mike, R., Rick, H.: Creating Continuous Flow. An Action Guide for Managers, Engineers and Production Associates. The Lean Enterprise Institute, Brookline (2001)

Mike, R., John, S.: Learning to See: Value Stream Mapping to Add Value and Eliminate Muda. Lean Enterprise Institute, Cambridge (1999)

Michael, S., Siegfried, W., Egon, M.: Factory Planning Manual: Situation-Driven Production Facility Planning. Springer, Heidelberg (2009)

Art, S.: A Lean Production-System Improvement Guide for Production-Control, Operations, and Engineering Professionals. Lean Enterprise Institute, Cambridge (2004)

Kiyoshi, S.: The New Manufacturing Challenge. Techniques for Continuous Improvement. Free Press, New York (1987)

Don, T., Tom, S.: Value Stream Management for the Lean Office: Eight Steps to Planning, Mapping, and Sustaining Lean Improvements in Administrative Areas. Productivity Press, New York (2003)

Hitoshi, T.: The Synchronized Production System: Going Beyond Just-In-Time Through Kaizen. Kogan Page, Philadephia (2006)

VDI-Guideline 5200, Part 1: Factory planning – Planning procedures. Beuth Verlag, Berlin (2010)

Hans-Hermann, W.: Turbulence Germs and their Impact on Planning and Control – Root Causes and Solutions for PPC Design. CIRP Annals Manufacturing Technology 56(1), 443–446 (2007)

Hans-Hermann, W.: Situative Konfiguration des Auftragsmanagements im turbulenten Umfeld. Jost-Jetter-Verlag, Heimsheim (2002)

Hans-Peter, W.: Load-Oriented Manufacturing Control. Springer, Heidelberg (1995)

Chapter 7
Appendix

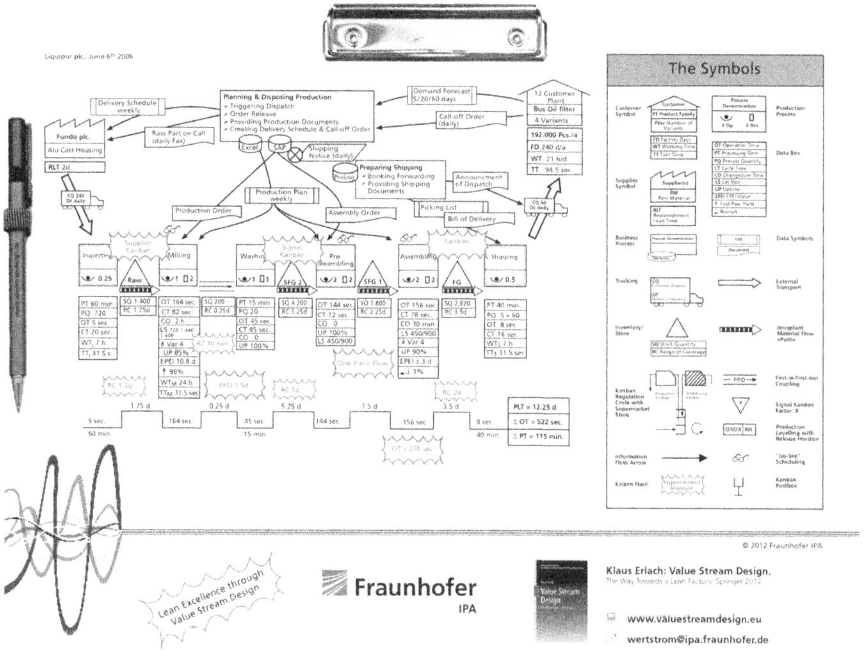

Fig. 7.1 The Value Stream Clipboard is sized to use ISO A3 and facilitates value stream mapping within production

7.1 Index of Formulae

7.1.1 Calculation Formulae for Value Stream Analysis

1. Calculation of Takt Time

1.1 General customer takt time for the whole value stream:

$$TT = \frac{\text{available working time per year}}{\text{customer demand per year}} = \frac{FD \times WT}{Pcs} = \frac{WT}{DD}$$

with: TT takt time [time unit/pcs.]
FD factory days [d/a]
WT daily working time [time unit/d]
Pcs annual piece number [pcs./a]
DD daily demand [pcs./d]

(2.1)

1.2 Process-specific takt time:

$$TT_P = TT \times \frac{Pcs}{Pcs_P} \times \frac{WT_P}{WT} = \frac{FD \times WT_P}{Pcs_P}$$

with: TT_P process-specific customer takt time [time unit]
WT_P/WT process-specific related to general working time [%]
Pcs_P/Pcs rate of process-specific annual piece number [%]
FD factory days [d/a]

(2.9)

2. Calculation of Cycle Times

2.1 Cycle time as net performance of a production process:

$$CT = \frac{OP \times \#P}{\#Res} \quad \text{oder:} \quad CT = \frac{PT \times \#P}{PQ \times \#Res}$$

with: CT cycle time [time unit]
OT operation time [time unit]
PT process time [time unit]
PQ process quantity for batch or continuous processing [pcs.]
\# P number of identical parts per finished product [pcs.]
\# Res number of same resources [pcs.]

(2.3)

2.2 Gross cycle time under consideration of time losses:

$$CT_{gross} = CT_{net} + CT_{CO} + CT_{UP} + CT_{Q} \tag{2.22}$$

with: CT_{gross} gross cycle time [time unit]
CT_{net} = CT; cycle time according to eq. 2.3 [time unit]

2.2.1 Additional time for time losses through changeover:

$$CT_{CO} = CT_{net} \times COF = CT_{net} \times \sum_{i=1}^{\#Var} \frac{CO_i}{LS_i \times OT_i} \times \frac{Pcs_i}{Pcs} = CT_{net} \times \frac{CO}{LS_{\varnothing} \times OT} \tag{2.20}$$

with: # Var number of variants
CT_{net} = CT; cycle time according to eq. 2.3 [time unit]
CT_{CO} time losses through changeover [time unit]
COF factor for losses through changeover [%]
Pcs. annual piece number [pcs./a]
OT operation time [time unit]
LS lot size [pcs.]
$LS_{\varnothing}$ average lot size [pcs.]
CO changeover [time unit]

2.2.2 Additional time for time losses through breakdown:

$$CT_{UP} = TT \times (1 - UP) \tag{2.21}$$

with: CT_{UP} time losses through breakdown [time unit]
TT customer takt time [time unit/pcs.]
UP uptime [%]

2.2.3 Additional time for time losses through quality defects:

$$CT_Q^m = CT_{net}^m \times QL^m = CT_{net}^m \times \left(\hookleftarrow_m + \sum_{pss=1}^{m} (1 - \uparrow_{pss}) \right) \tag{2.7}$$

mit: *pss* index for *m-1* subsequent processes after production process *m*
ZZ_Q time loss by quality defects [time unit]
CT_{net} = CT; cycle time according to eq. 2.3 [time unit]
QL rate of quality losses [%]
↑ first past yield [%] *often also as:* ↓ rejections [%]
↵ rework [%]

3. Calculation of EPEI Value for fixed lot sizes

$$EPEI_{fix} = \frac{\sum OT + \sum CO}{\# Res \times UP \times WT}$$

$$EPEI_{fix} = \frac{\sum_{var=1}^{n} LS_{var} \times OT_{var} \times (1 + \hookleftarrow_{var}) + \sum_{var=1}^{n} CO_{var}}{\# \mathrm{Res} \times UT \times \mathrm{WT}}$$

with: *var* index for *n* variants (2.8)

$EPEI_{fix}$	EPEI value for fixed lot sizes [d]
LS	lot size [pcs.]
OT	operation time per piece [time unit]
$\hookleftarrow$	rework [%]
CO	changeover time [time unit]
# Res	number of same resources [pcs.]
UP	uptime [%]
WT	daily working time [time unit/d]

4. Calculation of range of coverage

4.1 Range of coverage of inventory and buffer stock:

$$RC = \frac{SQ \times \uparrow}{DD \times \# P}$$

with: RC range of coverage [d] (2.13)

SQ	stock quantity [pcs.]
DD	daily demand [pcs./d]
$\uparrow$	first pass yield [%]
# P	number of identical parts per finished product [pcs.]

4.2 Range of coverage of order queue:

$$Q = \# O \times TT$$

with: Q queue [time unit] (2.14)

# O	Number of waiting orders [pcs.]
TT	customer takt time [time unit/pcs.]

4.3 Production lead time for the whole value stream:

$$PLT = \sum_{i=1}^{Process\ n} \frac{WIP_i \times \uparrow}{DD \times \# P} + \sum_{i=1}^{Stock\ m} RC_i = \sum_{i=1}^{m} \frac{(SQ_i + WIP_i) \times \uparrow}{DD \times \# P} = \sum_i RC_i$$

(2.16)

with: PLT production lead time [d]

RC range of coverage [d]

WIP work in process [pcs.]

SQ stock quantity [pcs.]

DD daily demand [pcs./d]

↑ first pass yield [%]

\# P number of identical parts per finished product

5. Identification of potentials for improvement and waste

5.1 Degree of flow as measure for dynamics of value stream:

$$DF = \frac{\sum_{Process} (OT + PT)}{\sum_{Store} RC \times WT} \times 100$$

(2.17)

with: DF degree of flow [%]

OT operation time [time unit]

PT processing time [time unit]

RC range of coverage [d]

WT working time per day [time unit/d]

5.2 Degree of capacity utilisation as measure for capacity availability in value stream:

$$DU = \frac{1}{n} \times \sum_{i=1}^{Process\ n} \frac{CT_i}{TT_i} \times 100$$

(2.18)

with: n number of production processes within value stream

DU degree of capacity utilisation [%]

CT cycle time [time unit]

TT customer takt time (process specific) [time unit]

7.1.2 Specification Formulae for Value Stream Design

1. Specification of capacity profile of value stream

$$\#\text{Res} = \text{ROUNDING UP}\left[\frac{OT}{TT \times (1-COF) \times UP \times (1-QL)}\right] \tag{3.1}$$

with: # Res resource requirements
OT operation time [time unit]
TT takt time [time unit]
COF factor for losses through changeover [%]
UP uptime [%]
QL rate of quality losses [%]

2. Specification of a continuous flow production

$$\#\text{OP} = \text{ROUNDING DOWN}\left[\frac{\sum OT_i}{TT \times 95\% \times \#\text{Res}}\right] \tag{3.2}$$

with: # OP number of operators in continuous flow production
OT_i operation time at station *i* [time unit]
TT takt time [time unit]
Res number of parallel lines

3. Specification of FIFO Buffer

$$ConWIP = \frac{(OT \times CQ) + MZ}{TT} + \frac{(OT_{max} - OT)}{TT} \times LS + \frac{CO}{TT} + \frac{DT_{max}}{TT} + PQ$$

(3.3)
(3.4)
(3.5)
(3.6)

with: ConWIP limited inventory level [pcs.]
OT operation time at preceding process [time unit]
CQ container quantity for conveyance [pcs.]
MT move time [time unit]
TT takt time [time unit]
OT_{max} operation time at preceding process [time unit]
LS lot size [pcs.]
CO changeover time [time unit]
DT_{max} maximal downtime [time unit]
PQ process quantity [pcs.]

4. Specification of production kanban

4.1 Estimation of necessary work in process (WIP):

$$WIP = RLT \times DD \approx 4 \times TCT \times DD \tag{3.8}$$

with: WIP work in process [pcs.]
RLT replenishment lead time [d]
DD daily demand [pcs./d]
TCT transport cycle time [time unit]

4.2 Determination of one variants maximum of supermarket stock:

$$SQ_{Var} = WIP + BS + SS$$
$$\approx 4 \times TCT \times DD_{Var} + 4 \times TCT \times \Delta DD_{Var} + RC_{SS} \times DD_{var} \tag{3.9}$$

with: SQ_{Var} maximal stock in supermarket of a variant [pcs.]
WIP work in process [pcs.]
BS buffer stock [pcs.]
SS safety stock [pcs.]
TCT transport cycle time [time unit]
DD_{Var} daily demand of a variant [pcs./d]
ΔDD_{Var} maximal additional daily demand of a variant [pcs./d]
RC_{SS} range of coverage of safety stock [d]

4.3 Deduction of required number of kanbans per variant:

$$\# K_{Var} = \text{ROUNDING UP}\left[\frac{SQ_{Var}}{CQ_{Var}}\right] \tag{3.10}$$

with: $\# K_{Var}$ number of kanban of a variant
SQ_{Var} maximal stock in supermarket of a variant [pcs.]
CQ_{Var} container quantity of a variant [pcs.]

4.4 Range of coverage of supermarket stock:

$$RC = \frac{1}{2} \times \frac{\# K \times CQ}{\# P \times DD} \quad \text{respectively:} \quad RC = \frac{1}{2} \times \frac{\sum_{i=1}^{\text{var}} \# K_{Var} \times CQ_{Var}}{\# P \times DD}$$

with: # K number of kanban in circulation

RC supermarkets range of coverage [d] (3.7)

CQ container quantity of kanban [pcs.]

DD daily demand [pcs./d]

P number of identical parts per finished product [pcs.]

$\# K_{Var}$ number of kanban of a variant

CQ_{Var} container quantity of a variant [pcs.]

5. Specification of signal kanban

5.1 Calculation of EPEI Value to determine lot sizes:

$$EPEI_{\min} = \frac{\sum CO}{\text{daily CO}} = \frac{\# Var \times CO}{(WT \times \# Res) \times UP - MLT}$$

$$= \frac{\# Var \times CO}{(WT \times \# Res) \times UP - \sum_{i=1}^{i=\text{var}} OT_i \times DD_i}$$

with: $EPEI_{\min}$ EPEI value to determine minimal lot size [d] (3.6)

CO changeover time [time unit]

Var number of variants

WT daily working time [time unit/d]

Res number of resources

UP uptime [%]

MLT machine loading time [time unit]

OT operation time [time unit]

DD daily demand [pcs./d]

5.2 Calculation of order point:

$$RL_{Var} = WIP + BS + SS = RLT \times (DD_{Var} + \Delta DD_{Var}) + RC_{SB} \times DD_{Va}$$

$$OP_{Var} = \text{ROUNDING UP}\left[\frac{RL_{Var}}{CQ_{Var}}\right]$$

with: RL_{Var} variant-oriented reorder level [pcs.]
OP_{Var} order point of a variant [number of containers]
WIP work in process [pcs.] (3.14)
BS buffer stock [pcs.]
SS safety stock [pcs.]
RLT replenishment lead time [d]
DD_{Var} daily demand of a variant [pcs./d]
ΔDD_{Var} maximal additional daily demand of a variant [pcs./d]
RC_{SB} range of coverage of safety stock [d]
CQ_{Var} container quantity of a variant [pcs.]

with an average replenishment lead time of:

$$RLT_{\varnothing} = 2{,}5 \times (OT \times S_{\varnothing} \times CQ + CO) + 2 \times TCT$$

with: $RLT_{\varnothing}$ average replenishment lead time [d]
OT operation time [time unit]
CQ container quantity of kanban [pcs.] (3.16)
$S_{\varnothing}$ average signal kanban factor
CO changeover time [time unit]
TCT transport cycle time [time unit]

5.3 Range of coverage of supermarket stock:

$$RC = \frac{1}{2} \times \frac{\sum_{i=1}^{var} (OP_{Var} \times CQ_{Var} + S_{Var} \times CQ_{Var})}{\# P \times DD}$$

with: RC supermarkets range of coverage [d] (3.15)
OP_{Var} order point of a variant [number of containers]
CQ_{Var} container quantity of a variant [pcs.]
S signal kanban factor
DD daily demand [pcs./d]
P number of identical parts per finished product [pcs.]

6. Specification of release unit

6.1 Release unit in serial and variant production:

$$RU_{Series} = TT \times b \times CQ$$

$$RU_{Variety} = TT \times \frac{\sum_{i=1}^{var} b_i \times CQ_i}{\#Var} \qquad \langle b_i \times CQ_i = const. \rangle$$

(3.20)

with: RU release unit [time unit]

TT takt time [time unit/pcs.] *or* [time unit/quantity unit]

b integer factor for containers to be released simultaneously

CQ container quantity [pcs.]

#Var number of variants

6.2 Release unit in single-item production:

$$RU_{Single-Item} = TT \times b \times PQ$$

$$RU_{OT} = CQ \times OT_{\varnothing}$$

with: RU release unit [time unit]

TT takt time [time unit/pcs.] *or* [time unit/quantity unit] (3.21)

b integer factor for containers to be released simultaneously

PQ package quantity raw material [quantity unit]

CQ container quantity: number of products in combination with defined size ranges

$OT_{\varnothing}$ operation time [time unit]

7.2 Symbols

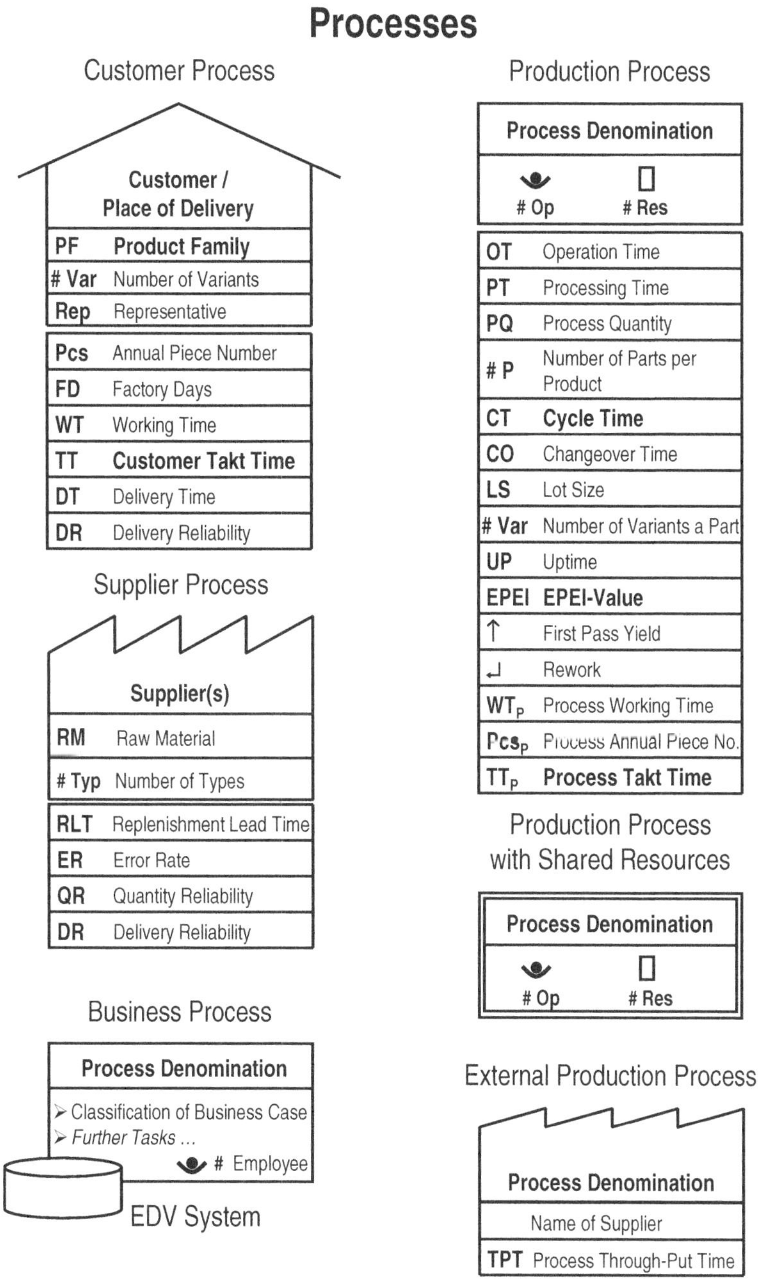
Processes
Customer Process
Customer / Place of Delivery
PF Product Family
Var Number of Variants
Rep Representative
Pcs Annual Piece Number
FD Factory Days
WT Working Time
TT Customer Takt Time
DT Delivery Time
DR Delivery Reliability
Supplier Process
Supplier(s)
RM Raw Material
Typ Number of Types
RLT Replenishment Lead Time
ER Error Rate
QR Quantity Reliability
DR Delivery Reliability
Business Process
Process Denomination
➢ Classification of Business Case
➢ Further Tasks ...
Employee
EDV System
Production Process
Process Denomination
Op
Res
OT Operation Time
PT Processing Time
PQ Process Quantity
P Number of Parts per Product
CT Cycle Time
CO Changeover Time
LS Lot Size
Var Number of Variants a Part
UP Uptime
EPEI EPEI-Value
First Pass Yield
Rework
WTP Process Working Time
PcsP Process Annual Piece No.
TTP Process Takt Time
Production Process with Shared Resources
Process Denomination
Op
Res
External Production Process
Process Denomination
Name of Supplier
TPT Process Through-Put Time

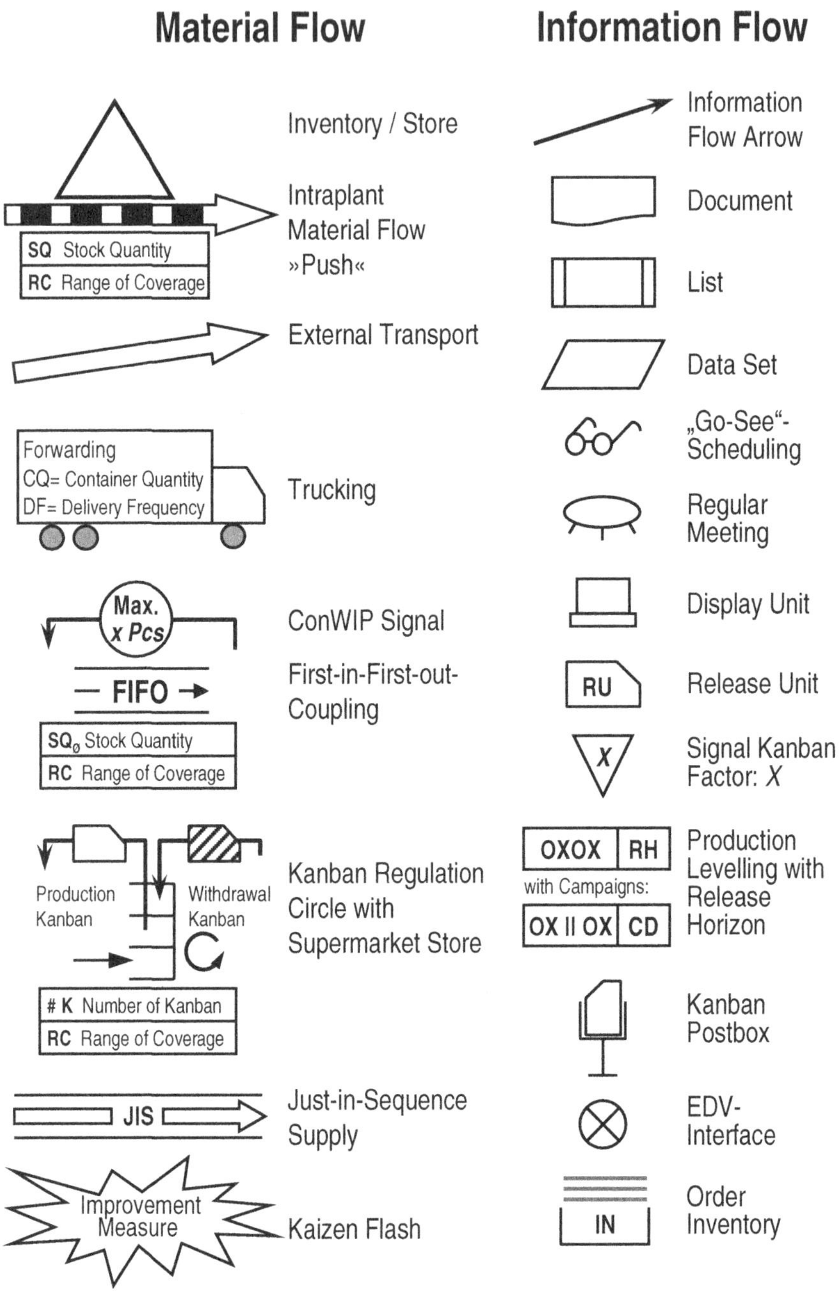
Material Flow
Information Flow
Inventory / Store
SQ Stock Quantity
RC Range of Coverage
Intraplant Material Flow »Push«
External Transport
Forwarding
CQ= Container Quantity
DF= Delivery Frequency
Trucking
Max. x Pcs
ConWIP Signal
FIFO
First-in-First-out-Coupling
$SQ_{\varnothing}$ Stock Quantity
RC Range of Coverage
Production Kanban
Withdrawal Kanban
Kanban Regulation Circle with Supermarket Store
K Number of Kanban
RC Range of Coverage
JIS
Just-in-Sequence Supply
Improvement Measure
Kaizen Flash
Information Flow Arrow
Document
List
Data Set
„Go-See“-Scheduling
Regular Meeting
Display Unit
RU
Release Unit
X
Signal Kanban Factor: X
OXOX RH
with Campaigns:
OX II OX CD
Production Levelling with Release Horizon
Kanban Postbox
EDV-Interface
IN
Order Inventory

Subject Index

5-W-method 226

A

ABC analysis 70, 108, 109
ABC classification 176, 202
allocation of resources 115
ancillary times 49
andon 129
assembly-to-order 185
autonomous automation 129
availability 14

B

batch 211
batch formation 152, 207
batch process 50, 53
benefit analysis 243, 246
bottleneck 90, 91, 147, 206, 208
bottleneck control 206–208
branching symbol 67, 153
buffer 150, 151, 152, 164, 195
buffer quantity 68
buffer stock 165
bullwhip effect 197
business process 26, 46, 72–74, 77

C

campaign 200, 201
capacity 51, 91
capacity control 225
capacity dimensioning 118, 121
capacity flexibility 135, 136
capacity profile 91, 206
caravan system 134
case study – Liquipur plc. 29, 33, 34, 36, 40, 42, 43, 63, 72, 82, 88, 92, 94, 114, 116, 120, 127, 155, 181, 190, 205, 216
changeability 226, 233, 244
changeover matrix 54
changeover portfolio 169
changeover reduction 174
changeover time 50, 54, 59, 93, 150, 170, 173, 200, 210
circular walk 134
clearance time 223
container quantity 67, 70
continuous flow 52
continuous flow production 121, 123, 128, 139
continuous improvement 9
continuous process 50
control board 180
conveyor 68
ConWIP 145, 147, 148, 154, 208, 209
customer decoupling point 113, 184
customer symbol 38
customer takt time 39, 40, 61, 62, 90, 91, 118, 132
customer's point of view 3, 23, 28, 92
cycle time 50, 52, 53, 61, 91

D

data box 39, 51, 52, 55, 57, 59, 60, 62, 68–70, 75, 160
degree of automation 128, 130
degree of flow 9, 84, 87, 90
delivery capability 17
delivery frequency 16, 68, 70
delivery reliability 17, 40, 70, 224
delivery time 40, 187, 222
demand-controlled disposition 79
design guideline 97, 118, 122, 145, 157, 184, 191, 196, 206, 238, 241
determination of lot size 171
dilemma of material planning 15
dilemma of production scheduling 14

dimensioning 117, 118, 120, 161
downtime 151, 166
drum-buffer-rope-control 208

E

EDP systems 74
efficiency 12
emergency kanban 162, 200
energy value stream design XI
enterprise resource planning 74
EPEI value 24, 59, 60, 61, 170, 171, 173
external production process 47

F

factory 4, 5
factory goals 11
factory layout 233, 237
factory planning 105
factory structure 138, 237
FIFO line 145, 146, 147, 148, 150
first pass yield 60
fixed station assembly 136
flexibility 11, 19, 51, 59, 224, 233
fluctuations in customer demand 41
frozen zone 223

G

'go-see' scheduling 77
goal conflicts 14, 20
goal dilemma 16, 19
goal dimension 11–13, 250
goals of production 3, 17, 248
gobalization XIII
golf ball signal 154

H

heijunka 24
heijunka box 197

I

ideal layout 241, 242
ideal planning 235
ideal planning principle 236
ideal production 121
implementation 215
improvement measures 215, 216
information flow 26, 46, 73, 82
information flow arrow 75, 77
inventory 99, 126, 145, 146, 148, 157, 160, 161, 164, 168, 174, 176, 178, 187, 195, 199, 212, 225

J

jidoka 129, 248
just in sequence 152

K

kaizen 215
kanban 164, 169
kanban board 168
kanban control 157, 159, 170, 180
kanban post box 178
kanban rules 162
kanban system 157, 162, 164, 169, 170
key figure 51, 63, 224

L

lead time 85, 86
lean factory 231
lean order management X
lean production 3
levelling of the production mix 196
line balancing 131,133, 134, 150, 210
line balancing losses 90
line production 144, 155
Little's Law 86
logical square of goals 20, 23, 246
logistic linkage 26, 73
logistician 239
lot production 156
lot size 14, 54, 55, 81, 167, 168, 170, 173
lot size calculation 55, 57, 102
lot size formula 55, 170
low-cost automation 130

M

make-to-order 184, 187
make-to-stock 187
manufacturing execution systems 74
material flow 26, 46, 66, 67, 72, 240

material flow arrow 70
material flow structures 242
milk run 178, 180
missing parts 212

O

operation time 49, 61
operational model 249
operator balance chart 24, 90, 91, 92, 93
order lead time 188
order point 168, 174
order processing 75, 102, 143, 188
order release 144, 145
overall equipment efficiency 57
overproduction 98

P

pacemaker process 184, 187, 191, 208, 209, 215
package quantity 193
paper-kaizen 128
performance figure 19, 250
performance measurement system 19, 219
pitch 191
planning horizon 79, 197, 221
planning principle 235
portfolio 111, 112
portfolio analysis 110
potential for improvement 27, 83
processing time 49, 87
product family 31, 32, 37, 38, 106, 111, 243
product family matrix 32
production audit 84
production kanban 158, 159, 163, 165, 167, 170, 180
production lead time 18, 86, 123, 188, 224
production optimization 3, 13, 23
production planning 77, 78, 79
production planning and control 73
production procedure schemes 34
production process 26, 46
production structure 105
production system 2, 3, 24, 247, 250
productivity 224
pull system 158
purchased parts 176
push production 78

Q

quality 12, 16, 23, 51, 61, 62, 93, 125, 244
queue 75, 194, 212

R

range of coverage 69, 85, 160
real layout 243
real planning 235, 236
rejection 58, 60, 101
relay system 134
release horizon 198, 201, 222
release interval 191, 196
release signal 145, 146, 206
release unit 191, 192, 196, 200, 221
reorder level method 80
replenishment lead time 16, 70, 165, 166, 168, 175
resource requirement 118
resources jointly used 48
restrictive bottleneck 209
rework 61
rework rate 60

S

safety stock 80, 165, 196
sales planning 221
scenario planning 236
scheduling and material planning dilemma 16
secondary bottleneck 210
segmentation 105, 106, 107, 108, 110, 113, 114, 116
separation principle 239
sequence 145, 168, 200, 209, 222
sequencing restrictions 125, 207, 210
sequential pull 145
shipping kanban 199, 204
signal kanban 167, 168, 170, 174, 175, 180
smoothing of production volume 191, 194
space analysis 237
space categories 237
space consolidation 240
space efficiency 240
space grid 137
space productivity 124

space quality 237
space requirement 238
space utilization 237
spaghetti diagram 173
standardization 9
stock quantity 69
storage symbol 66, 68
supermarket store 157, 158, 160, 161, 164, 165, 167, 174, 180, 194, 195, 202
supplier kanban 176, 178, 180
supplier symbol 70
supply chain 152
symbols 27

T

the material flow arrow 67
time losses 58, 93
time slice 202
timeline 86, 88
transition time 80
transparency 27, 84, 143, 212, 245
transport 67, 148, 165
transportation 66
trigger point 183
turbulence germs 141
turbulence profile 140, 142

U

uptime 57, 60, 93, 125
urgent order 199

U-shape 134, 241
utilization 18, 49

V

value 7
value stream analysis 26, 30, 246
value stream clipboard 307
value stream depiction 26, 27, 46, 86, 96
value stream design 246
value stream drawing 63
value stream management X, 219, 220, 222, 228
value stream manager 219, 233, 247
value stream mapping 27, 44, 84
value stream method 1
value stream performance 62
value stream perspective 9, 25, 243
value stream sections 215, 247
value stream-oriented factory planning 1, 232, 234
Value-added 238
value-added time 49
value-adding 9
variability 19, 24, 124
VDI-Guideline 5200 on factory planning X, 4, 234, 235
visualization 7, 27, 240

W

waste 61, 97, 98, 100
withdrawal kanban 159, 163
work in process (WIP) 66, 85, 165, 224
working time 39
working time model 57, 63

The manufacturer's authorised re EU is Springer
Nature Customer Service Centre itz 3, 69115 Heidelberg,
Germany. If you have any concer roducts, please
contact ProductSafety@springer

Printed and bound by CPI Group CR0 4YY
30/09/2
02225599